Calling for Help

Pragmatics & Beyond New Series

Volume 143

Calling for Help: Language and social interaction in telephone helplines
Edited by Carolyn D. Baker, Michael Emmison and Alan Firth

Calling for Help

Language and social interaction in telephone helplines

Edited by

Carolyn D. Baker†

Michael Emmison

University of Queensland

Alan Firth

Aalborg University

John Benjamins Publishing Company

Amsterdam / Philadelphia

∞™ The paper used in this publication meets the minimum requirements of American National Standard for Information Sciences – Permanence of Paper for Printed Library Materials, ANSI z39.48-1984.

Library of Congress Cataloging-in-Publication Data

Calling for Help : Language and social interaction in telephone helplines / edited by Carolyn D. Baker, Michael Emmison and Alan Firth.
p. cm. (Pragmatics & Beyond, New Series, ISSN 0922-842X ; v. 143)
Includes bibliographical references and indexes.
1. Social interaction. 2. Helplines--Social aspects. 3. Telephone--Social aspects. 4. Telephone calls. 5. Interpersonal communication. I. Baker, Carolyn D. II. Emmison, Michael, 1948- III. Firth, Alan. IV. Series.

HM1111.C35 2005
302--dc22 2005050837
ISBN 90 272 5386 2 (Hb; alk. paper)

John Benjamins Publishing Co. · P.O. Box 36224 · 1020 ME Amsterdam · The Netherlands
John Benjamins North America · P.O. Box 27519 · Philadelphia PA 19118-0519 · USA

In memory of Carolyn Baker
With affection and respect

Table of contents

Notes on contributors

Carolyn Baker was Associate Professor in the Graduate School of Education, the University of Queensland, Australia. In addition to her research on helpline interaction her interests included studies of language and literacy, the ethnomethodological analysis of interviews, talk and interaction in classrooms, meetings and other educational settings. She was the author of many articles and book chapters. Her books include *Children's First School Books: Introductions to the Culture of Literacy* (with Peter Freebody), Oxford: Basil Blackwell, and *Australian-Italian Courtship and Marriage Stories* (with Greer Johnson), Central Queensland University Press. Carolyn died in 2003.

Denise Chappell completed her PhD at the University of Southampton. She is Lecturer in applied linguistics at Anglia Polytechnic University, Cambridge, UK. Her current research focuses on patient empowerment in calls made to NHS Direct.

Susan Danby is Senior Lecturer in the Centre for Learning Innovation at Queensland University of Technology. Her research interests are the study of children and young people's talk-in-interaction, adult-child talk and early childhood education. Her most recent research projects include an Australian Research Council grant investigating young children's governance in home, school and community settings; children's troubles talk on helplines for children and young people; and friendships of twin children.

Michael Emmison is Associate Professor in Sociology, School of Social Sciences, University of Queensland, Australia, where he teaches courses on qualitative analysis and the media. His research interests are primarily in the field of language and interaction and he is currently examining the impact of technology (telephone, email and online web counselling) on troubles telling on a national children's helpline. His previous books include *Accounting for Tastes* (with Tony Bennett and John Frow), Cambridge University Press 1999, and *Researching the Visual* (with Philip Smith) Sage 2000.

Alan Firth is Associate Professor in the Department of Languages and Intercultural Studies, Aalborg University, Denmark, where he teaches applied linguistics and conversation analysis and leads a graduate programme on intercultural communication. He has published articles on conversation, helpline interactions, negotiations, second language acquisition, and English as an international language. He edited *The Discourse of Negotiation* (Pergamon, 1995) and is currently writing *Spoken Discourse and Social Interaction* (Edward Arnold).

Hanneke Houtkoop defended her PhD thesis on *Establishing agreement: an analysis of proposal-acceptance sequences* in Amsterdam in 1987. She was Associate Professor in the Communication Studies Department of Utrecht University, The Netherlands. Hanneke published numerous articles and books on conversation analytic subjects such as doctor-patient interaction, telephone conversations, and the design of telephonic survey interviews. Her last book was *Interaction and the Standardized Survey Interview*. Hanneke died in 2002.

Frank Jansen defended his PhD on syntactic constructions in spoken Dutch in 1981 in Leiden. He is Associate Professor in the Department of Communication and Information Studies of Utrecht University, The Netherlands. His current interests are politeness phenomena in business communication, and the effects of working with new media.

Wilbert Kraan is an educational technologist at the University of Wales, Bangor. He graduated with a BA in English and Education from the University of Wolverhampton, was awarded a Dutch teaching licence from Hogeschool Holland and an MA in Modern English Language from the University of Amsterdam. He taught general linguistics and sociolinguistics at Hogeschool Holland and the University of Wales, Bangor. His interests include cognitive semantics, pragmatics, sociolinguistics and multilingualism.

Håkan Landqvist defended his PhD on morality and conversational strategies in calls to the Swedish Poison Information Centre (2001). He is a research associate at FUMS, Uppsala University, Sweden. His interests are institutional discourse, interactional linguistics and the construction of knowledge.

Vesa Leppänen is a Researcher at the National Institute for Working Life in Malmö, Sweden. He received his PhD from the Department of Sociology at Lund University, Sweden. His research interest focus on work organization and social interaction with clients in human service organizations, especially health care.

Ged M. Murtagh is a Senior Lecturer in the Department of Applied Social Sciences at London Metropolitan University, UK. His research interests are in the analysis of communication. His work has focused on ethnomethodological and conversation analysis of professional–lay interaction.

Christopher Pudlinski is Associate Professor of Communication at Central Connecticut State University, USA. His areas of research interest include warm lines and the situated practice of advice, empathy, sympathy and other types of supportive communication.

Hedwig te Molder is Associate Professor in the sub-department Communication Science at Wageningen University, The Netherlands. She has published on a number of topics including government communicators' talk, online talk on depression and identity management in relation to food choice. Her work reflects a concern with both theoretical and applied aspects of discourse analysis. Together with Jonathan Potter, she edited *Conversation and Cognition* (CUP, 2005).

Brian Torode is a sociologist at Trinity College Dublin, Ireland. He sees Ethnomethodology, and the study of language games, as an everyday sociology, which can analyse consumerism, the mass media, popular culture, and politics in a way accessible to ordinary readers.

Anja Walstock studied communication studies from 2000–2004 in Utrecht, The Netherlands. She specialized in conversation analysis and graduated on the subject of the structure of the telephone conversations of helpdesk assistants. She is now applying her skills in the field of physiotherapy.

Jack Whalen is a Principal Scientist in the Knowledge Systems Area of the Systems and Practices Laboratory at Palo Alto Research Center (PARC – formerly Xerox PARC). His research focuses on the design and deployment of systems to support knowledge-sharing in work communities and the use of artificial intelligence applications in the workplace. His most recent projects include knowledge systems work with manufacturing engineers for a large American automobile company, customer support representatives for Telecom Italia, and Xerox service engineers. He is currently working on a project in

Japan. Whalen received his PhD from the University of California at Santa Barbara and joined PARC in 1997 after three years at the Institute for Research on Learning in Menlo Park, where he was Senior Research Scientist, and eleven years at the University of Oregon, where he was Associate Professor of Sociology and Department Head.

Don H. Zimmerman is Professor of Sociology and Affiliated Faculty in Communication at the University of California, Santa Barbara, USA. He has served as Associate Editor of *Sociological Methodology* and *American Sociological Review* and served on the editorial board for *Symbolic Interaction*, *Human Studies*, *Language and Society* and *Research on Language and Social Interaction*, of which he is currently editor. His research interests are in conversation analysis, talk in institutional settings, and the social life of very young children.

Preface

Help is only a phone call away
(The Gambling Helpline)

Helplines are out there in their thousands, a mere phone call away, offering help, advice and counselling, offering a sympathetic ear, offering someone at the end of the line who will listen to your woes, your fears, your worries, offering an expert verdict, offering a diagnosis of your ailments, instructing on how to connect or cancel your mobile phone, how to draw a vertical line in Word, how to dress a septic wound, how to plan a wedding, how to sue your brother-in-law, guiding on how to recognize and react to substance abuse, how to resist temptations of the flesh, the bottle, the casino and the dog track, informing of risks, dangers, tsunamis, shark attacks, rock falls, snowfalls, train delays and alternative routes. Helplines are variously staffed by volunteers, by dedicated amateurs, by highly-trained, often frustrated and invariably poorly-paid professionals. Helplines are located in office-block basements, in homes, in radio stations, in enormous, hangar-like call-centres on the edge of windswept industrial estates. Helplines are dedicated telephone-based services that provide assistance and/or guidance on a plethora of topics and concerns, ranging from emotional to legal, technical to financial, medical to strategic, mundane to bizarre. Helplines have grown exponentially over the last two decades and have become increasingly more specialized. Think of an illness, disease, pathology, phobia, addiction, machine, household appliance, the matter that kept you awake last night, and you will almost certainly find a helpline dedicated specifically to it.

A central feature of helplines is that helpline users – callers and call-takers, on either end of the line – talk to one another. Talk is the help of helplines. Help is sought, demanded, explored, resisted, rejected, challenged, accepted and provided through talk, talk that occurs in real-time, in the dialogic processes of focused social interaction. Here, on a collective basis, we pose the questions: how is helping undertaken through talk? How is the helpline-call setting and the talk of help-seeking and help-providing, mutually and reflexively configured? What are the skills and competences that callers and call-takers routinely draw upon when engaging one another in helpline calls? What do callers and call-takers do, what do they attend to, in talk, during helpline calls?

Helplines are socially ubiquitous, wide-ranging in scope and subject-matter, and at all times talk-saturated. It is surprising, then, that helplines have hitherto attracted little attention from scholars of spoken language and social interaction. In spoken language research terms, the helpline has been largely overlooked. Over the last two decades, as the research map of 'institutional' forms of interaction has become increasingly more detailed and multilayered, the helpline as an institutional context has essentially remained *terra incognita*. We thus have extremely scant knowledge of how, through talk, callers go about seeking help and how call-takers go about their institutional mandate of providing help – if this is, indeed, what actually happens in helplines.

In Aalborg, Denmark, in the autumn of 2000, a group of researchers had gathered to grasp the proverbial nettle and engage in a collective and detailed exploration of talk and social interaction in telephone helplines. In short, the meeting attempted to fill a research void. This book is the result of our forays into unchartered territory. In bringing together some of the leading as well as rising researchers of spoken language and social interaction, we here, in this collection of original studies, endeavour to cast light on language use and social interaction in a tenebrous though important area of modern living.

The symposium brought together researchers from Europe, the US and Australia who shared a deep interest in the analysis of help-seeking and help-giving behaviour conducted over the telephone. We believe that the symposium was the first occasion of its kind – one in which the methods of Conversation Analysis – with its high-powered, micro-analytic lens – were brought to bear on helpline interaction data materials. On the basis of the findings contained in the chapters in this collection, we believe that the helpline context has shown itself to contain an exceedingly rich vein of research findings for scholars committed to understand and explicate the social organization of spoken interaction in helplines in particular and work-related encounters more generally. What is offered here is thus only a beginning, a modest foundational stone that nevertheless hints at and leads the way to the potential rewards awaiting future research undertakings. We hope that others, after reading this book, will be inspired to build upon the foundations we attempt to lay forth.

The book has had a somewhat lengthy gestation period, due in large part to the illness and tragic death of our colleague, Carolyn Baker. Carolyn's death, in July 2003, overshadowed the usual trials and tribulations that frequently beset editors. With Carolyn's passing we have lost a dear personal friend and an inspirational colleague. We are proud that this publication will stand as Carolyn's final contribution in an illustrious career and we take great pleasure in dedicating the volume to her memory. We also, sadly, acknowledge the death of one of our contributors and colleagues, Hanneke Houtkoop. Hanneke died in late 2002, shortly after submitting her co-authored chapter on 'Collaborative problem descriptions in help desk calls'. With Carolyn's and Hanneke's passing, the Ethnomethodological and Conversation Analysis community has lost two exceptionally gifted scholars.

A book such as this one cannot come near to fruition without the support and collaboration of a number of individuals and institutions, and we would like to take this opportunity to acknowledge our debt of gratitude to them. First a heartfelt thanks to our contributors – for their commitment, dedication, and unfailing patience. Second to our supporting institutions: in Denmark, the Department of Languages and Intercultural Studies and the Faculty of Humanities at Aalborg University, and not least the Obel Family Foundation, all of whom have consistently supported Alan Firth's research sojourns to Australia, enabling the editors to collaborate in data-gathering activities and invaluable face-to-face meetings. In Australia we thank the School of Social Sciences, and the School of Education, University of Queensland, for their enduring commitment to and support of our international research collaboration. We have also benefited greatly from a generous Australian Research Council (ARC) grant and the supportive comments, challenging questions and collegiality offered by the transcript analysis group at the University of Queensland and Queensland University of Technology.

AF and ME
Aalborg and Brisbane, April 2005

Chapter 1

Calling for help

An introduction

Alan Firth, Michael Emmison and Carolyn Baker

1. Overview

This book is an edited collection of language and interaction-centred studies that explore what happens when people use the telephone to call for help. More specifically, the focus throughout this collection is on diverse aspects of spoken language and patterns of social interaction in calls made by members of the public to a variety of telephone *helplines*. Helplines are telephone-based services that offer callers help, advice or support in a wide range of areas, most commonly in areas relating to health and medicine, the law, finance, psychological wellbeing, interpersonal relationships, various forms of addiction, and computer technology. Prototypically, a helpline is a 'dedicated' service that provides help in a single, particularised area (for example breast cancer, AIDS, gambling addiction, computer software, bereavement, domestic violence, consumer rights). The help available in a helpline is expert advice or specialised knowledge and is delivered in the form of talk: callers engage, conversationally, with helpline call-takers in an attempt to obtain relevant information, instruction and/or guidance. In the vast majority of cases, helpline calls begin with the caller describing a 'problem' he or she is currently experiencing. Caller and call-taker will then discuss the problem, during which time the call-taker will offer advice or provide information or instructions, with the aim of resolving or alleviating the caller's 'problem(s)'. Once this has been achieved, the helpline call will be brought to a close.

Such specialised, telephone-mediated help has existed for almost over half a century (one of the oldest helplines being *The Samaritans* – a UK-based emotional crisis counselling service, see Farberow & Shneidman 1961; Fox 1968; Lester & Brockopp 1973), but it was not until the 1970s that helplines began to diversify and extend beyond the already well-established area of emotional crisis-counselling (WHO 2002:5). By the 1980s the number and variety of helplines were growing rapidly. This development has continued to the present day. Indeed, at present in Western Europe and other developed areas of the world, the telephone-based services sector is one of the major sources of employment. In the UK alone there are currently over fifteen hundred helplines in operation.

With the exception of a small number of notable studies,[1] helplines have yet to be investigated by scholars of language and social interaction. Thus we have scant understanding of what actually happens, in language and social interaction, when people call for help. Very few studies have hitherto attempted to describe and explicate in detail how helpline callers interact and use language in order to *seek* help, and how helpline call-takers *provide* help.

The helplines examined in this volume cover a range of issues with which callers – for a variety of reasons – have a perceived need for 'help', 'assistance' or 'support'. The scare quotes around these terms alert the reader to the view – one held throughout this book – that notions of 'helping', 'assistance' and 'support' are situationally defined and contingently achieved within the dynamics of spoken interaction. Like other social actions, seeking help and providing help are thus seen as interactionally negotiable and socially accomplished. One of the important goals of this book is to develop an understanding of *how*, through language and social interaction, *helping, assisting* and *supporting* are made manifest, situationally defined, contextually configured, and socially accomplished within helpline calls.[2]

The helplines under investigation here deal with primary healthcare, computer software, emotional crises, mobile telephones, poisoning, tourist accommodation, consumer rights, telecommunications, and emergency services.[3] The helplines are located in a number of different countries around the world, namely Australia, Sweden, the US, the UK, Ireland, and The Netherlands.

Although a diverse range of concerns is dealt with in the helplines examined in this collection, there are nevertheless a number of features that appear to transcend the various helplines. First, the encounters may be characterised by the participants' joint orientation to the overarching *interactional goal* of *seeking and providing help*. Second, and relatedly, the encounters are characterised by the participants' joint orientations to, and instantiation of, a *role and responsibility separation* between caller-as-help-seeker and call-taker-as-help-provider. In these respects, helpline encounters would appear to typify 'institutional' forms of interaction which, according to Drew and Heritage (1992:22), "involves an orientation by at least one of the participants to some core goal, task or identity (or set of them) conventionally associated with the institution in question". Third, help is sought and provided solely through *spoken interaction*. (In some helplines, additional 'hands-on' help may also be provided at a later date, depending on the resources available and/or the callers' requirements;[4] see, for example, Houtkoop et al. this volume, Zimmerman & Whalen this volume.) Fourth, interaction is *mediated by the telephone*. Helpline callers and call-takers must thus operate within the parameters of the telephonic medium. The medium impacts upon the character of the interaction in a variety of ways. To mention only one example: although the telephone allows for synchronous, real-time, conversational interactions, the participants do not have visual access to one another (see e.g. Rutter 1984; Hutchby 2001:85–89). This contingency must be dealt with and managed conjointly, within helpline calls. The nature of the demands posed, and the ways of meeting those demands, are contingent on the specific helpline in question and the type of interaction being undertaken in the specific helpline.

To summarise: across the range of helplines there are points of convergence that *may* have an impact upon the way language is used and interaction is structured and managed. These points of convergence pertain to the overarching goal of the encounter, role differentiation and responsibilities, communicative modality (speech) and communicative medium (the telephone). Detailed analyses – such as the ones presented in this book – will reveal whether and, where appropriate, how such factors impact upon helpline interaction.

The analytic foci of the studies contained here are diverse. They include the structure of helpline call openings, the discursive construction of social identities in helpline interaction, issues of caller competence and the notion of 'knowledge' in helpline interaction, modes of collaboration between caller and call-taker, problem-solving activities in helpline calls, the effect of the 'ecology' of the helpline worksite on patterns of interaction during calls, advice-giving and instructing in calls.

The studies are micro-analytic in methodological orientation and heavily influenced by the concepts, theories and analytic procedures developed within Conversation Analysis (hereafter CA; see e.g. Atkinson & Drew 1979; Atkinson & Heritage 1984; Schegloff 1968, 1986, 1992; Drew & Heritage 1992; Sacks 1992: Vol. 1&2; for methodological overviews, see Goodwin & Heritage 1990; Psathas 1995; Hutchby & Wooffitt 1998; ten Have 1999). This is serendipitous, for it is arguable that the contextual roots of CA lie in telephone helplines. Harvey Sacks, who is widely credited as the founder of CA, collected his first spoken-language data from a suicide-prevention helpline based in a psychiatric hospital (Sacks 1992 Vol. 1 [1964]: Lecture 1).[5] Indeed, one of Sacks' earliest publications (Sacks 1967) is based on data collected from the suicide prevention helpline. In his lecture from the Fall of 1964 (published in Sacks 1992 Vol. 1:3–11), Sacks describes how his interest in the details of talk emanated from the psychiatric hospital's concern that callers would not disclose their names when calling the helpline:

> The hospital's concern was, can anything be done about it? One question I wanted to address was, where in the course of the conversation could you tell that somebody would not give their name? So I began to look at the materials. It was in fact on the basis of that question that I began to try to deal in detail with conversations. (Sacks 1992 Vol. 1:3)

Sacks subsequently attended to some of the key features of crisis-counselling helpline interaction, including the delicacies of call openings and the notion and possibility of 'helping' someone with a mental problem or disorder. His analyses led to observations and methods of working that have since become paradigmatic within CA. For example he noted how certain types of talk are understood with reference to particular contextual roles, and that some expressions (such as 'May I help you?') are "manifestly organizational" (op. cit.: 10) in character. Talk, Sacks was beginning to demonstrate, is reflexively related to context: it is both 'shaped' by context and context 'shaping' (Heritage 1984: 242). Further, and perhaps most crucially for the development of CA, in attending to the way the helpline calls began, Sacks noted how one utterance 'shapes' or 'projects' a particular kind of next utterance. What Sacks brought to attention was that intersubjectivity is incrementally achieved and constructed across and within sequences of turns at talk (Heritage op. cit.: 254–260). This realisation – based as it was on analyses of recordings of naturally-occurring talk (which in itself was a relatively novel research procedure at that time) – was to have a major impact on the way researchers viewed notions of 'meaning', 'intersubjectivity' and 'context'. To Sacks, and CA researchers who have followed his lead, notions of 'meaning', 'intersubjectivity', and 'context' are framed as *interactional phenomena*, as phenomena that are dynamically and contingently constructed in the minutiae of 'talk-in-interaction' (Schegloff 1987), in accordance with the contextual relevancies that participants (rather than, say, analysts) attend or 'orient' to in their dealings with one another.

Seen in terms of the study of language and social interaction, Sacks' work on the suicide prevention helpline was momentous, for CA has emerged as one of the most influential and powerful analytic methodologies within the social sciences. CA research endeavours to describe and explicate the finely detailed ways in which talk-in-interaction is constructed, managed, and understood, within actual contexts of use. CA is thus quintessentially micro-analytic in orientation, and sees 'context' as something that is produced, ongoingly, within talk. In terms of analytic foci, CA is attuned to the participants' *displayed orientations* and the *interactional relevancies* that the participants conjointly make public, in each encounter. It is, throughout, based on analyses of audio- or video-recordings of naturally-occurring[6] interactions in a wide range of social settings.

Although relatively little work has since been done on helplines *per se*, Sacks' interest in the 'institutional' character of the suicide prevention helpline laid the foundations for a major development in CA research. For over the last two decades, the theme of 'institutional' forms of talk and interaction has come to occupy a central, if not *the* dominant, position within CA research. Settings such as classrooms, courtrooms, doctors' surgeries, business meetings, radio phone-ins, emergency services, and activities such as counselling, interviewing, playing, mediating, nursing, negotiating and plea-bargaining have been investigated by researchers working within the CA paradigm (see e.g. McHoul 1978; Atkinson & Drew 1979; Maynard 1984; Heritage & Greatbatch 1991; Greatbatch & Dingwall 1997; Drew & Heritage 1992; Boden 1994 and 1995; Firth 1995b; Peräkylä 1995; Hutchby 1996; Drew & Sorjonen 1997; Luff, Hindmarsh, & Heath 2000; Clayman & Heritage 2002; Maynard 2003).

The goal of this research is to study how, through talk and social interaction, work-based institutions and concomitant institutional activities are, as Heritage (1984: 290) phrases it, "talked into being" and thereby rendered recognisable. How this is achieved is viewed as an empirical question, answerable by dint of detailed descriptions of talk within natural settings. In an important overview of institutionally-focused CA research, Drew and Heritage (1992: 21–25) argue that institutional interactions are prototypically (1) goal-oriented encounters, (2) characterised by 'special constraints' on 'allowable contributions to the business at hand', and (3) that institutionally-specific 'inferential frameworks' may apply. However, because contexts are viewed as dynamically and contingently produced, the 'institutionality' of a particular piece of interaction cannot be taken for granted by analysts, for example due to its occurrence in a physical setting that is 'institutional' – such as a school or hospital. Rather, analysis must show how 'institutionality' is *made* relevant and produced by the parties involved (Schegloff 1992). This is achieved in the way people talk to and interact with one another.

The studies in this volume build upon such research and seek to extend existing knowledge in four ways: First, by establishing the telephone helpline context as an important and productive site of language and social interaction research in its own right; second, by reaching an improved understanding of the discursive and interactional character of telephone helpline calls; third by contributing to research undertaken on talk and social interaction on the telephone (see e.g. Hopper 1992; Luke & Pavlidou 2002); and lastly by extending the CA-inspired body of research investigating the detailed nature of talk and social interaction in 'institutional' settings (see Drew & Heritage 1992: Chap. 1; Drew & Sorjonen 1997; ten Have 2001).

At this juncture it may be helpful to point out to the reader that the studies in this volume are essentially *descriptive* in character. Critiques, admonitions, evaluative assessments of helplines and/or helpline call-takers, and recommendations for changes in working practices, are thus avoided. Because so little is known about what happens when people talk to each other in helplines, we proceed from the position that detailed description must, necessarily, precede moves to 'improve' upon the way helplines are organised and administered, including the efficacy (or otherwise) with which call-takers perform. Nevertheless it is hoped that helpline practitioners, professional call-takers, helpline trainers and educators will find the studies to be informative and useful, and that the studies will lead to an increased awareness of and sensitivity to issues of communication in telephone helplines. We suspect that most helpline organisations have not encountered this style of research before, that they are more familiar with impressionistic evaluation studies using checklists or other protocols, or with client satisfaction surveys. We hope that the work here will demonstrate to helpline professionals the benefits of detailed study of *how* talk is organised and managed in helpline calls, and as such inspire academic and practitioner-based research collaboration, and more generally provide the basis for reflection and discussion between scholars of language and social interaction and helpline professionals.

The remaining sections of this chapter are organised as follows: First we provide some background information on helplines, looking particularly at their exponential growth over the last decade. Second we reflect upon the notion of *help*; in doing so we consider Nikolas Rose's (1999) arguments relating to the role of 'experts' in modern societies. Third we consider reasons for the recent growth in the popularity of helplines and assess the interactional research implications of this growth. Finally we describe briefly the individual studies and outline their points of divergence and convergence.

2. Helplines: Some background

The last decade and a half has witnessed a dramatic growth in the number of telephone helplines. The social implications of this development are likely to be far-reaching – a topic much discussed in the media of late. In a 1998 article, *The Economist* reported that the proliferation of helpline and telephone services in general in the UK had been so rapid over the preceding decade that employment in telephone services now exceeded employment in heavy industries (e.g. coal mining, steel production, ship building) and, moreover, that such developments were revolutionising the nature of work.[7] Over five million jobs have been created in telephone help and assistance services in the US since 1990, and in Australia employment in telephone helplines has increased annually by 25% since 1996.[8] In the year 2000, *The Helpline Association* (THA) listed 'over 900' voluntary, telephone-based 'help and advice' services in the UK. By 2004 this figure had risen to 'over 1,200'.[9]

In tandem with the increase in the number of helplines is the number of calls made to helplines. For example, in 2004, *NHS Direct*, the UK's state-funded medical illness helpline, reported a 500% increase in calls since its launch in 1999. Presently *NHS Direct* receives a staggering half a million calls each month.[10, 11] *The Samaritans' Helpline*, which provides psychological crisis counselling in the UK, similarly reported a 28% increase in calls between the years 2001 and 2002, and presently receives two and a half million calls per year. Such statistics are illustrative of a general (and perhaps cyclical) trend of more helplines, more calls.

According to *The Helpline Association* website, the majority of helplines provide advice and counselling on matters relating to a wide range of *physical illnesses* and *emotional wellbeing* (or lack thereof). Included in the former category are dedicated helplines dealing with heart disease, respiratory illnesses, allergies, degenerative illnesses, viruses, sexually-transmitted diseases, and cancers; and from the latter category, helplines dealing with suicide prevention, numerous forms of addiction, divorce, marital guidance, child abuse, and domestic violence.[12] Pudlinski (this volume) notes the relatively recent emergence of so-called 'warmlines' in the US. These are emotional counselling helplines that provide what Pudlinski (op. cit.) refers to as a *pre-crisis* service. In cases where the callers' problems are adjudged to be severe (for example if callers appear to be suicidal), warmline call-takers will 'escalate' the call by transferring or referring the caller to a *hotline*. Overwhelmingly, such helplines (and 'warmlines') are operated on a non-profit basis, staffed by volunteers (who undergo professional training), and funded publicly or through charities.

Other types of helplines are also to be found – for example, within the state sector. Most public-utility companies (providing gas, water, and electricity) and local government offices in industrialised countries nowadays offer telephone-based assistance on a range of issues (typically public transport, housing, refuse collection, street lighting, pest control and billing), while national or local health authorities increasingly provide a dedicated helpline service (akin to the UK's *NHS Direct*) that deals with 'front-line', general health-related enquiries, functioning in much the same way as a 'general practitioner' or family MD. Additionally, the private or commercial sector is beginning to make extensive use of telephone-based services, as witnessed by 'hotlines' (as they are often called in this domain) provided by mobile phone companies, rail companies, internet providers, household-appliance manufacturers, detergent and food manufacturers, tour operators, and insurance companies.

3. Seeking and providing help

As the studies in this volume reveal, helplines provide help in a variety of ways – not only in terms of subject-matter (be it emotional, physiological, technical, legal, or factual), but also in character, for the help sought and offered will be critically dependent upon the nature of the problem(s) presented by the caller as well as the remit of the helpline.[13] To cite one transparently obvious example: the help required from and provided by a computer software technical-support helpline will invariably be of a quite different order from that sought from and provided by a psychological crisis counselling helpline. As Baker, Emmison and Firth (this volume) show, technical support calls are typically dominated, interactionally, by the call-taker giving the caller real-time *instructions* on how to rectify a software-related problem. However, as Pudlinski (this volume) demonstrates, in the counselling 'warmline' calls he has examined, call-takers are trained to resist 'instructing' or 'advising' callers on how they should act upon or perceive problematic events in their lives, even when callers request this (although, as Pudlinski shows, actually refraining from offering advice is a difficult task for call-takers). In the counselling context, then, the notion of 'helping' clearly takes on a quite different meaning, and interactional form, compared to the former (technical support) context.

Nevertheless, the research presented here provides compelling evidence that the notion of 'helping' transcends helplines, regardless of the subject area covered by the helpline. But in all cases, help has to be interactionally accomplished *in situ*, through the micro-dynamics of talk and social interaction.

What we can hope to begin to tease out are answers to the question of *how* this achieved.

Some insights are readily available. For example the 'helping' *role* of the call-taker is both instantiated and reflected in the way calls to many helplines are answered. Helpline calls routinely begin with the call-taker producing a phrase that contains the word 'help', as Baker, Emmison and Firth (this volume) show in their study of calls made to Microsoft's 'technical support' helpline (although see Danby, Baker and Emmison (this volume) for an example of a helpline where call-takers purposively refrain from asking how they may 'help' callers). Here calls almost invariably begin thus: *Welcome to Microsoft's technical support. My name is [first name]. How may I help you?* This opening format, which has become standardised not only in Microsoft helplines but also (with minor modifications) in many different kinds of helpline calls, is overwhelmingly oriented to, by callers, as 'normal' and appropriate and, moreover, as an invitation for callers to reveal the reason for their call.

Similarly, callers will often deploy the word 'help' in their problem descriptions; for example by producing utterances such as *well I need some help with [X]* or *I think I need you to help me with [X]*. But as we see in te Molder's (this volume) study of a psychological counselling helpline, in some cases helpline callers will challenge the presumption (encapsulated, for example, in call-takers' opening format) that they need to be helped. This finding harks back to Sacks' earliest work on the suicide prevention line (see Sacks 1992 [1964] Vol. 1:Lecture 1). To this end, te Molder shows, callers do interactional 'work' to emphasize that their need is for 'someone to talk to', rather than for someone to 'help' them. Thus for some callers, in some helplines, being 'helped' may be understood differently from the way call-takers understand the term. Our task here, and in future research, is to view the issue of how 'help' is requested, elicited, offered, challenged, or situationally defined, as essentially interactional in nature and as an issue that may or may not become *interactionally relevant* within helpline calls. Where such an issue does become relevant – as evidenced in and through the participants' orientations – the analytic task is to describe and explicate how it is rendered relevant, as it occurs, within the unfolding context.

While attention is given to explicating the detailed nature of helpline interaction, the research presented here also invites more general reflections upon the social implications and social functions of 'helping' and 'being helped' within the community, via telephone helplines. For example, what does it mean to call a helpline, and in so doing talk with an anonymous call-taker[14] about issues of concern in one's working or private life? When someone calls a helpline they commonly divulge their feelings, thoughts, troubles, and frustrations – whether these are mundane, prosaic or life-threatening, whether they relate to their home, school, workplace, their malfunctioning technological equipment and mechanical devices, their illnesses, relationships, their needs, their children, or myriad other possible sources and causes. In many cases, such feelings and concerns are intensely personal and private. Yet in increasing numbers, people are calling helplines and divulging their troubles to and seeking help from anonymous 'experts'. The point of doing so, we surmise, is to find solutions to specific and non-specific problems, to find solace and comfort, and to seek advice and support; that is, to somehow *manage* their actions, their 'intimate selves' and, ultimately, their lives. What, then, is the role of the anonymous experts who, as unpaid volunteers or professionals, act as call-takers on helplines? Might it be the case that when people call helplines they are consulting what Nikolas Rose (1999:3) calls "engineers of the soul": those specialists who are reconstructing "our ways of thinking about and talking about our personal feelings, our secret hopes, our ambitions and disappointments" (ibid.)? And might it be the case that helplines not only provide *prima facie* help and assistance on a multitudinous range of issues, but also in some important ways impact upon the public's perceptions of and reactions to 'problematic' or 'troubling' events that occur in their daily lives? Moreover, are helplines an instantiation of what Rose (op. cit.) describes as a *new form* of expertise, one that is predicated on the professional "management of subjectivity" (Rose op. cit.:2)? Such expertise, claims Rose (op. cit.), is somehow filling "the space between the 'private' lives of citizens and the 'public' concerns" of organisations and institutions (op. cit.:3). And lastly, are helplines both a product of our much-touted 'information age' *and* a cause of a particular kind of help-seeking behaviour – behaviour which is predicated on a 'manufactured' awareness of what is 'problematic' and 'solvable'?

These are trenchant and important questions, and in this volume we point towards possible answers. Here, though, we take the position that comprehensive and compelling answers to the questions posed above will be at least informed by, if not directly based upon, detailed examination and descriptions of the patterns of talk and interaction that are produced within helpline calls. For it is through the processes of talk and social interaction that telephone helplines – as institutions, as sources of help and support, and as sites of expertise – essentially come into being and make an impact upon people's lives (see Heritage 1984:290). With regard to Rose's (op. cit.) arguments, a number of questions thus arise. For example, *how* is subjectivity actually *'managed'* within the dynamics of helpline calls? If helpline call-takers, as 'experts' or 'engineers of the soul', are indeed 'reconstructing' ways of thinking and talking about feelings (etc.), then *how* is this being done? How, for example, do people talk about feelings, frustrations (etc.) when they call a helpline? In this volume we adopt the position that such 'how'-questions are best approached and answered *empirically*, through detailed analyses of social action, as it happens, *in situ*, in actual, real-life helpline calls.

4. The popularity of helplines and some interactional implications

Even a cursory search in the telephone directory or on the internet provides compelling evidence that helplines abound, while new helplines are emerging almost on a weekly basis – along with a subset of neologisms such as crisisline, warmline, quitline, kidsline, shelterline, crimeline, careline, AIDSline and debtline. Helplines are growing in number and becoming increasingly specific in terms of their areas of concern. Whereas thirty years ago the UK had *The Samaritans'* helpline as the sole source of psychological support or advice, prospective callers nowadays are offered – in addition to *The Samaritans'* helpline – an array of specific and dedicated helplines covering the gamut of social and psychological concerns (e.g. depression, addictions, bipolar disorders, suicide, substance abuse, phobias, self-assertion, trauma). The underlying expectation here is that callers display increasing sophistication in terms of self-diagnosis of their 'problems', prior to calling a particular helpline.

Perhaps the increasing specificity and exponential growth of helplines should hardly surprise us, particularly those of us inhabiting 'post-industrial societies' (see Bell 1976), where the desire for *knowledge* (Böhme & Stehr 1986), *information* and *specialisation* (Castells 1996) are claimed to underpin modern living. The 'desire', as Castells (op. cit.) points out, is at least in-part socially constructed – through marketing and advertising. And helplines are, without question, marketed products, their saleability being *accessibility* to specialised knowledge and information – for those who find themselves in a health/legal/emotional (etc.) predicament. Arguably, then, helplines have emerged from, and are simultaneously a vivid instantiation of, our consumer-driven 'information age' and its growing and apparently insatiable need for rapid access to specialised knowledge.

It appears to be self-evident that the need for telephone-based 'help' and 'support' *is* growing, regardless of the source or cause of that need. Furthermore it is difficult to escape the impression that telephone helplines are increasingly becoming the community's principal source and providers of information, 'help', 'support' and/or 'advice'. Whereas previously – e.g. less than a generation ago – someone might routinely seek consultation with an 'expert' (e.g. a lawyer, marriage guidance counsellor, local government official, careers advisor, physician, insurance agent and travel agent) solely in a face-to-face encounter, nowadays one is increasingly being encouraged (by helpline advertising) to *call* the relevant helpline when a 'problem' arises, when 'expertise' is required, or simply when a 'question' needs answering. This is clearly the case with the *NHS Direct* helpline, for instance, which was established principally in order to lighten the consultancy load of general practitioner MDs. The same trend is evident in the law firms, police 'crimelines', crisis counsellors, city councils, airlines, car hire companies, insurance companies, and others who, through advertising, openly express their desire and preference for telephone enquiries as the first (and often the only) point of contact.

Moreover, it appears that helplines are inexorably extending into spheres of life that were perhaps once the sole domain of the family or church. This is suggested in the emergence of helplines covering marriage guidance, bereavement, family planning, pregnancy, domestic violence, incest, child-rearing, nutrition and diet. On this basis we might reasonably venture that helplines are a *response* to a secular society's growing need for access to expertise – regardless of the fact that the expertise is almost invariably provided over the telephone by an anonymous individual.

However, taking a lead from Rose (1999), we might also postulate that helplines are simultaneously *creating* a here-and-now need for expertise, in the form of information, guidance, advice, solace, and/or emotional support, in a wide range of human affairs, in both private as well as professional spheres of life. Helplines market themselves as 'life-management' resources with a 'personal face': as 'someone to talk to', 'someone you can call', 'someone who will advise/help you'. They send out a general message that 'finding the solution [to your problem] may only be a phone call away'. This message is encapsulated in the following quotation extracted from the Telephone Information Support and Counselling Association of Australia website:

> When troubled by a particular problem, it is easy to feel isolated and quite alone if you don't know where to turn for help. You may know that help is 'out there' somewhere, but don't know how to begin to find it. Sometimes the answer is as easy as picking up the phone.[15]

The need, then, is not only for 'expertise', but *on-hand expertise*, provided by an *interactively-* or *conversationally-engaged* other. The fact that helplines are invariably staffed by individuals who are and will remain anonymous to callers seems to be immaterial. Thus it appears that despite (or in part perhaps as a result of) the plethora of text-based information widely available on the internet and in specialised publications, people retain a desire, or, indeed, experience an increased desire, to *talk to and engage with* someone in their search for help.

Now if such postulates hold, they would seem to do so principally for four reasons, namely because helplines – regardless of their remit or area of concern – provide relatively (1) *low-cost* (2) *access* to (3) *an anonymous expert* who is (4) *conversationally engaged*. These four factors, we submit, have contributed significantly to the prolific growth in the number of helplines over the last decade and a half. Moreover, both individually and collectively they are potentially significant in terms of the patterns of interaction that occur during helpline calls. Let us examine each of the four factors, briefly, in turn.

Low cost

Low financial cost is unquestionably a major factor in the growth and current popularity of helplines – for both helpline caller and provider – and is likely to be significant in terms of what happens, communicatively, during calls. For callers, the financial cost of contacting a helpline will in many cases amount to no more than the telephone charge (which is most often set at a 'local-call' rate), in that the majority of helplines are free (non-fee-paying) services, particularly those providing emotional and psychological counselling. Most 'hotlines' (i.e. telephone services provided by commercial companies) are free-calls, the telephone charges being paid for by the hotline providers. Even when callers are requested to pay for telephone-based help or assistance – as is typically the case with software manufacturers or law firms, amongst others – the charges are often significantly lower than those that would apply if the caller had sought or requested a face-to-face consultation.

In some cases, helplines provide a free 'first-tier' pricing arrangement, in the sense that callers are given a limited time-on-the-telephone free-of-charge initially, before a service charge will be required for continuation of the call. At the time of writing, Microsoft's 'Technical Support' helpline in Sydney, Australia, operates such a free 'first-tier' policy, and 'escalates' the 'problem' to a fee-paying one if the technical issue cannot be resolved within the first five minutes (or so) of the call.[16] Some of the interactional implications of this arrangement are described in Baker, Emmison and Firth (2001:54–55) and in more detail in Emmison, Firth and Baker (in prep.). This payment arrangement imposes itself on the way callers and call-takers interact with one another.

This is illustrated rather nicely in the following data extract, where the Microsoft helpline call-taker (CT), on realising that the caller (C) has not yet paid for the helpline service call, cuts herself off in mid-utterance (lines 14 and 16, arrowed) in order to refrain from revealing how the caller's particular software-related problem may be resolved:[17]

Baker, Emmison and Firth (2001: 54–55)

```
1   CT:     How can I help you?
2   C:      I've recently installed Microsoft office pro:
3           (0.4)
4   CT:     yeah
5   C:      and in the access part of the thing I've- I wanna
6           use the membership (.) but I've got those
7           Americanised dates an' phone numbers and erm
8           there's some form of ma:sking on them?
9   CT:     yes there's an input mask on them
10  C:      yeah I wanna- how do I get that bin to an
11          Australian standard?
12  CT:     you need to edit the ma:sk
13  C:      e[dit the-
14  CT:→     [edit the input mask=you just change it to:
15          (0.8)
16     →    what er whatever format you want (.) if you need me
17          to step you
18          through the procedure Benny I'll need to set up a
19          support contract for it
20          (0.5)
21  C:      o:h dear me (0.4) an' what's that gonna cost me,
22          Leena?
```

After the caller describes his 'problem' (lines 2, 5–8), and then produces a direct question (*how do I get that bin to an Australian standard?* lines 10–11), the call-taker, in line 9, begins to answer the question by revealing what must be done in order to solve the caller's problem: he has to *edit the mask* (line 12). In line 14 the call-taker then voluntarily embellishes the details by saying *edit the input mask*. In the same turn, without hesitation (the '=' symbolises 'latching', or talk produced without any perceptible pause) she adds: *you just change it to:* (line 14). Here she clearly cuts herself off in mid-utterance. This 'cut-off' is followed by a relatively long 0.8 second pause. In line 14 what the call-taker is doing is refraining from revealing to the caller what, exactly, he should do to 'edit the input mask'. Such information is evidently 'classified', and can only be revealed once payment has been made (note, incidentally, the call-taker's euphemistic

I'll need to set up a support contract (lines 18–19)). Payment, we see, is quite a delicate issue, both interactionally and pragmatically.

In other cases, helpline calls are 'pre-paid' by the callers: they are either purchased in advance as individual calls or as a 'package' of calls. Additionally, customers may purchase an arrangement where they are permitted to make an unlimited number of helpline calls within a stipulated period. This arrangement is common when hi-tech equipment or machinery has been purchased.

Helpline providers have become increasingly cost-conscious over the last decade, and, as Cameron (2000) describes in her study of communication in 'call centres', the communicative implications of this appear to be wide-ranging. For example, a growing number of banks, insurance companies, travel agents, utility companies, airlines, satellite television companies and mobile telephone companies (amongst many others) are transforming their regional or high-street in-person 'customer services' into centralised, telephone-based help/hotlines. Local offices are being closed in order to cut costs and centralise working practices, the theory being that when one single call centre deals with customer enquiries and/or complaints, as opposed to calls being taken at physically separate, local offices or branches, a company can make significant savings as well as impose uniform working practices, not least regarding the way call-takers talk to callers.[18] Recently, many larger companies have gone a step further and begun outsourcing their helplines to sub-contractors whose sole responsibility is to staff and run the helpline on behalf of the contracting company. In other recent cost-cutting moves, numerous helplines have been outsourced to foreign countries, where lower labour-costs apparently allow for even greater financial savings to be made. The communicative implications of such moves – for instance in terms of international and intercultural communication – have yet to be investigated.

Cameron (op. cit.: 99–106) describes the impact of uniform working practices and cost-conscious call centres on call-centre employees, who are increasingly subjected to greater time pressures and regimented working conditions. For example, in many cases, call centre call-takers are being ordered to follow prescribed dialogic 'scripts' or 'prompt sheets' when engaging with callers, and ordered not to exceed prespecified maximum time-limits for each call. In some call centres, greater rewards are given to call-takers who process the largest number of calls per working day and/or working week – a practice the present authors also witnessed during fieldwork at a large, Sydney-based helpline call centre, during the late 1990s. Such working conditions are being enforced in order to reduce what call-centre managers term 'handle time', that is, the amount of time each call-taker spends on calls.

Amongst the few language and social interaction researchers who have studied the implications of call-takers being institutionally obliged (by employers) to follow institutionally-prescribed work procedures (regardless of whether these were imposed for 'cost-cutting' or 'quality assurance purposes', or indeed both), such as the stipulation that call-takers follow dialogic 'scripts' and complete forms during interaction with callers, are Richard Frankel, Jack Whalen, Marilyn Whalen and Don Zimmerman. In a series of studies of '911' emergency calls in the US, the Whalens and Zimmerman (e.g. Whalen, Zimmerman, & Whalen 1988; Whalen & Zimmerman 1990; Zimmerman 1992) examine the interactional implications of emergency call-takers having to manage interactions with callers along lines dictated by the textual design of 'dispatch sheets' – an organisational requirement that frequently had a negative impact on how talk in 911 calls was conducted. Whalen, Zimmerman and Whalen (1988) also show how the institutionally-imposed 'agenda' of the call-takers – namely to elicit specific types of information in a particular order prior to confirming that the relevant service had been, or would be, dispatched – clashed with the 'agenda' of the callers, who were unaware of the strictures and limitations that 911 call-takers were working under. For the callers, the 911 call-takers were 'nit-picking' over unimportant details and/or asking irrelevant questions, while they should have been dispatching emergency services. This perception resulted in communicative difficulties, with fatal consequences. Frankel's work (1989) details similar difficulties experienced by poison control centre call-takers. Here too, call-takers are faced with the task of following institutionally laid down 'rules of procedure' relating to the ordered completion of forms. Such a practice is often the cause of miscommunication between callers and call-takers, as a result of the fact that both parties are pursuing conflicting goals: callers require immediate advice and instruction, while call-takers require detailed and specific information on the potential poisoning, before they advise or instruct.[19]

Accessibility

Telephone helplines commonly market themselves as being widely accessible. One of the gambling helplines in Australia carries a message that typifies the accessibility feature that helplines endeavour to communicate to the public through advertising and promotions, the message being that callers can *'Just call us. We're always here to help'*. In order to increase their accessibility, most helplines operate outside normal working hours; some operate on a twenty-four hour basis, seven days a week. But accessibility does not always extend to access to a human being – at least not before a sometimes lengthy waiting period has passed. Most larger helplines (size being gauged in terms of number of calls and call-takers) deploy automatic queuing system technology, where callers are placed in a queue and requested (by a recorded voice-message) to 'wait in-line for the next available operator who will be with you shortly'. In call centres and on the larger helplines, call-takers can observe computer screens that automatically monitor the number of 'calls waiting', and observe colour-coded bar graphs that change from green to amber to red, indicating how long callers have been waiting for access to a helpline call-taker. The red colour flashes out a warning to call-takers, signalling that 'if calls are not answered within the next two minutes, there is a ninety-percent likelihood that the caller will hang up'.[20] For some helplines, it appears, 'accessibility' is a major selling point, though it is not always easy to provide in practice.

Messages such as the one contained in the gambling helpline advertisement are typical of helpline advertising, and aim to give the public the impression that help is 'on tap' and, moreover, that their 'problems' are helpable and, indeed, helpable over the telephone. What is implied, then, is that using the telephone to seek help is a simple, straightforward matter. This viewpoint undoubtedly emanates from a culturally-shaped and culturally-shared perception that the telephone is part and parcel of everyday living, and that we can unproblematically carry out a wide range of *talk-based* tasks over the telephone – including problem-solving across a diverse range of concerns. Telephone companies too have sought, through vigorous advertising campaigns, to reinforce and promote this viewpoint, as witnessed clearly in *British Telecom's* campaign in the UK in the 1990s, where the slogan *It's good to talk* appeared in numerous television, radio and print commercials.

In many respects the telephone, like the automobile and the television, is emblematic of twentieth-century living, and today – not least as a result of the advent and popularity of cellular telephones – the telephone has become commonplace in virtually all spheres and contexts of daily life, be it the workplace, the home, the beach, the shopping mall, the street, the car, the train, etc. Overwhelmingly, we take its use for granted – helplines included. Yet the telephone has brought about major changes in ways of living and has led to a blurring of the distinctions between home, leisure and working life, as people have learned to exploit the advantages of communicating not only 'at a distance' (Hutchby 2001:81–85) but also 'on the move'. In some areas (for example, western Europe, Australia, and North America) no less than ninety percent of households are estimated to contain at least one telephone, while mobile telephones have resulted in an even greater accessibility to, and use of, telephone technology amongst the general public. Seen in this light, it is hardly surprising that telephone helplines have become widespread, numerous and diverse in their range of concerns. Yet notions of helpline accessibility have yet to be investigated from a language and social interaction perspective. Questions such as how the telephonic medium impacts upon problem-solving, how 'queuing' for access to a helpline call-taker impacts upon the way calls are managed, and whether and/or how mobile telephony influences the character of interaction in helplines, await investigation.

An additional matter in need of attention is the impact of accessibility pressures on the work of helpline call-takers. Often, notes Cameron (2000: Chap. 4), there is a clash of interests between the organisation financing the helpline and the caller, and in many cases the caller loses out by having his or her 'problem' or 'issue' rushed through in order that the call-taker can meet the stipulated target-number of calls. Ultimately such pressures will produce a sense of frustration for both caller and call-taker, and – for call-takers – result in 'chronic frustration' due to repeated exposure to the dilemma adumbrated above. Chronic frustration and stress are seen by many helpline professionals as the major reasons for the occurrence of the phenomenon known in the helpline profession as 'call-taker burn out' and the rapid turn-over of staff within call centres.[21] This problem appears to be an intriguing inversion of the problems described by Zimmerman and the Whalens, who observed that in the emergency calls they examined, it is callers, and not call-takers, who often feel that the 'help' required (the dispatching of fire, police, and/or ambulance services) is too slow in coming, and that it is callers who express feelings of frustration. Questions thus emerge: How, in language and social interaction, do call-takers deal with the apparently conflicting demands that Cameron (op. cit.) discusses? And how do institutional demands more generally manifest themselves in the forms of talk and social interaction that occur in helplines?

Anonymous expert

In an article appearing in *The Guardian* in 2001 it was reported that eight out of ten callers to a parenting helpline said that they found it difficult to discuss family problems with their partner and that they *preferred* discussing matters with a helpline.[22] Walther (1996) has coined the term *hyperpersonal* to describe interactions between anonymous parties where at least one of the parties experiences a heightened (and positive) sense of emotional closeness *as a result* of interacting with an anonymous and, crucially, visually inaccessible, other. Clearly in some cases, helplines offer 'hyperpersonal' experiences that may be preferred to the 'personal' interactions more commonly experienced in face-to-face encounters. Many helplines exploit this feature in their marketing and promotional materials, for example by informing prospective callers that 'help [by implication, 'help' of any kind] is *only* a phone-call away' or (as we saw is the case on *The Telephone Information Support and Counselling Association of Australia* website) that "[s]ometimes the answer is as easy as picking up the phone." As Hutchby (2001: 82) has pointed out, the telephone is nowadays unproblematically associated not only with 'business', but very much also with intimate and personal contact, and many helplines have exploited the telephone's extensive functionality, such that 'intimacy' has become a 'business' matter.

Helplines – particularly (though not exclusively) those providing emotional counselling – thus appear to offer what we might call *intimate anonymity*. This is perhaps in-keeping with the times, for as Lyn H. Lofland argues in her classic study *The World of Strangers* (1973), the post-industrial age is characterised by the individual's routine though transitory engagements with 'anonymous others' who act in some kind of 'official', information-giving capacity. These are the personally unacquainted individuals we meet with in the gamut of service encounters, in the tax or welfare office, make inquiries at 'information desks', speak (via telephone) to 'directory enquiries', emergency services, and so on. We are socialised into inhabiting communities of 'strangers', and see such an existence as 'routine' and 'ordinary'. Telephone helplines are in many ways an exemplification of this 'world of strangers', where the 'strangers' in this case regularly engage with one another over matters of an intimate or technical nature. Indeed, a general characterisation of helplines (particularly those concerned with physical and mental health, grief, crisis, divorce, addictions, phobias, trauma, and law) is their overriding tendency to emphasise the fact that call-takers and callers are, and will remain, personally anonymous to one another.

We need to improve our understanding of the implications of 'intimate anonymity' in terms of the way people socially interact and use language. To this end, a number of empirical questions spring forth: how do callers and call-takers 'manage' 'intimate anonymity' in their interactions with one another? How do call-takers, as anonymous 'professionals', elicit 'troubles talk' from callers-as-strangers? Do callers at times challenge the 'intimate anonymity' character of the talk? If so, how do they do this? In some helplines, callers and some call-takers may be 'known' to one another as a result of repeated and regular interactions; this 'knowing', however, will almost invariably be of a non-personal and non-reciprocal nature. This raises the question: What is the interactional character of helpline-based, 'non-reciprocal' *troubles talk* (see

Jefferson & Lee 1981), and how does it differ from 'troubles talk' that occurs in casual, everyday encounters between people who are personally acquainted?

The conversationally-engaged call-taker

One of the main attractions of helplines is that, once callers have progressed through the queuing and call-routing system, they experience that 'someone will be there to talk' to them. This message figures prominently in helpline advertising and promotional materials. The talk, however, will invariably be an amalgam of 'business-like' and 'casual' forms of interaction, with a shifting emphasis between the different helplines as well as within one and the same call (on this specific point, see Torode this volume). In general, helpline call-takers, and their trainers, are aware of the importance of establishing a *conversational footing* (Goffman 1979) in calls: greetings and 'how-are-yous' may be exchanged (although typically in sequential environments that differ from their occurrence in 'casual' telephone calls[23]), pleasantries such as talk about the weather regularly occur,[24] first names are often used, questions are asked and instructions are given by call-takers, but invariably in 'staggered', quasi-conversational ways (see e.g. Baker, Emmison, & Firth this volume; Leppänen this volume), perhaps in efforts to ensure that the call does not become overly formal or call-taker dominated, and hence manifestly non-conversational in tone.[25] In many helplines (particularly those dealing with emotional difficulties), intimacy and openness is encouraged (see e.g. Danby, Baker, & Emmison this volume; te Molder this volume; Pudlinski this volume). This is almost invariably achieved through the artful deployment of conversational devices that display empathy and involvement (for example by the call-taker producing empathetic receipt tokens during callers' descriptions of their 'problem', by the use of appropriate – e.g. non-technical – lexical items, by both parties taking turns in cooperative ways, and by call-takers' skilful and interactionally appropriate use of questions, comments and suggestions). In all cases, callers and call-takers draw upon culturally-learned knowledge of interaction in 'everyday' conversations and apply it, in modified forms, to helpline interactions.

The reason for the modifications is that various 'institutional' laminations will characteristically impose themselves on the forms of talk in helpline calls. The anonymity of the call-taker is one obvious example. Others include the goal-driven and goal-focused nature of the interaction; the stringent time constraints that the call-taker may be working under; the fact that callers are informed that the call 'may be (audio) recorded for quality control purposes'; the cataloguing or inscription work of the call-taker, who may be required to collect and ongoingly enter (via the computer keyboard) information relating to the caller and his or her problem; the payment requirements that are a feature of some helplines and, of course, the ubiquitous knowledge differentials that are invariably orientated to and thus interactionally realised within helpline calls. Calls to helplines feature 'engaged' interactants, but the interactions are 'conversational' only in certain respects.

The challenge for analysts, then, is to address questions of how callers and call-takers conjointly construct 'conversational' engagement, while also attending to the 'institutional' contingencies, demands and tasks that underpin helpline interactions. This question – of how and where 'institutional' forms of interaction diverge from or converge with the 'bedrock' form of talk that is purportedly casual conversation – is a live one within CA research (see e.g. ten Have 1999: 167–170) and provides the theoretical motivation for a growing number of empirical studies of talk and interaction in 'institutional' settings.

Another important issue is how conversational engagement is maintained in the face of different role responsibilities and knowledge differentials within calls. Helpline callers routinely engage with call-takers who have (or are assumed by the caller to have) expertise or training in the particular area covered by the helpline. We may reasonably assume that callers and call-takers, in the ways in which they interact, display at least some cognizance of knowledge differentiation, and that this will impact upon the language used and the unfolding patterns of interaction emergent in helpline calls. Clearly though, we need to know precisely how such 'cognizance' is made manifest within calls and thus how the roles of 'expert' (or 'counsellor', 'technician', 'nurse', etc.) and 'novice', 'layperson' (etc.) are constructed and negotiated, through interaction, within calls. Furthermore, since engagement between helpline callers and call-takers revolves around an attempt to address, and resolve, a problem (or set of problems) raised by the caller, and is in this sense 'goal-driven' (see Hymes 1974; Drew & Heritage 1992), how are the goals of problem-solving and the 'expert-novice' (and the various gradations of 'novice') relationship accomplished through the deployment of (modified) 'conversational' resources? The work contained in this book sheds much-needed light on such questions, but the task of providing comprehensive answers lies beyond our remit and awaits investigation in future research undertakings.

5. The studies

Each of the studies in this volume presents detailed analyses of talk on a telephone helpline. There is a specific analytic focus in each chapter, where the authors investigate aspects of the interaction that they find salient. In all cases, then, analyses have been generated by working with the data to find a dimension of activity that appears to be significant in the specific helpline. We have organised the chapters according to the type of help sought and provided. The sections thus deal with (1) technical assistance, (2) emotional support, (3) healthcare, and (4) consumer assistance; the final section of the book (Part V) is concerned with aspects of call-management.

Part I: Technical assistance

The chapter by Baker, Emmison and Firth, dealing with calls to a computer software support helpline, focuses on an issue endemic to such helplines, namely the requirement for call-takers to monitor the caller's reporting of his or her trouble so as to provide the appropriate level of technical information in their 'real-time' trouble-shooting of the caller's problem. The question of 'talking at the client's level' as the organisation describes it, or of 'calibrating' for the perceived level of caller competence, as Baker, Emmison and Firth conceptualise it, thus becomes a pressing and ongoing concern for call-takers. Baker, Emmison and Firth show that this process of calibration begins with the caller's initial description or formulation of their problem which the design of the helpline call-taker's opening question 'How can I help you?' obliges them to produce. Call-takers closely monitor the displayed ease – or uncertainty – which characterises these initial caller turns, for this will provide evidence of the caller's prowess or ability in handling technical advice. Although the work of calibration would seem to refer more to the activities of the call-taker, Baker, Emmison and Firth suggest that there may also be a reciprocal process in which callers can be seen as actively producing conditions for calibration through various and successive indications of what they know and what they can (or cannot) do.

The theme of callers' contributions to problem solving is also raised in Houtkoop, Jansen and Walstock's study of callers to a Dutch bank's communication centre and the call takers who handle these inquiries. A crucial difference in this context is that the call-takers are 'non-technical staff' and therefore must refer problems they cannot answer themselves to technicians who work for the Dutch Telecom company, KPN. For KPN to solve the problems referred to them they need clear and unambiguous descriptions of the problems. However the KPN technicians frequently complain that the bank's call-takers fail to describe the reported problems fully or correctly. In an interesting intervention, Houtkoop, Jansen and Walstock report on the results of a small-scale field experiment in which they provided the bank's communication centre call-takers with a procedural script designed to regulate the interaction during the initial calls. In essence the script required call-takers to read back to callers what they had entered as the problem description, thus providing callers with an opportunity to verify what had been entered. Houtkoop, Jansen and Walstock describe the ways in which the new procedure resulted in collaborative work between the parties in ensuring that the problem description was efficiently and accurately recorded. For example the design of the questions required by the script meant that callers were frequently able to anticipate up-coming questions regarding routine information matters and were able to supply such information to the call-taker in advance of the question being posed. Moreover, call-takers began to 'work aloud' when undertaking the business of entering the problem description on their keyboards and this 'commentary' became a resource which the callers were able to monitor to determine if their initial problem-reports had been correctly interpreted. Callers, in turn, designed their talk with an orientation to the eventuality that the information they provided was recorded, in writing, in electronic format.

Kraan's chapter on the use of spatial metaphors by call-takers and callers reporting computer problems provides additional insights into the social organisation of technical support. Kraan's principal argument is that talk about computers and computer systems is particularly susceptible to images or metaphors of space. That is, users of computer systems frequently describe their actions by resorting to a number of devices in which movement, space or journeys are recurring tropes. Kraan distinguishes between what he calls, first, the 'agent-trajectory model', in which the user envisages the computer system as a space occupied by its constituent components – servers, programs, files etc. – which are encountered in some sort of serial fashion by the user; second the 'direct-interface' model which places more emphasis on the close proximity of the user and the various computer peripherals; and third the 'personification' model, in which the computer takes on an agentive or human-like role. Kraan shows how callers to a university computer help-service typically drew upon these various metaphors at different stages of the call. Openings to calls characteristically exhibit a shift from an 'agent-trajectory' model to a 'personification' model. Overall, these spatial metaphors play an important role in mediating the interaction between the technical experts and the lay callers. For example, novice users are more likely to talk about their computers in terms of the 'direct-interface' model; this can become a resource taken up by helpdesk personnel in structuring their replies.

Part II: Emotional support

Pudlinski's chapter examines three different consumer-run (peer-to-peer) 'warm lines' located in the northeastern United States. Warm lines differ from helplines in that their main objective is listening and supporting, not referring or advising, as is the case with helplines. However, Pudlinski calls into question this particular characteristic, and finds that giving advice is one of the activities working consumers, like other social support providers, consistently perform. Warm lines, as support services for people with mental disabilities, foster client decision-making through a web of what Pudlinski calls 'contrary themes', namely connectedness, nondirectiveness, and problem solving. As part and parcel of this dilemmatic ideology that characterizes consumer-run warm lines, circumstances arise in which working consumers must put forth advice. Putting forth safe, uncontroversial options and mitigating these options are ways to be consistent with contrary themes. It was in the warm line's interests for working consumers not to take responsibility for client solutions yet it was also in the warm line's interests to help clients solve problems. These features arose when working consumers encouraged adoption of a solution. Encouraging adoption of a solution involved formulating a course of action and either positively evaluating it or using a form of obligation. Pudlinski finds that this occurs in more urgent circumstances and seems particularly fitted to such circumstances. Where a client's problem is especially serious or urgent, working consumers seemingly assume more responsibility for the client. Contrary themes of problem solving and nondirectiveness are addressed when this strategy is used. Working consumers tend to deal with these dilemmas by seeking that elusive middle ground between taking responsibility for client welfare and respecting clients as decision-makers.

Danby, Baker and Emmison's study of the Australian children's counselling service – *Kids Help Line* – examines the interactional implications of an opening format which differs from the standard helpline opening of *How may I help you?* Call-takers at *Kids Help Line* purposively refrain from deploying such a format. Instead, in their initial turn they provide the caller with an organisational identification, namely *Hi, Kids Help Line*. The authors argue that the particular design of this turn offers callers greater interactional leeway than would be the case than if an explicit offer of help was produced, and that such a talk-based practice serves important organisational functions for this particular helpline, whose organisational philosophy is enshrined in the motto 'we listen, we care'. The authors show that, by withholding their first name (in contrast to the practice on the software helpline where first names are routinely exchanged), call-takers do conversational work that will reduce the likelihood that callers will feel obligated to reveal their identities. This finding resonates with one of Harvey Sacks' (1992 Vols. 1 & 2) earliest concerns in his research on a suicide prevention helpline. Another important difference between the calls to *Kids Help Line* and those in the technical helplines examined by Houtkoop, Jansen & Walstock, Baker, Emmison & Firth, and Kraan (see Part I), is that the caller typically does not have a 'problem' which requires resolving. Rather, they may simply wish to talk with the counsellor about an 'issue' that has arisen in their personal life. Counsellors are alert to this possibility and, as Danby, Baker and Emmison show, closely monitor the callers' opening turns for the specific reason for the call at this point in time, as opposed to the more general background problem.

Te Molder's chapter explores the ways in which callers and call-takers on an emotional counselling helpline in the Netherlands categorize themselves and each other, and reveals the interactionally sensitive business being performed through these identity ascriptions. Drawing upon a single case, te Molder examines how the helpline caller undermines the institutional nature of the interaction by constructing a 'deviant identity'. It is shown how the caller turns the interaction into 'everyday talk' by portraying herself as an 'ordinary person'. The caller's 'ordinariness' deprives the call-taker counsellor of her right to take the initiative in the introduction and the shaping of conversational topics. In a subtle way, however, the call-taker re-claims the initiative, after which the caller can again be defined as a potential help-seeker. Te Molder's chapter combines an interest in the detailed analysis of institutional talk with a strand of recent work in psychology which aims to respecify cognition from an analyst's category into a participant's resource. The analysis provides insight into some of the social and rhetorical activities performed through the subtle management of identities and counter-identities in helpline talk.

Part III: Healthcare provision

Leppänen's chapter deals with callers' help-seeking behaviour in telephone calls to a Swedish primary care helpline. The calls are often the first steps that people take when they experience medical problems that require professional help. Usually nurses answer and listen to the callers' problem presentations, ask diagnostic questions, make decisions about what measures need to be taken, suggest their decisions to the callers in the form of advice, and finally sum up the new cases in medical files that often are sent to doctors or others who do the actual physical examinations of the patients. Leppänen analyses the first few seconds of these calls, during which callers' subjectively experienced illnesses are formulated as possible medical conditions. The chapter describes the different ways in which callers present their problems and discusses the interactional consequences of using different problem-description formats, namely requests, questions, or narratives. The three formats have different interactional consequences for the talk that follows. Leppänen also focuses on the following interactional issue: How do callers present their 'problem' as serious enough to be given attention from the medical institution (for instance, as a problem requiring examination by the doctor)? This concern seems to motivate many callers to present their conditions as deteriorating. But, as Leppänen shows, there is a possible risk in the caller being seen as too focussed on the problem, in which case the nurse may question the caller's veracity. Therefore callers are faced with a second issue: How do I present myself as a trustworthy witness of the problem I describe?

Landqvist's chapter details ways in which advice-giving is achieved in calls to the Swedish Poison Information Center (PIC). Landqvist concentrates on the seriousness of the incidents, and the callers' demonstrated willingness (or lack thereof) to co-operate. Landqvist demonstrates that the caller in the calls to the PIC is not merely a passive recipient of advice. Rather, s/he contributes actively in the construction of advice. This active co-construction of advice is shown in the way the caller, during advice-giving, contributes information that makes the pharmacist interrupt his advice and modify it so as to fit the re-evaluated situation. This finding resonates with the 'calibrating for competence' phenomenon identified by Baker, Emmison and Firth (this volume). More than one advice sequence normally appears in a call. The prototypical position for the main advice is after the interrogative series and the temporary break, during which the pharmacist consults a database or archive. In many calls where there is only a slight risk of poisoning, this is also the place where advice is first offered. Where the risk of intoxication is greater, a first instance of advice is often positioned in the interrogative series and before the break. Another common advice sequence in the less severe calls is what Landqvist calls 'just-in-case advice', where the pharmacist explains what to do if the incident turns out to be more serious than expected. Giving just-in-case advice seems to be a way for the professional and the institution to handle possible liability issues. This sequence is normally positioned after the main advice, at the very end of the call, just before the closing sequence.

Part IV: Consumer assistance

Chappell's study examines interaction in an accommodation bureau that caters for the tourist consumer market. The study focuses particularly on the bureau's mediating work in providing assistance. This involves receipting tourists' requests in face-to-face exchanges, and making telephone calls to potential accommodation providers. In her analysis of these exchanges, Chappell shows how formulations and accounts are critical interactional resources in meeting tourists' requirements. 'Formulations' summarize an informant's earlier statement; 'accounts' are explanations for contextually unwanted occurrences. Chappell identifies the locations in which formulations and accounts play a role in problem-solving and negotiation work. These devices are deployed by bureau personnel at strategic moments in the interaction, notably at the interface of the closure of the call with the accommodation provider and the continuing face-to-face interaction with the tourist. Drawing on foundational work by Jefferson and Lee (1981), Chappell demonstrates that accommodation bookings differ in the degree to which they might be termed 'troubles-tellings' and 'problem-solving' service encounters and identifies the location at which interlocutors' perceptions of the ongoing talk change from 'troubles-telling' to 'problem-solving'. It appears that bureau personnel view the encounter as a problem-solving task only where difficulties arise when seeking suitable accommodation. For the tourist, however, it may continue to be viewed as a 'trouble', in that their role changes to one of 'troubles-recipient' when problems arise.

Torode's chapter provides a single-case analysis of a telephone complaint made to the Office of Consumer Affairs in Dublin, Ireland. The recent debate concerning the ability of conversation analytic work to adequately document the precise characteristics of 'institutional talk' provides the context for Torode's study. In contrast to those who maintain that institutional talk can be identified by the presence of certain – non conversational – turn-taking and turn-organizational features, Torode argues that the institutionality of any talk is best described through the idea of 'language games', most famously exemplified in Wittgenstein's work but more recently in the writings of Garfinkel (1967). In the examination of the complaint call Torode identifies a series of different 'games' played between the caller and consumer affairs advisor in relation to how the caller's problem (a pair of defective working boots) can be rectified. Torode maintains that the 'institutional' and the 'everyday' should not be seen as separate 'species' of talk but rather as a continuum which are invoked or appealed to in the various language games comprising the overall call. In this way, both 'everyday' and 'institutional' social structures – feelings, entitlements, rights and redress – are accountably accomplished.

Part V: Aspects of call management

One of the constituent features of almost all helpline interaction is a requirement for the exchange of information. That is, before any help or advice can be offered, call-takers must record or document much in the way of routine or preliminary information: callers' names, their telephone or fax numbers, the type of software they are experiencing problems with, the names or descriptions of chemicals accidentally ingested and so on. In most cases there is an organizational requirement that such matters be accurately recorded. Murtagh's paper, dealing with calls made to a UK mobile phone help centre, places the investigation of the ways in which such 'instructions transfers' are given and acknowledged as its central concern. In doing so he focuses in depth on the properties of the instruction/receipt adjacency pair. Instruction giving is characteristically a sequential phenomenon which requires several turns for its full completion. Murtagh finds that instruction transfer is very much a collaborative matter; typically instruction givers (for example callers parsing mobile phone numbers) will not complete the delivery of their number unless a receipt of each of its component units is heard. Of course, instruction transfer is not always successful, but situations in which ambiguity and misunderstanding arise do not pose major interactional problems, for instruction sequences readily lend themselves to the inclusion of 'repair slots' which are locally sensitive to the precise conversational environment in which they are invoked.

The theme of collaboration is continued in the chapter by Whalen and Zimmerman, whose study examines calls to an emergency dispatch centre. The authors examine ways in which call-takers collaborate informally, 'behind the scenes', during interaction with callers, and as such work as a *team* of call-takers. The focus of their chapter, then, is the collective work undertaken by the various personnel who manage calls at an emergency dispatch call centre. Typically in this setting, although only one organisational member may be conversationally engaged with a caller, the other co-present personnel will contribute in closely coordinated ways to the processing of the call. Whalen and Zimmerman argue that the work of emergency telecommunications cannot be adequately grasped unless this 'multiparty' involvement is understood. The 'ecology' of the dispatch centre, with its close alignment of consoles and work stations permits, and indeed requires, the almost continuous mutual visual and aural monitoring of each other's conduct. Whalen and Zimmerman describe how the dispatch centre personnel artfully oscillate between their 'not-work' and 'work' actions with the former providing a unique interactional scaffolding upon which the latter activities can be assembled.

Notes

1. Notably Whalen and Zimmerman's work on '911-emergency' services in the US, e.g. J. Whalen and Zimmerman (1987), M. Whalen and Zimmerman (1988), M. Whalen and Zimmerman (1990), Zimmerman (1992), Frankel's work (1989) on calls to poison-control centres, and Torode's research on calls to a consumer rights helpline, e.g. Torode (1995).

2. This view of *helping* – as being something that is situationally defined and interactionally accomplished – follows ethnomethodological emphases on the occasioned and achieved nature of social phenomena, and the concomitant methodological requirement to describe and explicate *how* social phenomena are, in practice, situationally defined and interactionally accomplished. The original source of ethnomethodology is Garfinkel (1967). For overviews of ethnomethodology, see Heritage (1984), Livingston (1987), Firth (1995b).

3. Telephone *emergency* services are *mediated* services in that normally *help* – in the form of police, paramedics, the fire service, or coastguard – is dispatched on the basis of information provided by the caller. In the majority of helplines examined in this book, and discussed in this particular chapter, help or support is provided directly, *over the phone*, through the medium of speech.

4. Many helplines also provide additional help in the form of written materials (brochures, etc.), which are posted to callers following helpline interaction.

5. It should be noted that Sacks (1992 Vol. 1: Lecture 1) does not use the term 'helpline'. In the lecture he talks of "telephone conversations collected at an emergency psychiatric hospital" (op. cit.: 3). Nevertheless he implies that callers are in search of some kind of help or assistance. Thus "[Callers] can be either somebody calling about themselves, that is to say in trouble in one way or another, or somebody calling about somebody else" (ibid.).

6. 'Naturally-occurring' means that the interaction has arisen for 'natural' reasons (rather than, say, for experimental, researcher-induced purposes) and as such its occurrence is not conditional upon, nor caused by, the interests or presence of the researcher.

7. *The Economist*, May 14, 1998.

8. These figures are taken from http://www.callcentres.com.au, July, 2004.

9. This information is extracted from The Helpline Association website (http://www.helplines.org.uk) in October, 2004. In addition to the 1,200 helplines covered by *THA* there are over five hundred other telephone help services ('hotlines', 'crimelines', commercial helplines, etc.) in the UK.

10. Taken from the NHS Direct website (www.nhsdirect.org.uk).

11. The record number of calls in one day was set on 26th December 2003, when 28,000 calls were made to *NHS Direct*, an increase of 5% compared with the same day the previous year (reported in *The Guardian* newspaper, 29 December, 2003).

12. The Helpline Association: http://www.helplines.org.uk

13. It is also important to mention here that the term 'help' is best seen as differentially serving the interests of both the caller and the companies or institutions providing the helpline service. For example, the call to a city council helpline may resolve the caller's specific problem regarding planning permission, while also fulfilling important bureaucratic and public-relations functions for the city council providing the helpline.

14. Call-takers may not, strictly speaking, be completely anonymous, but it is our experience that exceptionally few helplines operate where call-takers disclose their full names or other distinctive markers of their individual identity. In most cases, it appears, helpline call-takers will use their first names only, when engaging with callers, and will resist callers' requests to reveal their full names or other identity markers.

15. http://www.helplines.org.uk

16. The five-minute time period is a 'rule-of-thumb' guideline, used by Microsoft's call-takers, who are trained to assess, at very early stages of a call, the amount of time approximately required in order to resolve a caller's problem. If the estimation is that the call (problem) will demand more than approximately five minutes' call time, then call-takers are expected to inform the caller that the call will be 'chargeable'. This leads to some interesting interactional issues, as described in Emmison, Firth and Baker (in prep.).

17. In this particular case, Microsoft was operating a different payment policy from the free 'first-tier' arrangement already described. In the data segment reproduced here, callers were transferred (by switchboard-operators-cum-quasi-technicians) to technicians *free-of-charge* if it appeared that the caller's problem was due to either installation difficulties or a computer 'bug' or 'virus'. If callers' problems were deemed to be caused by other factors (for example the caller's lack of 'software competence'), then payment was obtained by the switchboard operators, prior to transfer to a technician. Where callers were transferred free of charge, technicians were faced with the task of assessing whether the actual problem was in fact due to one of the two factors that warrant a free call. If it subsequently proved not to be the case that either a virus/bug or installation were the causes of the caller's problem, then the call-taker was obliged to reveal this to the caller, at which point payment for the call was required. This is what we see in operation in the data segment. Microsoft (Sydney, Australia) have now abandoned this payment arrangement and replaced it with the free 'first-tier' format. Needless to say, the issue of payment remains a live and delicate issue – for both callers and call-takers.

18. Cameron (2000: 93) reports that there were 5,000 call centres in the UK in 1998. Niels Kjellerup, editor of http://www.callcentres.com.au, estimates that by 1999 Australia had between five to six thousand call centres.

19. Whalen and Vinkhuysen (2000) also describe the (negative) impact of a requirement that call-takers follow a 'script' when diagnosing callers' technical problems with photocopiers (although it is unclear whether the imposition was made for cost-cutting or work-

quality reasons). The authors show how call-takers commonly deviate from the strict 'script' format when attempting to diagnose callers' problems. The irony here is that, while the scripts were developed for the purpose of improving the levels of service and competence of the call-takers, call-takers broke the institutional rules by deviating from the scripts *in order* not to appear incompetent.

20. This information is included in the training manual of a large Australian helpline dealing with mobile telephony.

21. See the articles on problems with call centre staff turn-over and employee 'burn out', on the helpline managers' website http://www.callcentres.com.au

22. *The Guardian*, 4 July, 2001.

23. Schegloff's (1986) work on openings of casual conversations in the US posits a four-sequence structure as the 'canonical opening'. The four-sequence structure consists of (1) a summons-answer sequence, (2) an identification/recognition sequence, (3) a greeting sequence, and (4) an initial inquiry sequence (e.g. 'How are you?'). In our work on Microsoft's technical support helpline (see Baker, Emmison, & Firth 2001), we describe how Schegloff's 'canonical' sequence is radically modified in the helpline in question. Elsewhere (Firth, Emmison, & Baker in prep.), we show that 'how-are-yous', if they do occur in helpline calls, emerge in so-called 'dead-air' segments of the calls; for an explanation of 'dead air', see Note 24.

24. Research undertaken by Firth, Emmison and Baker (in prep.) on Microsoft's technical support helpline discovered that such 'pleasantries' (typically on the topic of the weather) occur in conversational environments that may appear marked or unusual in 'casual-conversational' calls. In the technical support calls, such pleasantries invariably occur during what call-takers refer to 'dead air' moments, that is, during phases of interaction where the interactants are waiting for the computer to be rebooted, or whilst software programmes are being reloaded or re-installed. In these cases, 'work-in-progress', relating to the caller's problem, has been temporarily suspended. During such 'dead air' moments, pleasantries frequently arise, initiated by either caller or call-taker. They were not seen to arise at other times during calls – for example in closings or call openings.

25. An obvious exception to the requirement that a casual 'conversational-footing' be established are emergency-service calls, which are largely devoid of 'conversational' characteristics. See Zimmerman (1992), Whalen and Zimmerman (this volume).

References

Atkinson, J. M. & Heritage, J. (Eds.). (1984). *Structures of Social Action: Studies in Conversation Analysis*. Cambridge: Cambridge University Press.

Atkinson, J. M. & Drew, P. (1979). *Order in Court: The Organisation of Verbal Interaction in Judicial Settings*. London: Macmillan.

Baker, C., Emmison, M., & Firth, A. (2001). 'Discovering order in opening sequences: calls to a software helpline'. In A. McHoul & M. Rapley (Eds.), *How to Analyse Talk in Institutional Settings: A Casebook of Methods* (pp. 41–56) London: Continuum.

Bell, D. (1976). *The Coming of Post-Industrial Society: A Venture in Social Forecasting.* Harmondsworth: Penguin.

Böhme, G. & Stehr, N. (1986). *The Knowledge Society.* Dordrecht: Kluwer.

Boden, D. (1994). *The Business of Talk: Organizations in Action.* Cambridge: Polity Press.

Boden, D. (1995). 'Agendas and arrangements: everyday negotiations in meetings'. In A. Firth (Ed.), *The Discourse of Negotiation: Studies of Language in the Workplace* (pp. 83–100). Oxford: Pergamon.

Boden, D. & Molotch, H. L. (1994). 'The compulsion of proximity'. In R. Friedland & D. Boden (Eds.), *Now Here: Space, Time and Modernity* (pp. 257–286). Berkeley: University of California Press.

Cameron, D. (2000). *Good to Talk.* London: Sage.

Castells, M. (1996). *The Rise of the Network Society.* Oxford: Blackwells.

Clayman, S. E. (1988). 'Displaying neutrality in television news interviews'. *Social Problems, 35*, 474–492.

Clayman, S. E. & Maynard, D. W. (1995). 'Ethnomethodology and conversation analysis'. In P. ten Have & G. Psathas (Eds.), *Situated Order: Studies in the Social Organization of Talk and Embodied Activities* (pp. 1–30). Washington, DC: University Press of America.

Clayman, S. & Heritage, J. (2002). *The News Interview: Journalists and Public Figures on the Air.* Cambridge: Cambridge University Press.

Drew, P. & Heritage, J. (1992). 'Analysing talk at work: an introduction'. In P. Drew & J. Heritage (Eds.), *Talk at Work: Interaction in Institutional Settings* (pp. 3–65). Cambridge: Cambridge University Press.

Drew, P. & Heritage, J. (Eds.). (1992). *Talk at Work: Interaction in Institutional Settings.* Cambridge: Cambridge University Press.

Drew, P. & Sorjonen, M.-l. (1997). 'Institutional dialogue'. In T. A. van Dijk (Ed.), *Discourse as Social Interaction: Discourse Studies 2 – A Multidisciplinary Introduction* (pp. 92–118). London: Sage.

Emmison, M. (1988). 'On the interactional management of defeat'. *Sociology, 22*, 233–251.

Emmison, M., Firth, A., & Baker, C. (in prep.). 'Calling for help, charging for support: Some features of the negotiation of payment on a commercial software helpline'.

Farberow, N. L. & Shneidman, E. S. (1961). 'A survey of agencies for the prevention of suicide'. In N. L. Farberow & E. S. Shneidman (Eds.), *The Cry for Help* (pp. 136–149). New York: McGraw-Hill.

Firth, A. (1995a). 'Ethnomethodology'. In J. Vershueren, J.-O. Östman, & J. Blommaert (Eds.), *Handbook of Pragmatics* (pp. 269–278). Amsterdam: Benjamins.

Firth, Alan (Ed.). (1995b). *The Discourse of Negotiation: Studies of Language in the Workplace.* Oxford: Pergamon.

Firth, A., Emmison, M., & Baker, C. (in prep.). 'After "How may I help you": problem descriptions in calls to a software helpline.'

Fox, R. (1968). 'The Samaritans'. In H. L. P. Resnik (Ed.), *Suicidal Behaviors: Diagnosis and Management* (pp. 405–417). Boston: Little Brown.

Frankel, R. M. (1989). "'I wz wondering uhm could *Raid* uhm effect the brain permanently d'y know?": some observations on the intersection of speaking and writing in calls to a poison control center'. *Western Journal of Speech Communication, 53*, 195–226.

Garfinkel, H. (1967a). *Studies in Ethnomethodology.* Englewood Cliffs, NJ: Prentice-Hall.

Garfinkel, H. (1967b). 'Practical sociological reasoning: Some features in the work of the Los Angeles Suicide Prevention Center'. In E. S. Shneidman (Ed.), *Essays in Self-destruction* (pp. 171–287). New York: Science House.

Garfinkel, H. (Ed.). (1986). *Ethnomethodological Studies of Work*. London: Routledge & Kegan Paul.

Goffman, E. (1979). 'Footing.' *Semiotica, 25*, 1–29.

Goldberg, J. (1975). 'A system for transfer of instructions in natural settings.' *Semiotica, 14*, 269–296.

Greatbatch, D. (1988). 'A turn-taking system for British news interviews'. *Language in Society, 17*, 401–430.

Greatbatch, D. & Dingwall, R. (1997). 'Argumentative talk in divorce mediation sessions'. *American Sociological Review, 62*, 151–170.

Have, P. ten (1999). *Doing Conversation Analysis: A Practical Guide*. London: Sage.

Have, P. ten (2001). 'Applied conversation Analysis'. In A. McHoul & M. Rapley (Eds.), *How to Analyse Talk in Institutional Settings: A Casebook of Methods* (pp. 3–11). London: Continuum.

Heritage, J. (1984). *Garfinkel and Ethnomethodology*. Cambridge: Polity Press.

Heritage, J. & Greatbatch, D. (1991). 'On the institutional character of institutional talk: the case of news interviews'. In D. Boden & D. Zimmerman (Eds.), *Talk and Social Structure* (pp. 93–137). Cambridge: Polity Press.

Hester, S. & Francis, D. (2000). 'Ethnomethodology, conversation analysis and "institutional talk"'. *Text, 20*(3), 391–413.

Hopper, R. (1992). *Telephone Conversation*. Bloomington: Indiana University Press.

Hutchby, I. (2001). *Conversation and Technology: From the Telephone to the Internet*. Cambridge: Polity Press.

Hutchby, I. (1996). *Confrontation Talk: Argument, Asymmetries and Power on Talk Radio*. Hilldale, NJ: Erlbaum.

Hutchby, I. & Wooffitt, R. (1998). *Conversation Analysis: Principles, Practices and Applications*. Cambridge: Polity Press (UK and Europe), Blackwell Publishers Inc.

Hymes, D. (1974). *Foundations in Sociolinguistics*. Philadelphia: University of Pennsylvania Press.

Jefferson, G. & Lee, J. R. E. (1981). 'The rejection of advice: managing the problematic convergence of a "troubles telling" and a "service encounter"'. *Journal of Pragmatics, 5*, 399–422.

Lester, D. & Brockopp, G. W. (Eds.). (1973). *Crisis Intervention and Counseling by Telephone*. Springfield, IL: Charles C. Thomas.

Livingston, E. (1987). *Making Sense of Ethnomethodology*. London: Routledge & Kegan Paul.

Lofland, L. H. (1973). *The World of Strangers*. New York: Basic Books.

Luff, P., Hindmarsh, J., & Heath, C. (Eds.). (2000). *Workplace Studies: Recovering Work Practice and Informing Systems Design*. Cambridge: Cambridge University Press.

Luke, K. K. & Pavlidou, T.-S. (Eds.). (2002). *Telephone Calls: Unity and Diversity in Conversational Structure across Languages and Cultures*. Amsterdam: John Benjamins [Pragmatics & Beyond New Series 101].

Maynard, D. W. (1984). *Inside Plea Bargaining: The Language of Negotiation*. New York: Plenum.

Maynard, D. W. (2003). *Bad News, Good News: Conversational Order in Everyday Talk and Clinical Settings*. Chicago: University of Chicago Press.

Maynard, D. W. & Clayman, S. E. (1991). 'The diversity of ethnomethodology'. *Annual Review of Sociology, 17*, 385–418.

McHoul, A. (1978). 'The organization of turns at formal talk in the classroom'. *Language in Society, 7*, 183–213.

McHoul, A. W. (1982). *Telling How Texts Talk: Essays on Reading and Ethnomethodology*. London: Routledge & Kegan Paul.

Peräkylä, A. (1995). *AIDS Counselling: Institutional Interaction and Clinical Practice*. Cambridge: Cambridge University Press.

Psathas, G. (1995). *Conversation Analysis: The study of Talk-in-Interaction*. Thousand Oaks: Sage [Qualitative Research Methods 35].

Rose, N. (1999). *Governing the Soul: Shaping of the Private Self*. London: Free Association Books Ltd.

Rutter, D. R. (1984). *Looking and Seeing: The Role of Visual Communication in Social Interaction*. London: John Wiley.

Sacks, H. (1967). 'The search for help: no one to turn to'. In E. S. Shneidman (Ed.), *Essays in Self-destruction* (pp. 203–223). New York: Science House.

Sacks, H. (1992). *Lectures on Conversation*. 2 vols. Edited by Gail Jefferson with introductions by Emanuel A. Schegloff. Oxford: Basil Blackwell.

Schegloff, E. A. (1968). 'Sequencing in conversational openings'. *American Anthropologist, 70*, 1075–1095.

Schegloff, E. A. (1979). 'Identification and recognition in telephone conversation openings'. In G. Psathas (Ed.), *Everyday Language: Studies in Ethnomethodology* (pp. 23–78). New York: Irvington.

Schegloff, E. A. (1986). 'The routine as achievement'. *Human Studies, 9*, 111–152.

Schegloff, E. A. & Sacks, H. (1973). 'Opening up closings'. *Semiotica, 8*, 289–327.

Schegloff, E. A. (1987). 'Between macro and micro: contexts and other connections'. In J. Alexander et al. (Eds.), *The Micro-Macro Link* (pp. 207–234). Berkeley and Los Angeles: University of California Press.

Schegloff, E. A. (1992). 'On talk and its institutional occasions'. In P. Drew & J. Heritage (Eds.), *Talk at Work: Interaction in Institutional Settings* (pp. 101–134). Cambridge: Cambridge University Press.

Torode, B. (1995). 'Negotiating "advice" in a call to a consumer helpline'. In A. Firth (Ed.) (1995b), *The Discourse of Negotiation: Studies of Language in the Workplace* (pp. 345–372). Oxford: Pergamon.

Turner, R. (Ed.). (1974). *Ethnomethodology: Selected Readings*. Harmondsworth: Penguin.

WHO (2002). *HIV/AIDS Counselling, Just a Phone Call Away: Four Case Studies of Telephone Hotline/helpline Projects*. Unaids, World Health Organization, Geneva.

Walther, J. B. (1996). 'Computer-mediated communication: impersonal, interpersonal, and hyperpersonal interaction'. *Communication Research, 23*(1), 3–43.

Whalen, J., Zimmerman, D. H., & Whalen, M. R. (1988). 'When words fail: a single case analysis'. *Social Problems, 35*, 333–362.

Whalen, J. & Zimmerman, D. H. (1998). 'Observations on the display and management of emotion in naturally occurring activities: the case of "hysteria" in calls to 9-1-1'. *Social Psychological Quarterly, 61*, 141–159.

Whalen, J. & Vinkhuyzen, E. (2000). 'Expert systems in (inter)action: diagnosing document machine problems over the telephone'. In P. Luff, J. Hindmarsh, & C. Heath (Eds.), *Workplace Studies: Recovering Work Practice and Informing Systems Design* (pp. 92–140). Cambridge: Cambridge University Press.

Whalen, M. R. & Zimmerman, D. H. (1987). 'Sequential and institutional contexts in calls for help'. *Social Psychology Quarterly, 50*, 172–185.

Whalen, M. R. & Zimmerman, D. H. (1990). 'Describing trouble: practical epistemology in citizen calls to the police'. *Language in Society, 19*, 465–492.

Zimmerman, D. H. (1992). 'The interactional organization of calls for emergency'. In P. Drew & J. Heritage (Eds.), *Talk at Work: Interaction in Institutional Settings* (pp. 418–469). Cambridge: Cambridge University Press.

Part I

Technical assistance

Chapter 2

Calibrating for competence in calls to technical support*

Carolyn Baker, Michael Emmison and Alan Firth

1. Introduction

This chapter discusses the ever-present consideration in answering telephone calls for assistance on a software technical support helpline, namely that of assessing and responding to the diversity of knowledge and competence that callers display when they present their computer software problems. Since anyone can purchase the software, anyone can call for assistance and this means that any next caller might be vastly different from any prior caller not only in the problem presented, but also in the fluency and precision of their description of the problem. Almost by definition, a call to this technical support service indicates that the caller is experiencing difficulties; and the very presence of an advertised technical support telephone number indicates to the purchasers/users of the software that in their work with the software they may encounter problems. Encountering some difficulties is therefore understood by the helpline call-takers (hereafter CTs) and by callers to be a normal kind of thing to happen and that it can happen to anyone. A caller, then, may take it that a call for technical assistance is *not* in itself an indication of incompetence.

Yet the competence of the caller is a consideration for the technicians who answer the calls. Occasionally, a caller is made a source of amusement within the workplace. Although there is little spoken interaction between CTs during their work shifts, the ecology of their open-plan work environment means that CTs are easily observable by their colleagues. During our fieldwork observations at the research site we have witnessed technicians conspiratorially making silent fun of incompetent callers and the frustration of dealing with them through exaggerated facial expressions, by pacing impatiently or perhaps kneeling on the floor; by silently pounding the desk, or by making pretend strangling actions, and so on. None of this is available to the callers, of course. It is worth noting at the outset that some of the calls made by these less competent callers are very long – a few in our corpus last for over three hours – and that the CTs handle all of these with the patience, calm and friendliness to callers that also characterise much shorter calls.

However, our interest is not in the fun or frustration that an incompetent caller can generate in the workplace. Our interest is in explicating the more mundane activity of what the organisation describes as "talking at the client's level", which means adjusting the way a call is dealt with according to the heard competence of the caller. We are interested in unpacking this gloss, and do so by examining transcribed segments of calls to show what this CT activity consists of, interactionally. We show that this form of "recipient design" (Sacks & Schegloff 1979) attends to much more than the technical competence of the caller – it includes hearing and responding to 'identity matters' (lay and expert), to a preferred pace of information-giving and receiving, and even to 'extraneous' considerations such as what is at stake for both caller and CT in trying to resolve the problem (cf. Johnson & Kaplan 1980).

The focus of this chapter, then, is the ways in which various dimensions of competence are displayed by caller and CT, including both technical competence and what we designate as social-interactional competence. Our use of 'competence' as the key term rather than, say 'knowledge' is meant to imply 'knowledge-in-use', 'know-how' rather than a repository of information held by either party to the talk. Our analysis is of how the work of the call is accomplished. We show how the technical competence of the caller is displayed to the CT and then oriented to, and accommodated for, by the CT. This is one aspect of this project's interest in the social organisation of expert-lay relations. Another dimension of competence that we attend to is the social-interactional competence of both caller and CT to hear and make adjustments to what the other is saying about the problem and indirectly about each other's understanding of the problem.

2. Accounting for the call

An additional dimension of caller competence that we do not focus on in this paper is the ways in which callers account for the call in the first place. In a study of the openings of these calls (Baker, Emmison, & Firth 2001), we show that callers routinely account for the call in at least two ways. First, they describe the particular problem that is occurring (the reason for the call). Second, they describe the efforts they have made to sort the problem out for themselves prior to calling for technical assistance. Thus, in the first place, they routinely present themselves as competent *users of a technical assistance resource.* More specifically, they have demonstrated a willingness to attempt to help themselves, prior to asking for help from a technician.

With remarkable regularity, callers use a format with the following components:

> [I have installed/I am working with (software) product X]
> and/or [I'm trying to do Y]
> followed by [and/but]
> followed by [something is happening that should not happen]
> and/or [something is not happening that should happen]

We have observed that this format shows the caller's orientation to the fact that the CT might be asked about any one of many products and, within any one product, a vast number of locations. Thus the callers specify the focus of attention for the CT in their very first turns. Their description of the problem is usually infused with commentaries on what the caller has tried to do in the way of diagnosis *prior to* calling, often in the past progressive tense as in "I've been trying to…". Hence a further dimension of competence with the computer software service that callers display is the knowledge of *when* a call to technical support would be heard as warranted – that is, after they have done some double-checking or even detective work on their own. Such behaviour is common is everyday life, of course. For example, if one woke in the morning with a (mild) pain in the shoulder, one would not (normally) hurry to the doctor; one would wait to see whether the pain continued or disappeared before asking for help (see, e.g., Leppänen this volume, data Extract 2, lines 2–3, where the patient, calling a medical helpline, reports on having a headache and feeling dizzy 'all day', prior to making the call). Likewise, callers to the software support service indicate that they are calling not at the very first hint of a problem, but after they have established that the problem is, indeed, stable or recurring. Callers are obliged to pay for the technical assistance they may receive, so it is doubly commonsensical not to telephone before establishing the necessity to do so.

We have noted also that their accounts of 'why this call now' make reference to the time that has passed since the problem has been detected and established as an outstanding problem rather than a quirk. Callers use such terms as 'today', 'now', and 'suddenly', which has the effect of representing them as regular and serious users of the technology who need their software to be working all the time. Thus most callers open their inquiries such that they are hearable *as competent users of the technical assistance resource*. It is almost taken for granted that callers need to be at their 'problem machines' when they make the call – anticipating the work that might be done for and with them. We have one example of a caller who was not at his computer when his call was transferred to a technician (he was making his call while driving a car), and he was advised to call back when he was "in front of the computer" where the problem resided.

Thus, competence as a user of the technical support service is normally established in the opening phase of calls, and provides the fundamental grounds upon which the rest of the call can proceed. The accounting done by callers is predominantly a social rather than technical matter, although the details they provide about what they may have tried to do to fix the problem themselves may contribute technical information to the CT. Although this competence, *to know how to describe why one is calling with this problem, now*, is not the main focus of attention in this chapter, it is nevertheless an element of caller competence that is, (possibly) 'heard but unnoticed' by CTs.

3. Calibrating for competence

Technical competence

In this chapter we work with the three extended transcripts and trace the trajectory of the interactional display and recognition of levels of technical competence with the computer or computer software. Our first sense of 'competence', then, is technical competence as displayed to and heard by the CT. Without a deep knowledge ourselves of the technical details of how software programmes work, we cannot be certain how competent some caller-description of a problem might sound to a technician. And we are unable to be certain about how technically-difficult-to-solve the problems are (that is, simple or complicated problems). However, we can observe the CT's apparent ease in handling the technical problem, the speed with which the CT arrives at a diagnosis, the CT's need to consult screens or data bases, or, on the other hand, the ability to solve the problem without reference to such tools. This, of course, does not equate to how difficult the problem is, since the CTs are themselves specialists. The difficulty of the problem needs to be seen as relative to the analytic skill and/or experience of the technician. In our analyses, then, we work with a lay notion of 'technical competence', drawing on our knowledge as moderately competent software users, and observing how the technician responds to the caller, compared to responses throughout our corpus of recorded calls.[1]

Social-interactional competence

A second dimension of competence that we attend to in this chapter is social-interactional. This encompasses a range of activities that have been studied within ethnomethodological Conversation Analysis (CA) (e.g. Hutchby & Wooffitt 1998; Psathas 1990) that are deployed in the sequential organisation of talk-in-interaction. Within this dimension, we presume commensurate competence on the part of callers and CTs.

As a metaphor for the interactional work that we have observed in the corpus, we propose that CTs *calibrate for competence*: that is, they respond to the caller initially through first receipts of the caller's description of the problem, and then they make successive adjustments to how they speak with the caller depending on the perceived level of caller's competence that they hear in successive turns. This notion of calibration is meant to convey the CT's 'fine-tuning' as the interaction progresses. However we can also see in places a reciprocal process in which callers effectively send signals through their turns that invite (further) calibration-for-competence work by the CTs. This might be thought of not just as recipient design, but in some cases as a process of 'redesigning the recipient' – a notion that may have application beyond our corpus of calls to a technical support helpline.

The caller's first description of their computer/software competence

The work of diagnosing and resolving the caller's problem begins with the caller's initial description of the problem. As we have shown in our analysis of openings (Baker, Emmison, & Firth 2001), the CT's opening turn is very general – for example, "How can I help you?". This leaves it up to the caller to 'design' the way that the problem is presented. The work of diagnosing the technical problem is closely related to the work of assessing the caller's competence in describing it. That is, the resolution of the problem is both technical and social at the same time, and from the very beginning of the call.

In the following example (Extract 1) in lines 1 to 22 the caller presents himself as a consultant to other software users while he is, in this call, a client of the technical support service.

Extract 1

```
CT  how can I help you?
C:  er problem with a (.) system running er power point in actual
    fact uhm the customer's trying to do a pack an' go
    (1.0)
    a:nd
    (2.0)
    it's coming up with a message towards the end of it saying
    insufficient space
    (1.5)
    regardless whether we send it to the hard drive or the floppy
    disk, y'know it does not even check the floppy drive when it
    comes up with this error message
    (1.0)
    .hh also we cannot install office ninety seven again over the
    top .hh
    (1.0)
CT: you can't install it over the to:p?
C:  no it comes up with an el zed ((L.Z.)) thirty two dee el el
    ((D.L.L.) file is corrupt or damaged or missing
    (0.5)
CT: el zed? ((L.Z.))
C:  el zed thirty two? (.) dot dee el el ((L.Z.32.D.L.L.)) now
    this file resides in the windows system directory? I have
    removed it=renamed it also (.) taken it back off of the see
    dee ((C.D.)) .hh a:nd (.) put it back in the original
    location (.) it refuses to accept it
```

The CT listens without comment until line 17, even though there are interactional spaces in which she could have issued receipt tokens (e.g. at lines 4, 6, 9 and 13). This form of uninterruptive listening is common in the calls we have studied. In some openings, however, and as will be shown in the other two transcripts we analyse in this chapter, CTs do issue various receipt tokens such as 'mhmm' or 'yeah', during the caller's initial presentation of the problem. We might venture that the amount of receipt work is proportional to the heard needs of the client, and we intend to investigate this possibility in future publications. That is, some callers may 'ask' for feedback during their problem description more strongly than others through turn design and intonation. In the transcript above, the caller presents himself as having considerable software expertise himself. Yet he appears to have left spaces for CT feedback, but these spaces are unfilled. When the CT does speak, it is with a very strong uptake of the last point the caller made in lines 14 and 15. It seems as if the CT has found some important clue and she asks twice, at lines 17 and 21, for confirmation of the problem and the name of the error message. Beginning at line 22, the caller gives a further detailed account of what he has done prior to calling technical support, a description that presents him as having carried out appropriate or at least plausible efforts to fix the problem by himself.

Extract 2

```
    (1.0)
CT: okay I'm just having a look on the on the data base here to
    see exactly what that file i:s
    (1.0)
C:  .hh I did go on to the internet site yesterday and had a good
    look around to see if you had any information about it but
    there's nothing up there (.) through your knowledge base so:
    (4.0)
CT: okay
    (6.0)
CT: and this is power point ninety seven that you're using?
C:  yep power point ninety seven (.) now I have another system in
    the same building (.)exact specifications to this one (0.5)
    and it's actually wo:rking on it
CT: it's working okay is it?
C:  on the other machine yeah
    (2.0)
    it's just this one system it's not working on that I cannot
    get around this LZ thirty two DLL file
CT: that- LZ thirty two file comes up when: you reinstall?
    (2.0)
CT: [when you try      ] try to reinstall
C:  [when you run setup]
C:  yeah soon as you run setup it goes to the LZ thirty two DLL
    file
    (1.5)
    it appears to be an expansion (.) file something to do with
    the expanding of compressed files or something
    (3.5)
CT: thee- exact error that you're getting in power point when
    you're trying to pack and go?
C:  exact error is insufficient space
    (6.0)
    to do a pack and go it's just insufficient space it doesn't
    matter where we send it
    (1.0)
    floppy drive or hard drive
    (2.0)
CT: okay I'll have to-
    (3.5)
```

After CT, in line 28, announces that she needs to look up the file on the data-base, C provides yet more evidence of his know-how with software problems and their resolution (lines 31–33). Whether or not the 'knowledge-base' he refers to is the same base that the CT is currently looking up is not clear. However, we note that his reference to this activity is placed immediately after the CT's reference to looking on the helpline's data-base. We can conclude from this opening, then, that C presents himself as someone familiar with software and familiar with solving software problems, that is, as someone with technical skills .

These characterisations of his competence continue in response to various factual and diagnostic questions put by CT in lines 37 to 56. It is notable that the format of his answers to these questions is first to confirm the detail that CT is checking out or, at line 49, answering CT's question, and second, in each case to extend his answer to add more about his working knowledge of the problem. We see this in his answer turns beginning at lines 31, 38, 44 and 50. The caller uses as a contrast in presenting his problem, that he has another system in the same building with the exact specifications but which does not have this current problem. That is, his identity as a person with technical competence is threaded through his replies to diagnostic questions as well as in his opening description of the problem. CT accepts this extra information in line 41, but after that does not pick up on it again, as with the caller's proffered comments about what kind of file the problem seems to be resident (in lines 53–54). In fact, after a 3.5 second pause, she changes topic, and the caller elaborates again on what he has discovered through his own investigations. CT is presumably busy with her own diagnostic work, as she does not respond until line 65.

After a 3.5 second silence, at line 67 CT issues a further question in the form of a check on which product this is. At line 68 caller confirms that it *is windows ninety five*. This begins a sequence taken up primarily by the CT working on the keyboard, with no comment from the caller.

Extract 3

```
65  CT: okay I'll have to-
66      (3.5)
67      this is windows ninety five you're using?
68  C:  yes it i:s
69  CT: [okay
70  C:  [( )
71      ((keyboard sounds 9.0))
72      I'll just have a look on the data base (.) for that error
73      message
74      (17.0)
```

We arrive at a turning point in the call. After seventeen seconds of silence (line 74) CT begins another line of questioning: "*Now did you have a previous version of office on there?*" The caller's answer is definite: "*no we did not*" (see lines 75–76, below). Up to and including this point the caller has been definite about what he knows about the system he is calling about. CT pursues the matter by delving further into possibilities connected with the history of Office programs on the computer, which brings forth the first hesitations from the caller. This is also the first point where there are any hesitations or turbulence in the caller's issuing of questions. We note first that the CT pursues the matter of previous versions of Office despite the caller's answer. It seems that in lines 77–80 she is searching for words to put the question another way, as if she has tracked down the problem, and then she gives some suggestions for applications that might have been installed:

Extract 4

```
74       (17.0)
75  CT: Now did you have a previous version o:f (.) office on there?
76  C:  no we did not
77  CT: is it- (0.5) y- did you have a previous version of any of the
78      office applica:tions? (.)
79  C:  u[-
80  CT:  [wo:rd exce:l po:werpoint?
81      (2.5)
82  C:  I don't think so (1.0) these were new systems that were
83      suppli:ed um they had previous actually they woulda had a
84      previous version
85  CT: they would have?
86  C:  they woulda had yeah cause they used to run uh windows three
87      one one on their old network (.) an' then they purchased new
88      systems an' (.) also office ninety seven upgrade to go over
89      the top
90      (3.0)
91      so the data was transferred from their old system (.) to the
92      new (.) but I didn't actually think they transferred o: (.)
93      thee um (.) old office (.)[across
94  CT:                           [.hhh
95  C:  and then did- I- just put in the
96      diskette and told it to run it
97  CT: okay (.) um but you you're not too sure=
98  C:  =not [a hundred per cent=
99  CT:      [how they've done it
100 C:  yeah not a hundred per cent on how they actually (.) they ran
101     the installation
102     (2.0)
```

```
103    the original installation although I'm quite sure that it was
104    run exactly the same (.) on the other system
```

Line 82 is a crucial point in the conversation where the caller, in offering more knowledge about the system, 'discovers' a new possibility that CT had already proposed. "Actually", in line 83, marks a form of 'change of state of knowledge' token (cf. Heritage 1984) by the caller. The CT responds to this information with great interest: "*they would have*?" The CT has led the search for the solution to the mystery, and the search has led to a revision of the claimed breadth of knowledge of the caller. This revision is followed by an account by the caller of his possible error – "*but I didn't actually think…*" – in lines 92–93. Again he is threading his own narrative into the developing plot of what the problem is.

In the final few lines of this transcript segment, the CT picks up explicitly on the matter of how sure the caller is of his knowledge ("*but you're not too sure*", line 97), which the caller matches with the alternative assessment "not a hundred per cent sure". Note that this second assessment is overlapped, and is then *re-issued* by the caller, which ensures its inclusion in the story-so-far, via its hearability by the CT. The caller then finds an item about which he *is* "*quite sure*" to add to his account. This issue of how sure the caller is of his facts is probably important to CT only because it is part of the relevant technical information, the history of this particular software programme.

Our first extended transcript, then, shows a case where a relatively high degree of technical competence is demonstrated by the caller, and where the CT works with this relatively high degree of competence to resolve the problem. The CT in this case seems to focus upon the caller's statement that a prior version of Office had not been installed, which she does by a sideways move of suggesting various Office *applications*, which keeps the topic open without directly challenging the caller's position. In this first transcript we see 'calibration' around the matter of how sure the caller is of his facts (not really sure, one hundred per cent sure, quite sure), and we see this as an activity in which both participate. We also see in this example the caller *accounting for* the difference between what he initially knew and what he now concedes could be the case, which suggests that he takes it that the competence he initially displayed has come under some question. There is no equivalent evidence that his competence in this is any real problem or issue for the CT. She is not concerned with assessing his competence, only with estimating it as evidence with which to solve the problem. A significant part of her competence is to hear and respond to the technical information and to work with and around the caller's personal story, so to speak.

CT calibrations as orientation to the heard 'competence' of the caller

In the second transcript we encounter another form of calibration. In this case, the caller has made a previous call to the helpline relating to his particular problem. If callers need to call on more than one occasion regarding a given problem, they will likely not reach the same technician. Notes are made on each call such that any next technician that the customer reaches can find a written (screen-based) outline of the problem and proceed from there.

This call begins with CT's check of the customer number and her acknowledgement of the unresolved problem. The caller then reports the problem number, which is assigned when a problem cannot be resolved during a first call:

Extract 5

```
CT: Rob, can I have your customer number
C:  yes, three six nine, five eight one
    (2.0)
CT: okay Rob I'd just like to let you know that this call may
    be recorded for a quality (.) [control purposes
C:                                [that's not a problem=
CT: =o:kay, thank you very much .hh okay you've got a problem
    here with windows ninety eight
    (0.9)
C:  yes I certainly do:
CT: okay
C:  it's a continuation of problem number two two two
CT: yeah
C:  eight one double two
CT: okay
C:  I'm in all sorts a' bother
    (0.5)
CT: [uh y(h)e(h) a(h) h
```

In lines 10 and 16 the caller offers two successive self-descriptions that indicate his need for help. The first could be heard as a description of the technical problem *per se* ("*I certainly do* [have a problem]") but the second is hearable more as a description of his current condition ("*I'm in all sorts a' bother*"). These announcements at the beginning of the call are hearable as invitations for sympathetic treatment by the CT, and indeed we hear her make an affiliative receipt through laughter at 18. By reacting as she does here, CT is displaying her understanding of C's 'problem' as essentially non-serious. The CT, then, has already acknowledged the personal story that is woven into the call from the beginning.

Extract 6

```
18 CT: [uh y(h)e(h) a(h) h
19 C:  [.hh I've actually bin out of the loop in hospital for
20     some ti:me
21 CT: right
22 C:  I came back and em
23     (1.2)
24     what I did is I went through and I've (0.7) actually
25     (0.6) reloaded we- (.) I-I-I've formatted the disk
26 CT: yeah
27 C:  got rid of everything unfortunately (0.8) uhm (0.3)
28     reloaded
29     windows ninety five an' then I've (0.5) loaded windows
30     ninety eight
31     (0.8)
32 CT: y[eah
33 C:   [a:nd em what's happening today=I took- took the second
34     drive out I've only got the one drive in the machine now
35     (1.1)
36 C:  .hh an' what's happening now: (0.5) is (0.6) it's (0.5)
37     only loading up into safe mode=an' I get an unusual
38     message coming up which sez quit one or more .hh emri-
39     resident programmes or remove .hh unnecessary utilities
40     from your config sis or something=an' I can't- I haven't
41     got (.) time quite to read it
42 CT: ah okay uh is that- is that up in safe mode or is that in
43     normal mode that happens?
```

Note that the caller at line 19 offers an account for some lapsed time since his last call on this same technical problem – he has been "*out of the loop in hospital*". This may reflect an orientation to the matter of when a *second* or *next* call for assistance is accountably due. However, this reference to being in hospital is also hearable as part of the personal story. The sequence that follows looks like the enactment of perspective display (Maynard 1991) where as in transcript Extract 1, the CT grants the caller an extended turn in effect. In this case CT does provide some receipt tokens. In these turns the caller is giving an account of what he has recently done, presumably since his last call to technical support and since coming out of hospital. This is consistent with our analysis of the formatting of the openings of calls. The problem itself is not arrived until lines 34–38. The prior narrative is hearable as leading up to the point of the call – what is actually wrong. Notice also that there is an account embedded in this narrative for why this call *today* and *now*. It is clear that this caller is describing himself as 'out of his depth' with what is happening with his software. When we examine the design of his turn in lines 34–38 we sense that he describes himself as someone unable to understand or follow the software message, that "*sez quit ... from your config sis or something*":

> .hh an' what's happening now: (0.5) is (0.6) it's (0.5) only loading up into safe mode=an' I get an unusual message coming up which sez quit one or more .hh emri- resident programmes or remove .hh unnecessary utilities from your config sis or something=an' I can't- I haven't got (.) time quite to read it

He states that interpreting the message is too difficult for him to do by himself. At line 42 the CT begins with an "okay" to issue a question about where the problem occurs. The caller continues to explain what has happened with his software:

Extract 7

```
42  CT: ah okay uh is that- is that up in safe mode or is that in
43      normal mode that happens?
44      u:hm well it won't- load in ordinary mode now
45  CT: right ok[ay
46  C:          [it's going safe mode so- it was working okay
47      .hh uh (.) through most of the day
48  CT: yeah
49  C:  an' now all of a sudden this message has started to come
50      up an' I really don't know why
51      (0.2)
52  CT: okay, are you in safe mode at the moment?
53  C:  yeah=I'm just reloading in safe mode now, [yes
54  CT:                                           [okay (0.2)
55      just let me know when it's all loaded up
56  C:  okay thank you so what was your name?
57  CT: Jenny
58      (0.4)
59  C:  thanks Jenny
60  CT: that's okay
61      (4.0)
62  C:  it's just (.) loading in safe now:
63      (2.8)
64  C:  I donno what did- what happened to make it change its
65      mind, it was working quite well
66  CT: yeah what I'd say ah have you loaded all your software
67      programmes on there yet?
```

At line 42 the CT asks the caller whether the problem occurs in safe mode. The caller continues with a description of the problem and of his confusion in response to it: "*all of a sudden this message has started to come up an' I really don't know why*". At 52 the CT asks the caller whether he is in safe mode now. This question is not a reply to the caller's most recent description of his confusion. Instead it is straightforward and practical. The caller hears this question as the beginning of concrete help, since he reports that he is reloading. The CT asks him at line 55 to "*just let me know*" when he has completed the procedures. She has implied in this statement that she will wait until he is ready for a next step in resolving the problem. It is sensitive to the 'bother' that the caller feels himself to be in. It is a form of calibration for competence.

This seems to be heard by the caller as a turning point in the call, for we see the caller checking on the CT's name. We see this particular caller activity happening occasionally at turning points like this, or at points where the CT begins to suggest that they cannot go further together in the current call, and the caller requests the name of the CT indicating that they want to reach that same person again. In the latter case, callers are told gently that they may not reach the same CT again, but that notes are kept that any CT can proceed from. In this case, the check for the CT's name appears to anticipate lengthy or very important work about to happen for the caller. It comes immediately after the CT has implied that she will help this caller at his own pace. "*Thanks Jenny*" at line 59 following a brief silence is hearable as "thanks in advance for what you will do for me" more than "thanks for telling me your name". Jenny's reply, "that's okay", is then hearable as her further agreement to work further with this caller on his self-described terms. A kind of social contract seems to have been signed between a desperate caller and a kindly helper.

In line 66 the CT appears to begin giving an explanation in reply to the caller's musing about "*what happened to make it change its mind*", as if the computer has been a stable person who has suddenly begun behaving erratically. We submit that such *anthropomorphic* characterisations of the computer and the caller's relation with it display the caller's relative lack of sophistication in talking about the technology. However, the CT stops mid-turn to check that the current task of loading the software programmes has been completed, as if the caller needs to concentrate on one thing at a time:

Extract 8

```
66  CT: yeah what I'd say ah have you loaded all your software
67      programmes on there yet?
68  C:  ye:s
69  CT: yeah .hh what I'd say has happened is one of those
70      software programmes has put something in one of the
71      startup folders
72  C:  that it needn't do
73  CT: uh well not necessarily but that's just having problems
```

```
74        running? (.) so=
75   C:   =oh all right
```

C's turn at line 72 continues the anthropomorphism he attributes to the computer, but at the same time it can be heard as a possible 'bid' for the CT to upgrade his perceived level of competence by offering a candidate completion – albeit mistaken – to her diagnosis of the problem:

Extract 9

```
76   CT: when you start the computer these uhm .hh these
77        programmes
78        in the startup folder
79   C:   y[e:s
80   CT:   [all start up
81   C:   yeah
82   CT: and um an' more than likely that's causing the problem
83        an'
84        that's why the machine won't start
85   C:   oh I see
86   CT: see what safe mode does is that doesn't load anything (.)
87        any (.) like programmes running in the background or any
88        dri[vers or anything ] like that
89   C:      [r i g h t o k a y]
90   C:   [okay
91   CT: [so that's why you can start the safe mode but you c[an't
92   C:                                                         [all
93        right
94   CT: start the normal the normal mode
95   C:   okay it's just taking a while to load here Jenny
96   CT: that's okay it usually does when you start in s[afe mode
97   C:                                                   [yeah
98   C:   yeah
99        (4.8)
```

The CT's provision of an account of what went wrong, beginning at line 76, 'chunks' the information into several small packages each of which is receipted by the caller: yes, yeah, oh I see, right okay, all right. This is a form of turn calibration – measuring out the doses of information – that takes into account her hearing of the caller's competence to understand what the problem is. The sequence concludes with the caller commenting on the state of the loading process, "taking a while", which CT replies to with "*that's okay it usually does*", which is an assurance that this time taken is normal. This is followed by a further long silence after which the CT *herself* offers an account for the caller of the time it is taking, more assurance for the caller. Another measuring-out occurs soon after:

Extract 10

```
99         (4.8)
100 CT:  it always takes longer while you're sitting here waiting
101        for it doesn't it
102 C:   hm [hm
103 CT:      [h(h)u(h)h(h)um(h)m
104 C:   hm hm
105        (7.8)
106 C:   okay I'm just startin' to get a screen now?
107 CT:  that's okay
108        (2.6)
109 C:   okay now it sez windows is running in safe mode special
110        diagnostic mode enables you to fix the problem [that's
111 CT:                                                  [yeah
112        click
113 C:   keeping you from starting
114 CT:  okay on that message
115 C:   click okay?
116 CT:  yeah
117        (0.3)
118 C:   okay
119        (0.5)
120 C:   my icons should be coming up any minute now
121        (2.0)
122        okay the eye- desktop icons are here
123 CT:  okay .hh if you click on the start button
124        (0.2)
125 C:   ye:s?
126 CT:  and choose run?
127        (0.6)
128 C:   run
129 CT:  a:n' type in em ess ((M.S.)) con fig (.) that's em ess
130        ((M.S.)
131        see oh enn ((C.O.N.)) (.) e[ff eye gee ((F.I.G.))
132 C:                                [yeah
133 C:   okay I'll just take out what's in here (0.3)
134        em ess (.) no space?
135 CT:  no space, one word
136 C:   con, see oh enn ((C.O.N.))
137 CT:  yep
138        (0.2)
139 C:   eff eye gee? ((F.I.G.))
140 CT:  that's correct
141 C:   yes?
142 CT:  an' the:n click okay
143 C:   oka:y
144        (2.2)
```

In this instance of calibrating, what is measured out is instructions for what the caller should do. What we see here is quite common practice: A sequence of operations needs to be performed on the caller's screen, which of course the CT cannot see. It is only through the talk that the two speakers can ensure that the operations are being performed as required. In this case the CT issues the instructions one step at a time, and the caller replies in some way to indicate that each step is being attended to. There is a surface parallel in routine talk by pilot and co-pilot in preparing for take off. However, pilots need to check and double check each item in a standard list audibly, for the record, as a safety procedure, even though they are sitting side by side (Nevile 2001). In the case of calls to technical support, the CT's use of this step-check, step-check format for instructions is a contingent, situated use, and varies from call to call. We have called the very close checking format as used here with the caller, Rob, 'baby steps'. It is important to recognise how Rob effectively called on the CT to speak this way. See, for example, line 128, where Rob voices for her what he has just done on his computer: "run", meaning that he has chosen what he was designated to choose. At the same time, and prospectively, this is an indication that he will keep her informed as he performs each step, and that he wants to continue to proceed in this way.

In this call, the CT also puts the instructions in *exactly* the order that the caller will encounter them: that is, for example, instead of saying 'choose run' from the start menu, the caller is first asked to click on the start button, *then* choose run. This is similar to preformulations used in infants' classrooms to direct the attention of children to exactly where the solution to a question can be found (French & MacLure 1981). This is a discourse format *designed to bring the novice* to the correct location so that a competent performance can be attempted. In the case of this call, like many others, the caller participates fully in the discourse format by confirming or "okaying" every step in some way, so that the CT can be sure he has done it, rather than taking it for granted.

This, then, is another form of calibrating through turn design and turn placement, undertaken to take into account the heard level of competence of the caller – this time to work competently on the screen that the CT cannot see.

We have identified at least three orders of competence to which CTs and/or callers appear to be oriented: (1) competence as a user of the software technical support service, as shown in accounting for the call, particularly why this call now, after what other efforts have been made to resolve the problem; (2) competence with understanding or explaining the workings of programs, startup folders and other software dimensions; and (3) competence in making one's way around the keyboard with or without instructions, and if with instructions, how finely ordered the instructions need to be.

4. The contingent use of a pedagogical format

The third transcript we analyse in this chapter is a call that has been transferred to Dieter, the CT, from another telephone, which is why the transcript we work with begins at line 17. In the caller's first turn we observe again many of the features of openings that we have commented on previously. There is a narrative presented about the history of the problem, what the caller has tried to do, and an account for why this call is being made now:

Extract 11

17 CT: er good morning Hank is it?
18 C: yes Ha-
19 CT: my name's Dieter I'm from the operating system support team
20 how are you?
21 C: hello Dieter very good er erm I'm er having a lot of
22 difficulty with er (.) my pc getting a line to Optus internet
23 and I've had about three attempts through them after they
24 sent me a nice cd rom an' all that sort of stuff and I sa-
25 (.) put this cd rom in there follow the instructions and
26 you'll be on line with the internet well (.5) huh here we are
27 a couple of weeks later and we're not getting anywhere at all
28 CT: really? okay erm-
29 C: so basically today we've- I've been going through the
30 technician who in the end suggested I give you guys a call
31 because we have a problem apparently (.5) in adding erm to
32 the TCP slash IP protocol (1.0) and er when I get up for
33 example select network component type (2.0) and er I click on
34 the protocol box and try an' <u>add</u> (.) we just get exactly the
35 same screen coming back
36 (3.0)
37 CT: okay erm
38 (2.0)

The caller's problem description uses a mixture of technical vocabulary, for example "*adding erm to the TCP slash IP protocol*", sarcasm or colloquialism – "*they sent me a nice cd rom an' all that sort of stuff*", and a statement that he has not been successful in following the instructions: "*here we are a couple of weeks later and we're not getting anywhere at all*". The CT receives this description first with interest ("really?") and also twice with what could be a preliminary to starting some kind of diagnostic work ("okay um-"). The silences at lines 32 and 33 suggest that the Dieter has to do some thinking on how to approach the solution to this problem. At line 37, after what appears to be a moment's thought, the CT begins diagnostic questioning. The first question is about what was on the CD:

Extract 12

```
37 CT: okay erm
38     (2.0)
39 CT: let's see the cd that they sent you what was-what was on that?
40 C:  it's just a- just a int- inter- it's the Optus um(.)internet
41     er cd
42 CT: and that's- what has windows er internet explorer version 4
43 C:  correct yep yep (     ) s'got all the specs on the back of it
44     here
45 CT: sure (.) the operating system that you're using at the moment
46     what is that?
47     (1.5)
48 C:  .hhhhh
49 CT: windows 95?
50 C:  it's windows 95 yeh I don't know what version though
51 CT: okay no that's- that's no problem at all have you- you've
52     never connected to the internet?
53 C:  never- never before I was on li::ne (.) the only thing I was
54     on line to was fax via er (.) which I had to organise through
55     IBM because er we had er problems initially getting fax er
56     online and er calling IBM they ended up partitioning my hard
57     drive (.) but never been on the internet before
```

As in the call in transcript Extract 2, the CT here uses preformulations, as in "*the cd that they sent you what was- what was on that*?". This format first directs attention to some object and then asks a question about it. The answer in line 40 is halting and almost incoherent: "*it's just a- just a int- inter- it's the Optus um (.) internet er cd*". At line 42 the CT suggests what might be on the CD: "*and that's- what has windows er internet explorer version 4*". The CT issues another similarly-formatted question in line 45. This question is not answered, and the CT again offers a suggestion. During this diagnostic phase, then, the CT is learning as much about the caller's familiarity with the software and therefore technical competence, as he is about what software the caller is working with. In line 51 the CT guesses that the caller has not been on the internet before. His turn design suggests that he was beginning to ask: have you ever been on the internet? In mid-turn he then produces the alternative design, thus: "*have you- you've never connected to the internet*?". In these subtleties of turn design, we see quite clearly the CT calibrating for the caller's competence.

There is a further feature of the CT's interactional work with this caller that indicates calibration in a slightly different way. This involves the CT's use of an evaluation component similar to that used in formal instructional talk in classrooms (Mehan 1979). This consists of an assessment of the quality of a student's answer to a teacher's question. We observe this phenomenon starting up in the transcript segment above, and it continues immediately after the caller has said that he has "*never been on the internet before*":

Extract 13

```
CT: okay no that's- that's no problem at all have you- you've
    never connected to the internet?
C:  never- never before I was on li::ne(.)the only thing I was
    on line to was fax via er(.)which I had to organise through
    IBM because er we had er problems initially getting fax er
    online and er calling IBM they ended up partitioning my hard
    drive (.) but never been on the internet before
CT: excellent okay(1.)erm (2.5) let's see I just want to double
    check your er your settings can you go into control panel
    for me-
C:  >yeh won't be a second I'll just get rid of this other stuff
    here on the screen<
CT: sure
    (15.0)
C:  (get a reading) cd (.) rom=won't be a sec I'll just get rid
    of this
CT: sure no problem (4.0) how's your day been?
C:  please enter for okay (.) escape cancel
    (3.0)
CT: how's your day been Hank?
C:  not good hhchhh
CT: not good (       ) frustrating by the sounds of it
C:  very frustrating I'm- I don't profess to be the world's expert
    on pcs but I was- had no end of problems with this for some
    reason er please insert cd rom nnnnrrrr cd rom drive away
    please to escape okay I'll just get escape(1.5)still trying
    to get rid of the erm Optus screen here at the moment won't
    be a sec(8.0)come on(10.0)okay setti::ngs(.5) control panel
CT: that's the one er if you can (1.0) double click add remove
    programs for me
C:  won't be second just coming up now (4.0) double click add
    remove programs yeh correct
CT: that's the one (.) just tell me what version of internet
    explorer is currently listed er under installed components
    if you could
    (5.0)
CT: should be microsoft internet explorer
```

88 C: mic- yeh microsoft internet explorer version 4
89 (1.5)
90 CT: excellent okay then just cancel there for me if you could
91 C: 'kay
92 CT: er what I'll do is get you to er (1.0) double click the er
93 network icon for me
94 C: 'kay won't be a sec (5.0) okay figuration identification
95 access control
96 CT: okay if you can er starting at the top of that er list of
97 software can you er (1.5) remove (.) each of those?
98 C: yeah I'm a client for microsoft networks I click on that and
99 I- I remove
100 CT: yeh
101 C: now I've got dial up adaptor (.) click on that and remove?
102 CT: yeh please
103 C: and then last one - hello it's removed everything now we've
104 got- we've got a blank white sc- white panel now
105 CT: excellent, just click close for me (3.0) you'll be prompted
106 to restart the machine, don't do that-

The CT's receipts of the caller's turns include various qualitative assessments such as "*that's no problem at all*", "*sure*", and even "*excellent*". These turns have some similarity to the 'high-grade assessment sequences' which Antaki et al. (2000) have identified in their analysis of quality of life interviews. Antaki et al. speculate that the use of terms such as 'brilliant' or 'excellent' by an interviewer is not oriented to the informational content of a respondent's previous turn but rather serves as markers or indicators that the particular task at hand has been successfully completed. Similar considerations appear to apply in the case of the CT's contributions in this transcript extract. That is, it could not really be "excellent" that Hank has never been on the internet, given the problem to be solved in this call, or at line 90, "excellent" that the caller can read something on his screen. However they seem to be a nice way in which CT can mark the successful completion of each of the 'baby steps' – which the CT deems necessary to lead Hank through to the resolution of his problem.

In other analyses we have produced on calls to technical support, we have occasionally imported the metaphor of medical consultations, which include diagnostic questions, but which do not have the evaluation component found in classroom talk. We propose that this call and others like it are analysable as pedagogical interaction, as in a classroom lesson, and possibly as a remedial reading lesson, in which, given the weakness of the student, the teacher will be especially encouraging of any sign of competence. There are further pedagogical analogies to be drawn. At line 83 the CT asks the caller to read what is listed on the screen. After a five-second wait, the CT offers a suggestion of what should be there, just like a (remedial) reading teacher might offer a proposal such as "could the word be house?" The caller confirms the suggestion, and then receives an "excellent" for his work. There are also numerous places where the CT asks the caller to do things "*for me*" and "*if you could*". "For me" is a ploy that teachers use to get students to do things that they might otherwise not be interested in doing, and is a form that other professionals such as physiotherapists or nurses may use when some effort is needed from the other person.

It is important to note how readily the caller inserts himself into the pedagogical format as the CT issues instructions step by step. The use of such a format is contingent on the caller in some way asking for such close supervision. In this case, the CT is aware of the mistakes the caller might make, as in his warning in lines 105–106:

```
CT: excellent, just click close for me (3.0) you'll be prompted to
    restart the machine, don't do that-
```

This is similar to what teachers might say to students on the basis of their prior experience of what students may do incorrectly, such as "don't staple the pages, use a paper clip".

5. Conclusion

We have drawn on three transcripts of calls to technical support to show the various ways in which calibration for competence is done. The main general points arising from our analyses are:

- CTs open calls with a general inquiry (*How may I help you?*), thus leaving it to callers to describe, in their own terms, the reason for their call.
- Callers display, in their initial turns, aspects their computer/software competence.
- Callers invite CTs to recognise some degree of computer competence, either through the design of their turns or through explicit statements.
- CTs calibrate their next turns in orientation to the heard competence of the caller.
- CTs' calibration work is shown in their successive turn designs and placements.
- CTs' formulations of instructions to the caller are a reflection of the caller's hearable competence.

We have shown that calibration incorporates both technical and social-interactional dimensions. On the part of the CTs, this calibration work demonstrates a sensitivity to the social-interactional resolution of technical problems and is therefore a central workplace competency. However, callers are very much involved in the calibration activity, in that they produce conditions for calibration through various and successive indications of what they know and what they can do (or cannot do). We have shown how active the callers are in generating these conditions, such as by inviting or even instituting pedagogical formats.

We have drawn on the resources of Conversation Analysis to specify some of the activities that constitute important workplace competencies often conveyed through abstractions such as "talk at the customer's level" or "listen to the customer". These abstractions do not explicate what is listened for, how the hearings are conveyed, or how a customer's level is found in the first place. Nor do they offer any recognition of the importance of sequence and turn design in the performance of the work of talking and listening. Such abstractions miss entirely the "haecceity" (Garfinkel 2002) or the "just-thisness" of the work that is being done. Our specifications in this chapter draw attention to an order of competence that allows the technical problems to be solved and for the customer to feel satisfaction with the service they have paid for. We have presented a technical analysis of talking and listening that may be useful for other studies of helpline talk where the interactional organisation of problem-solving, instruction- or advice-giving, or even "merely listening" is of analytical interest.

Notes

* The data considered in this chapter were collected as part of an inquiry into the organization of expert–lay interaction as exemplified by the phenomenon of technical advice giving on a multinational computer software helpline. Data for the project were collected by the research team at the helpline's headquarters in Sydney, Australia, on several visits between 1997 and 2000 and comprise over 120 hours of audio and videotaped interaction between callers to the helpline and the software technical support personal who answer the calls. Preliminary versions of the present chapter were presented at the Aalborg symposium 'Calling for Help', September 2000, and at RC25 of the XV World Congress of Sociology in Brisbane, July 2002.

1. There are a few cases in our corpus where the technical competence of the CT is called into question by the caller, or where the CT acknowledges that they are beyond their depth – for example, by escalating the call to someone senior – but we have not included these cases

in this chapter. Our analysis, then, for the most part presumes the technical competence of the CT while looking for variation in the competence of the callers. This asymmetry of doubt is consistent with the very purpose and organisation of the technical support service, so is not a violation of the local rationality of the work that we study.

References

Antaki, C., Houtkoop, H., & Rapley, M. (2000). "Brilliant. Next question...'. High-grade assessment sequences in the completion of interactional units'. *Research on Langauge and Social Interaction, 33*(3), 235–262.

Baker, C., Emmison, M., & Firth, A. (2001). 'Discovering order in opening sequences: Calls to a software helpline'. In A. W. McHoul & M. Rapley (Eds.), *How to Analyse Talk in Institutional Settings: A Casebook of Methods* (pp. 41–56). London: Continuum.

French, P. & MacLure, M. (Eds.). (1981). *Adult-Children Conversation.* London: Croom Helm.

Garfinkel, H. (2002). *Ethnomethodology's Program: Working out Durkheim's Aphorism.* Edited and introduced by A. W. Rawls. Lanham: Rowman and Littlefield.

Heritage, J. C. (1984). 'A change-of-state token and aspects of its sequential placement'. In J. M. Atkinson & J. C. Heritage (Eds.), *Structures of Social Action: Studies in Conversation Analysis.* Cambridge: Cambridge University Press.

Hutchby, I. & Wooffitt, R. (1998). *Conversation Analysis.* Cambridge: Polity Press.

Johnson, F. G. & Kaplan, C. D. (1980). 'Talk-in-the-work: aspects of social organization of work in a computer centre'. *Journal of Pragmatics, 4*, 351–365.

Maynard, D. (1991). 'The Perspective display series and the delivery and receipt of diagnostic news'. In D. Boden & D. Zimmerman (Eds.), *Talk and Social Structure: Studies in Ethnomethodology and Conversation Analysis.* Cambridge: Polity Press.

Mehan, H. (1979). *Learning Lessons.* Cambridge, MA: Harvard University Press.

Nevile, M. (2001). 'Understanding who's who in the airline cockpit: Pilots' pronominal choices and cockpit roles'. In A. W. McHoul & M. Rapley (Eds.), *How to Analyse Talk in Institutional Settings: A Casebook of Methods* (pp. 57–71). London: Continuum.

Psathas, G. (Ed.). (1990). *Interaction Competence.* Washington, DC: University Press of America.

Sacks, H. & Schegloff, E. A. (1979). 'Two preferences in the organisation of of reference to persons in conversation and their interaction'. In G. Psathas (Ed.), *Everyday Language: Studies in Ethnomethodology.* Hillsdale, NJ: Erlbauum.

Chapter 3

Collaborative problem description in help desk calls*

Hanneke Houtkoop, Frank Jansen and Anja Walstock

1. Introduction

In this chapter we deal with audio recorded telephone calls to the communication centre help desk of a Dutch bank's branch office. Agents at the help desk answer phone calls regarding telecommunication problems in the bank. If possible, they solve these problems immediately. However, if the agents cannot solve the problems themselves, they relay these calls to Dutch Telecom (a company known as "K.P.N."), mediating between the persons reporting the problems and K.P.N.'s technicians. Our research focuses on troubles-reporting calls that are eventually relayed.

In these calls a caller (in this chapter referred to as male) reports some telecommunications problem that he or someone else in the bank is experiencing. We refer to this report as the *problem report*. If the call taker (hereafter CT and referred to as female) decides to relay the problem to K.P.N., she makes what is called a "ticket," opening a form in a computer program and entering the relevant information, including her description of the reported problem. We refer to this formulation as the *problem description*. The call is then closed, and the ticket is electronically relayed to K.P.N.

In order to solve these technical problems, K.P.N. needs clear and correct problem descriptions. Thus the problem description formulated on each ticket is an important interface between the problem as it is experienced and reported by (nontechnical) callers and the mental model of the problem, or the *problem definition*, as it is constructed by technicians. An adequate problem description ultimately corresponds with both the caller's problem report and the technician's problem definition. However, a recurrent complaint expressed by the technicians is that call takers fail to describe the reported problems correctly and/or fully.

A close look at these calls reveals that there is no systematic way for callers to know what CTs record as the problem description, and thus what CTs relay to K.P.N.'s technicians. For example, prior to the fragment below caller has explained his problem, and the CT has asked for and recorded the relevant names and numbers. We then get the following:

(1) CALL 01419

```
CT  ((typen))
    Goed (.) e:hm (.) na volgens mij weet eh weet ik voldoende.
C   ja?
CT  we gaan een eh (.) dat aanmelden en hopelijk wordt
    dat eh opgelos[t dan
C                  [o:kay:. Hartstikke fijn.
CT  goed=
C   =dank je wel.
CT  ja d[a:g
C        [da::g
.........................................................................
CT  ((typing))
    Alright (.) u:hm (.) well, I'd say I know enough.
C   yes?
CT  we're going uh (.)to make a ticket, and hopefully
    it will be uh solv[ed
C                      [o:kay: Great
CT  alright=
C   =thanks very much.
CT  yah. B[y:e.
C         [By::e
```

In this call, the CT enters the problem description into her computer without displaying to the caller what she is typing. She then says that a ticket will be made, so that the problem can be relayed. However, caller does not know whether the CT has recorded his problem correctly and completely.

In an attempt to improve the quality of these problem descriptions, we carried out a small-scale field experiment. We provided CTs with a procedural script designed to regulate their verbal and nonverbal actions during these calls. Specifically, the script directed CTs to read back what they recorded as problem descriptions to callers, thus enabling callers to verify what was entered into the computer, and ultimately, what was relayed to K.P.N. In the script it says: "The report is registered under number XXX", immediately followed by "I have noted as the trouble: 'read back content free field'". The next step in the script says "Add possible additions of client in ticket", followed by the final step "Close the call." We expected that this script would prevent some of the troubles with problem description experienced by K.P.N.'s technicians. Also, we were interested in observing how callers and (especially) CTs would cope with these new interactional requirements.

We transcribed 17 calls in order to examine whether the use of the script would indeed lead to more adequate problem descriptions. We do not deal here with the question of whether CTs follow the script, and to what effect. Rather, we look at these calls in a slightly more general fashion in this chapter, focusing on how call taker and caller collaborate to generate the completed electronic data entry form.

2. Overall organization of the calls

We shall first give an outline of the overall structure of these calls. After the *identification sequence*, the *greeting sequence*, and sometimes a *how-are-you-sequence* and some social talk (Schegloff 1986), callers provide narrative accounts of their problems (cf. Frankel 1989; Baker et al. 2002).

(2) CALL 01585

```
C       kijk ik-ik heb te maken met het volgende. de e::h
        medewerkers van de O en I in de UOZ in eh Veendam
        die verhuizen naar Groningen en die nemen hun fax ↑mee,
CT      mhm
C       en wij schakelen dan ook de eh fax vanuit Veendam
        door naar Groningen. althans dat was onze bedoel↑ing,
        met [sterretje eenentwintig
CT          [mhm
        mhm=
C       =nou (.) het is een analoge lijn ik heb er een
        analoog toestel aangehangen
CT      °ja°
C       ik doe wat in de telefoongids staat. sterretje
        eenentwintig sterretje telefoonnumer h↑ekje=
CT      =mhm=
C       =maar da' werkt niet.
        (0.8)
        .hhh e:h bij i.s.d.n. gebruik je geen sterretje
        eenentwintig, maar sterretje tweehonderdtien
        (0.2)
CT      mhm
C   →   maar dat werkt net zo min.
```

```
34          (0.2)
35          weten jullie daar wat meer van?
.................................................................................
12  C       look I-I have to do with the following. The u::h staff
13          members of the O and I with the U.O.Z. in uh Veendam
14          they'll move to Groningen and take their fax with th↑em,
15  CT      mhm
16  C       and so we put the uh fax indeed from Veendam
17          to Groningen. At least that is what we wanted,
18          with [star twenty-one
19  CT           [mhm
20          mhm=
21  C       =well(.) it's an analogue line I have hitched an
22          analogue appliance to it
23  CT      °yes°
24  C       I follow what's in the telephone directory.
25          star twenty-one star phone number h↑ash=
26  CT      =mhm=
27  C       =but tha' doesn't work.
28          (0.8)
29          .hhh u:h for I.S.D.N. you don't use star twenty-one,
30          but star twohundredandten
31          (0.2)
32  CT      mhm
33  C   →   but that doesn't work any better.
34          (0.2)
35          do you know a bit more about this?
```

As we can see in the first part of this call, the talking is done in an *information delivery format* (Silverman 1997:41), in which the caller and the CT are respectively aligned as the speaker and the recipient. This communication format then changes to an *interview format* (Silverman 1997:41), which occurs in this call after caller allocates the turn-at-talk to CT in line 35. The conversational roles thus change to those of questioner and answerer, with the CT asking questions in order to roughly diagnose the problem, and the caller answering these questions.

(2a) CALL 01585 continued (# indicates a key stroke)

```
37  CT      sterretje eenentwintig sterretje, he?=
38  C       =ja?
39          (0.2)
40  CT  →   [vanaf een meridian centrale
41  C       [wa-
42          .hh e::h
43          (0.8)
44  CT      ja ik weet niet of ze dat ook op de meridian zo- (.)
```

45 ja k- volgens mij wel
46 (.)
47 d[at kan gewoon
48 C [ie-iedereen roept dat altijd tegen mij.
49 zo moet je dat doen.
50 CT ja:
51 C hm (ha) maar dat werkt niet. ha ha
52 CT → stond die al doorgeschakeld, of niet?
53 C nee nee-nee
54 CT ah- .hh
55 CT → # # en wa-wat doet ie dan niet?
56 (0.5)
57 → of wat doet ie wel=
58 C =tijdens het intoetsen van de telefoonnummer
59 v'waar dat gesprek naartoe m↑oet=
60 CT =ja=
61 C =in dit geval dan een fax,=
62 CT =ja=
63 C =ehm geeft ie eh (.) ja hoe moet ik dat zien te vertellen
64 .hh dan begint ie e:h t-e:h tonen te produceren, eh net
65 hetzelfde als dat je een verkeerd telefoonnummer toetst.
66 CT [.hh
67 C [ik noem maar eens wat.=
68 CT → =ja eh even kijken eh toetst u wel een eh nul om naar
69 buiten te komen en dan (0.3)n[u:l
70 C [ja eerst een nul en dan
71 begin ik met dat sterretjesver↑haal=
72 CT =mhm
73 C en de l-lijn of het nummer wat ik dan kies da[t maakt
74 CT [mhm
75 C geen verschil uit of ik dat vooraf laat gaan door
76 een (0.3) twee nullen of een nul
77 CT → okay. e:hm::: loopt die lijn over de centr↑ale.
78 (1.0)
79 → die faxl↑ijn
80 C ja
81 CT → loopov- ehm heeft die lijn wel toestemming om naar
82 buiten te forwarden?
83 (0.6)
84 C Dat is een goeie vraag

....

37 CT star twenty-one star, right?=
38 C =yes?
39 (0.2)
40 CT → [from a meridian exchange
41 C [wha-
42 .hh u::h

```
43          (0.8)
44  CT      yes I don't know if they're on the meridan too- (.)
45          yes I- think they do
46          (.)
47          th[at's possible yes
48  C         [E-e-everyone tells me you should do it like this
50  CT      ye:h
51  C       hm (ha) but that doesn't work. ha ha
52  CT →    was it put through then, or not?
53  C       no no-no
54  CT      Ah- .hh
55  CT →    # # and wha-what does it not do then?
56          (0.5)
57     →    or what does it do=
58  C       =when keying in the phone number
59          fro' which that call has to go t↑o=
60  CT      =yes=
61  C       =a fax in this case,=
62  CT      =yes=
63  C       =uhm it gives uh (.) well how should I ever explain it to
64          you. .hh Next it begins u:h t-:h to make tones,
65          uh just the same as when dialling a wrong phone number
66  CT      [.hh
67  C       [I'm just saying things=
68  CT →    =yes uh let's have a look uh you do key in uh nought in
69          order to get outside and then (0.3) n[ou:ght
70  C                                            [yes first a nought
71          and then I start with this star stor↑y=
72  CT      =mhm
73  C       and the l-line or the number that I choose next tha[t makes
74  CT                                                         [mhm
75  C       no difference if I first choose
76          a (0.3) two noughts or one nought
77  CT →    okay. u:hm::: does that line run through the exch↑ange.
78          (1.0)
79     →    this faxl↑ine
80  C       yes
81  CT →    run throu- uhm does that line have permission indeed
82          to forward outside?
83          (0.6)
84  C       that's the question
```

One reason for asking these questions is that K.P.N. is not responsible for all communication equipment failures, and the call takers must assess who needs to deal with the reported problem (cf. Meehan 1989, on call takers assessing the 'police-worthiness' of citizen complaints to the police; Bergmann 1993, for calls to the fire department; and Whalen 1995, for 9-1-1 emergency calls). Another reason for asking these diagnostic questions is that the problem is best solved on the spot, if possible, which the organization refers to as "real time" problem solving.

Ticket announcement

Once it is clear that the problem cannot be fixed in real-time and must be solved by K.P.N., call takers usually announce that they are going to make a ticket, as we can see below in line 72:

(3) Call 01810

```
68  C       Ze is d'r achter gekomen die die- nou thh dat
69          die gesprekken niet op de band eh komen.
70          (2.3)
71  CT      Dat is <nogal een> behoorlijk probleem.=
72     →    =Eh ik-ik ga daar een melding van aanma[ken=
73  C                                                [ja.
74  CT      =en het was in kant↑oo:r
75  C       eh het heet eh [NAAM KANTOOR]
.....................................................................
68  C       She found out these these- well thh that
69          these calls will not uh go on tape
70          (2.3)
71  CT      That's <rather a> quite a problem=
72     →    =Uh I-I will make a ticket of [it=
73  C                                      [yes
74  CT      =and it was in offic↑e:
75  C       uh it's called uh [NAME OFFICE]
```

In order to make an electronic ticket, CTs have to start up a computer program, which usually takes some time, as is alluded to in line 124 of the fragment (4):

(4) Call 01586

```
90  CT →  Nou ja ik zal eh bij de storingdienst een navraag
91          ehm (.)of een aanvraag indienen
92          (3.1)
93          of dat klopt dat dat niet kan en zo ja eh of ze het
94          dan even willen opheff↑en
95  C       ja en dat het liefst met eh de grootst mogelijke spoed
96          (.)
97          .hh want ik ben nu dus per fax niet bereikbaar
            .
            .
            .
```

```
122 CT     mhm
123        (2.0)
124 CT →  nu is- het programma op aan het starten.
125        (.)
126        °even kijken (.)daar is tie.°
127        even kijken welk kantoor gaat het om?
..............................................................................
90  CT     well yes I will uh inquire with the fault-clearing service
           uhm (.) or make an application
91         (3.1)
92         if it's correct that that's impossible and so yes uh
93         if they just will discontinue
94  C      yes and that preferably at uh the greatest possible speed
95         (.)
96         .hh because I can't be reached by fax now
           .
           .
           .
122 CT     mhm
123        (2.0)
124        Now (.) the programme is starting up
125        (.)
126        °let's have a look (.)there we are°
127        Let's have a look, which office is concerned?
```

As CT announces in line 90 that she is going to make a ticket, she uses her mouse to start up the computer program. After a long silence in line 92 she tells caller what she is going to ask K.P.N. to do. The co-participants exchange some further talk, and in line 124 call taker informs caller that they have to wait for the program to appear on the screen. Right after her "*let's have a look* (.) *there we are,*" CT reads a question from the form: "*Which office is concerned?*".

In other cases this technical procedure does not take up any interactional time. In the fragment below CT states that she is going to report the trouble to K.P.N. by telephone, after which she immediately moves on to requesting the information she needs to fill out her form:

(5) Call 01604

```
58  CT     en eh wij kunnen gewoon als een storing aannemen,
59         en dan melden wij het meteen door en eh (.)dan
60         wordt het eh door KPN verholpen,
61         en dat kunnen we gewoon telefonisch doen.
62  C      ja?
63  CT     en dat ging om: (.) de:                         → QUESTION
64  C      om toestel eh zesnulzeven
..............................................................................
```

58 CT and uh we may take it as an interruption
59 and we'll report it straight away and uh (.)next
60 it will be repaired by K.P.N.
61 and we'll just do so by phone.
62 C yes?
63 CT And it concerned (.) the: → QUESTION
64 C uh extension uh sixnoughtseven

The completed electronic ticket contains all the information the technician needs in order to solve the problem. To assist CTs in acquiring the relevant information from callers during the telephone interaction, the help desk uses a *computer-aided dispatch system* (*CAD*) and an *electronic data-entry form*. The form looks like this:

Received Account Description
Office
Position notifier
Name
Phone/fax
Type of application
Issue
District
Address
Phone/Fax

After announcing that they will make a ticket, CTs turn to the electronic data-entry form, organizing their talk in accordance with the fields on the left of the form, from top to bottom. Some fields are automatically filled in for them by the system itself, and CTs may verify the correctness of this information with callers. CTs also ask questions of caller in order to fill in the fields that are left blank by the automatic system. In the fragment below we find both ways of generating information.

(6) Call 1408

43 CT Amsterdam Orlyplein UO[Z → VERIFYING INFORMATION
44 C [jaja:::::.
45 Ja::::::: inderdaad
46 (1.0)
47 Nah jah
48 (.)
49 CT E::n u bent de M.O.I. H[BD?
50 C [.hh ja nee: M.O.I.=
51 CT =M.O.I.
52 (1.0)

```
53  CT   E::n wat is uw naam?             → REQUESTING INFORMATION
54  C    Ronny van der Kraak
55       (4.0)
56  CT   °de Kraak°
57       [(2.5)
58       [#
59       °<Er, p[unt, [van der Kraak>°
                [#   [###
.................................................................................
43  CT   Amsterdam Orly square UO[Z.      → VERIFYING INFORMATION
44  C                           [right
         .
         .
         .
53  CT   A::nd what's your name?          → REQUESTING INFORMATION
54  C    Ronny van der Kraak
55       (4.0)
56  CT   °de Kraak°
57       [(2.5)
58       [#
59       °<R, f[ull stop, [van der Kraak>°
               [#         [###
```

Frequently there is some confusion about whose names and numbers are wanted (caller's or the person's having the telecom problem), and thus this exchange may take up to twenty turns-at-talk. Once the fields on the left are dealt with, CTs turn their attention to the right side of the form to enter the problem description.

By orienting to the items on the screen in order to complete this electronic ticket, CTs indicate that an organizational agenda is in operation (Zimmerman 1984, 1992; Meehan 1989; J. Whalen, Zimmerman, & M. R. Whalen 1988; Frankel 1989; M. R. Whalen & Zimmerman 1990; Heritage & Sorjonen 1992; J. Whalen 1995). As Jack Whalen points out (1995: 189), such an agenda is "in one sense anterior and 'external' to the conversational encounter even as it surely must be accomplished in and through it."

An important aspect of filling in a form (either electronically or not) is that CT produces a written document in an interactional context. That is, the written document is actually produced during social interaction. We thus find a close relationship between writing and talking (Frankel 1989; Whalen 1995; Longman & Mercer 1993; Houtkoop-Steenstra 2000). In the next section we deal specifically with this intersection of speaking and writing, and how speaking and writing influence one another.

When CT completes the ticket, she usually reads the problem description as it is now formulated on the ticket back to caller. We see this in the fragment below in lines 195–198.

CT then gives caller a reference number for the report she has made (lines 214–218), and she announces that the report will be relayed to K.P.N. (line 219), which opens the closing of the call (cf. Schegloff & Sacks 1973).

(7) CALL 1586

```
195 CT  .hh ik heb hier genoteerd faxnummer                 ┌READING
196     nulvierzevenvijf drievijfzevenzevennulnegen         │BACK
197     is niet door te schakelen met sterretje             │PROBLEM
198     eenentwintig sterretje                              └DESCRIPTION
        .
        .
        .
210 CT  ik meld het aan (.) bij de BSD.              ANNOUNCING RELAY
211 C   mhm
212 CT  Ehm en daar krijg ik vanmiddag vast en zeker nog
213     reactie op (.) ze zijn vrij snel
214     .hh en het is bij ons na te vragen onder            ┌REFERENCE
215     nummer nul vijftien zesentachtig                    │
216 C   Nul vijftien zesentachtig.                          │
217 CT  voor het geval u een college aan de lijn (0.5)      └NUMBER
218 C   hmm Ik heb het gedoneerd- genoteerd
219 CT  Okay[ik ga hem aanmelden                     ANNOUNCING RELAY
220 C       [goed zo. I:k ben benieuwd
221 CT  Prima(ha)=
222 C   =Okay
223 CT  Dag=
224 C   =Dag
.........................................................................
195 CT  .hh I have noted down fax number noughtfour-        ┌READING
196     sevenfive threefivesevensevennoughtnine             │BACK
197     cannot be put through with star                     │PROBLEM
198     twenty-one star                                     └DESCRIPTION
        .
        .
        .
210 CT  I'll report it (.) to B.S.D.                 ANNOUNCING RELAY
211 C   mhm
212 CT  Uhm and I'll get a reply definitely this
213     afternoon (.) they're fairly quick
214     .hh and it can be inquired from us under            ┌REFERENCE
215     number nought fifteen eightysix                     │
216 C   Nought fifteen eightysix.                           │
217 CT  in case you'll speak to my colleague (0.5)          └NUMBER
```

```
218 C    hmm I donated- noted it down
219 CT   Okay[I will report it                          ANNOUNCING RELAY
220 C        [good. I: wonder
221 CT   Grea(h)t=
222 C    =Okay
223 CT   Bye=
224 C    =Bye
```

3. The collaborative construction of the computer-aided ticket

We shall now discuss in detail how these computer-aided tickets are collaboratively constructed. The CT's task is to fill out the form and to enter the data, and caller's task is to provide the information that call taker needs. As we shall come to see, CT and caller closely and elegantly collaborate in performing these various tasks.

CTs production of incomplete and-prefaced statements as questions

When creating a ticket, CTs first fill in the fields on the left of the electronic form. These fields are labeled: office, position notifier, phone, etc. If the information is not already stored in the computer, CTs must ask the callers to provide the information. CTs mark these questions as *agenda-based questions*, in that they preface them by the connective "and." Heritage and Sorjonen observe:

> *And*-prefacing is primarily used by professionals to establish and maintain an orientation to the course-of-action character of their talk across sequences of question/answer adjacency pairs. Associated with this activity linkage [...] is the maintenance of an orientation by both parties to the questions as *agenda-based*, i.e., as members of a series that are in some way routine elements of an activity or as elements the questioner has anticipated or has "in mind" – or commonly, as externally motivated components of a bureaucratic task or other agenda which is being managed by the professional questioner as part of the "official business" of the encounter. (Heritage & Sorjonen 1994: 5–6)

Given the fact that the CT has already made clear that she is now going to make a ticket, *and*-prefacing the questions to come may contribute to caller's understanding that the answers to these questions will be included as relevant elements of the ticket.

CTs not only *and*-preface these questions; they also use grammar to organize the social interaction (cf. Schegloff, Ochs, & Thompson 1996; Lerner 1996), that is, they put these questions in a specific syntax. Rather than producing questions in interrogative form (e.g. 'What is your phone number?'), CTs tend to phrase questions in declarative form ("Your phone number is?"). If speakers put a wh-question (what, where, when, who and how-questions) in the syntactic format of a declarative sentence, they produce an incomplete statement, which ends just before the focus position where the wh-element is to be produced. By producing the turn-final element with upward intonation, CT invites caller to complete the unfinished turn-constructional unit by providing the relevant wh-element. For example, when asking for the identification of the problematic machine, we find CT producing an incomplete statement, and caller providing the missing element.

(8) Call 23

CT And this concerned (.) t↑he:
C extension uh six nought seven

CT stops her utterance at the point where the requested element is to be filled in. The final element of the utterance ("*th↑e:*") is pronounced with upward intonation and a lengthened vowel, signaling that caller is now to provide the lacking information (cf. Lerner 1996: 243 on how the syntactic environment of a pause or a sound stretch may inform the action it performs). Caller indeed treats this incomplete statement as a request to fill-in-the-blank: "*extension uh six nought seven*" (cf. Lerner 1995 on teachers employing incomplete turn-constructional units as a resource for getting students' participation).

The same procedure is to be found in the fragments below. Again CT invites caller to complete the incomplete statement:

(9) Call 1419

106 CT → en die meneer ↑i:s VELD: FUNCTIE MELDER
107 C hoofd bankdiensten
108 CT de HBD
.
.
.
122 CT → en het faxnumm↑e:r VELD: FAX
123 C eh is nulveertig tweedrie tweezeven
124 driezes vier

....

106 CT and this gentleman ↑i:s FIELD: POSITION REPORTER
107 C head bank services
108 CT the H.B.S.

```
          .
          .
          .
122 CT →  and the faxnumb↑e:r                 FIELD: FAX
123 C     uh is noughtfourty twothree twoseven
124       threesix four
```

(10) Call 1810

```
72  CT    =Eh ik-ik ga daar een melding van aanma[ken=
73  C                                            [ja.
74  CT →  =en het was in kant↑oo:r                        VELD: KANTOOR
75  C     eh het heet eh [NAAM KANTOOR]
.................................................................................
72  CT    Uh I-I will make a ticket of [this=
73  C                                  [yes
74  CT    And it was in off↑ice:                          FIELD: OFFICE
75  C     Uh it's called uh [NAME OFFICE]
```

We may say that CT and caller collaboratively produce the text to be recorded on the form. CT mentions the item to be dealt with in the form of a not-yet-complete statement, and caller provides the remaining part of this statement, which is then recorded by CT.

Caller's orientation to the information to be recorded by CT

Looking at the ways in which callers frequently phrase their answers to these questions, it is clear that callers orient to the fact that their information is being recorded by CTs. When asked for his name, the caller below does two things in succession. He first provides his given name plus family name. He then spells his name, thereby facilitating CT's writing task.

(11) Call 1582

```
52  CT    En (.) uw n↑aa:m
53  C     Gerard Kops (.) Kaa oo pee es.
.................................................................................
52  CT    And (.) your n↑a:me
53  C     Gerard Kops (.) Kay Oh Pea Es
```

Note that when caller spells his name, he does so for his family name only. Along with the spelling, this shows his orientation to the recording task.

This orientation becomes especially clear when callers provide numbers. In spoken Dutch, the expression of compound numbers is somewhat confusing. For example, the number 63 is 'three-and-sixty'. In fragment 8 above, caller provides the number of the problematic machine in such a way that he facilitates CT's ability to record it correctly. Rather than saying "six hundred and seven," caller provides the number digit by digit.

```
(re8) CT  And this concerned (.) t↑he:
      C   extension uh six nought seven
```

And number 040-2327364 is provided by caller as:

```
(re9) 122 CT en het faxnumm↑e:r
      123 C  eh is nulveertig tweedrie tweezeven driezes vier
      ..........................................................
      122 CT and the fax numb↑e:r
      123 C  uh is noughtfourty twothreee twoseven threesix four
```

So, if we study the interactional production of an electronic ticket by telephone, it is clear that there is a close collaboration between CT and caller. In indicating that she is going to make a ticket, and in *and*-prefacing her questions, CT signals that she is seeking information to be filled in. In formatting the questions as incomplete statements and producing the turn-final element with a rising intonation, CT projects what type of information she wishes caller to provide. Caller then provides the requested information in such a way that it is easy for CT to write it down.

CTs working aloud while typing

As mentioned previously, we provided CTs with a script that directed them to read their problem descriptions back to callers after recording them. The help desk supervisor had the impression that the use of this script had an interesting effect. Not only did CTs read back what they had entered as the problem description, but they also became more talkative with callers in general. The CTs tended to work aloud more, informing callers about what they were doing on their end. For example, they would tell callers when they were waiting for the computer program to appear on the screen, as well as telling them what they were typing when they worked on their keyboards, rather than performing these activities in silence, as they had done previously. We shall provide examples of this practice.

After the left side of the CAD form has been filled out, CTs turn their attention to the right side of the form to enter the problem description.

(12) Call 01695

```
81  CT   Weet je wat, ik maak toch even gewoon een melding aan.
         (...)
90  CT   Z- o[kay en de fax die eh waar het hier om gaat
```

```
91  C         [Da-
92  CT     i[s zes één vier vijf één drie vij[f.
93  C       [ee-                             [.hh ja ja[ja
94  CT                                                [en
95       → die werkt niet. en wa- geen lij[n of ↑eh
96  C                                    [E:h hij-
97         nee hij eh kan wel verzenden (.) maar niks ontvangen.
98         (.)
99  CT     [ Okay <niet (.) kan (.) ontvangen> ]   → WORKING ALOUD
100        [ ((......typen .................)) ]
101        ((2 seconden typen))                    → WORKING SILENTLY
102        okay
103        (1.0)
104        [ <wel (.) verzenden>. Punt ]           → WORKING ALOUD
105        [ ((.......typen ........)) ]
106        En faxapparaat staat goed ingesteld?
107 C      Ja..
108 CT     o↑kay
109        (3.0)
110 CT     °Even kijken°. Hij staat hier ge-re:gistreerd onder
111        nul zestienvijfennegent↑ig
112        E:n ik geef 'm door aan de storingsdienst
113        v[an de KPN
113 C       [o-
114        Okay
.....................................................................
81  CT     You know what, why don't I just make a ticket of this
           ((some lines ommitted))
90  CT     S- o[kay and the fax that's uh concerned he:re
91  C          [Tha-
92  CT     i[s six nought four five nought three fi[ve.
93  C       [A-                                    [.hh yes yes[yes
94  CT                                                        [And
95       → that doesn't work. And wha- no li[ne or ↑uh
96  C                                       [U:h it-
97         no it uh will send alright (.) but receive nothing.
98         (.)
99  CT     [ Okay <re:ceive (.) nothing> ]         → WORKING ALOUD
100        [ ((..... typing ..........)) ]
101        ((2 seconds of typing))                 → WORKING SILENTLY
102        okay
103        (1.0)
104        [ <can (.) send>. Full stop ]           → WORKING ALOUD
105        [ ((......typing ........)) ]
106        and fax appliance is set correctly?
107 C      yes.
108 CT     o↑kay
```

```
109        (3.0)
110 CT     °just a moment°. It's registered here under
111        nought sixteenninetyfi↑ve
112        a:nd I'll hand it down to the fault-clearing service
113        o[f K.P.N.
113 C       [o-
114        Okay
```

In the first part of this call, not shown here, caller mentions as his reason for calling: "*I'd like to report a trouble with a fax appliance*." After some joking about caller's recurrent troubles, CT announces her next action: "*You know what, why don't I just make a ticket of this*" (line 81). Upon CT's request, caller provides various names and numbers that are entered by CT. CT then moves on to the problem description.

In line 95 CT invites caller to formulate the problem with the fax machine: "*And that doesn't work. And wha- no line or uh*". Rather than using an open invitation to report the problem (e.g. "what seems to be the problem," CT provides a candidate problem ("*no line*"), followed by "*or uh*," produced with a rising intonation (cf. Houtkoop-Steenstra 1990, 2000, on this "is it X, Y or uh?" question format). Caller then indicates that the malfunction is not the candidate problem proposed by CT ("*no*"), but that the fax machine: "*will send alright* (.) *but receive nothing*". Having heard the problem, CT enters the problem description into her computer. While typing, CT says in a dictating tone of voice "*<Okay <re:ceive* (.) *nothing>*." By working aloud, CT displays to caller what she has made of his problem and what will eventually be relayed to K.P.N. In this specific instance, caller is informed that CT has (at the least) recorded that fax machine 6145135 can receive nothing, but it can send. If this formulation is to caller's satisfaction, he accepts CT's problem description.

4. Reading back the problem description

In most cases CTs read the problem description back to callers, as instructed by the scripts with which we provided them. Sometimes CTs use explicit announcements, like "I have noted here" or just "I have" to introduce their reading back of what they recorded. In other cases they preface the reading back by using markers, such as "Okay::" or "u:::h." As the two fragments below show, caller monitors CT's text.

(13) Call 1408

```
145 CT     E::::::h ik heb toe:stel vier twee één defec[t.
146 C  →                                                [ja
147 CT     Dat is een uh zesentwintig zestien dee.
148 C  →   Jah=
149 CT     =Kan mogelijk aan de lijn en toestel liggen.=
150 C  →   =Ja
151 CT     We gaan eh eh de KPN d'r naartoe stu[ren die ...
152 C                                          [Oh
.................................................................................
145 CT     U::::::h I have applia:nce four two one defec[t.
146 C  →                                                 [yes
147 CT     That's a uh twentysix sixteen dee.
148 C  →   Yes=
149 CT     =Could be the extension possibly and the appliance.=
150 C  →   =Yes
151 CT     We will uh have K.P.N. pay you a vi[sit they ...
152 C                                         [Oh
```

(14) Call 1667

```
55  CT     Ok↓ay::. Nul zeventig vijf één vijf nul één zes drie
56         is een analoge lijn. is dood sinds begin februari.
57  C  →   Ja
.................................................................................
55  CT     Ok↓ay::. Nought seventy five nought one six three
56         is an analogue line. is dead since the beginning of
           February.
57  C  →   Yes
```

Problem description – acceptance/non-acceptance

In acknowledging the text read aloud by CT, caller implicitly accepts this text as a correct formulation of his reported problem.

(15) Call 1582

```
CT     Okay:: [ °<telefoniste (.) hoort (.) niets>° ]
              [ ((....... Typen..................)) ]
       Okay, bedienpost kraakt lijnen vallen soms weg,
       en (.)telefoniste hoort n[iets.
C  →                            [Jah
CT     Okay. Staat genoteerd.=
       =Ik ga ernaar kijk↑en
C      J↑a::
.................................................................................
CT     Okay, operator cracks,lines fall away sometimes
       and (.) operator not not hear any[thing.
```

```
C  →                                       [Yes.
CT      Okay. I made a note.=
        =I'll have a look at ↑it
```

Due to the fact that CT and caller have collaborated so closely in the formulation of this text, it does not come as a surprise that caller usually accepts it. In the fragment below, however, caller adds something to CT's problem description:

(16) Call 01606

```
170 CT      'K meld 'm aan bij de KP↑N.=
171         =Ik heb hier staan toestel zevennnultwee van
172         de UOZ in Roermond zowel intern als extern
173         doorschakelen
174         niet meer mogelijk.
175 C       mhm
176         (1.0)
177 CT      'Kay
178 C   →   en een conference gesprek kan niet meer worden opgebouwd
179 CT      [ <tevens geen conference opgebouwd> ]
180         [ ((............typen.............)) ]
181         [ (4.0)    ]
182         [((typen))]
183         punt
184         (.)
185         goed ik heb hem hier geregistreerd.
186         hij gaat door naar de KP↑N.
187 C       ja:?
..............................................................................
170 CT      I'll report it to K.P.↑N.=
171         =It says here appliance sevennoughttwo of
172         U.O.Z. in Roermond both internal and external
173         switching no longer possible.
175 C       mhm
176         (1.0)
177 CT      'Kay
178 C   →   and a conference call can no longer be set up
179 CT      [ <also no conference set up> ]
180         [ ((........typing.........)) ]
181         [ (4.0)       ]
182         [ ((typing)) ]
183         Full stop
184         (.)
185         good.I've registered it here.
186         it will go further to K.P.↑N.
187 C       ye:s?
```

In line 175, caller withholds acceptance of CT's problem description. He produces a continuer ("*mhm*"), signaling that he expects CT to continue talking. When CT keeps silent for 1 second (Jefferson 1989), caller reminds her of the second problem he has reported: "*and a conference call can no longer be set up*." In fact, caller not so much "reminds" CT of the second problem, that is, he does not say something like "I also mentioned a second problem." In beginning his turn with the connective "*and*," and in using the rather formal passive construction typical of written language (Chafe 1982), he may be seen as dictating additional information that CT should write down: "*and a conference call can no longer be set up*." Without acknowledging that caller's contribution can be heard as a correction, CT continues her problem description, using telegraphic style: "<*also no conference set up*">. After some more typing CT markedly and hearably closes off her writing by saying "*full stop*." (line 183). She then states that the job of entering the problem description is done ("*good, I've registered it here*").

Below caller uses the reading back of the problem description to correct the number of the troubled machine. The fragment starts with CT typing aloud. In line 200 CT starts reading back what she has written down. Caller then notices that CT has the phone number wrong, and he correct this (lines 205–207). In line 208 CT accepts this other-initiated repair (Schegloff, Sacks, & Jefferson 1977; Schegloff 1992) by saying: "*Yea::: extension hundred fifty nine, indeed*." In the transcript the text that CT reads back is printed in bold.

(17) Call 01414

```
CT   ((2.2 seconden typen))
     <Na het bellen van toeste:l (.) honderd- [(.) blijft
                                              [#
     ((4.6 seconden typen))
     [de
     [#
     ((1.9 seconden typen))
     [Ja nou ik ben wel aan het rommelen
     [## #
C    [hmhm
     [## #
CT   [ <lijn van op het toestel waar vandaan gebeld is ]
     [ ((...............typen.......................)) ]
     (2.0)
     Pu[nt.
       [##
     [ toestel eenvijfnegen heeft nu geen signaal meer punt> ]
     [ ((................typen............................)) ]
```

200 **Na het bellen van toestel honderdvijfennegentig**
201 **blijft de lijn hangen op het toestel waarvan (.)**
202 **w[aarv-**
203 C → [>na het bellen van toestel< hondervijfennegentig?
204 (.)
205 → honderdnegenenvijftig
206 (1.4)
207 → [eenvijfnegen
208 CT [Ja:::: toestel honderdnegenenvijftig inderdaad.
209 .hhhh **<na het bellen van toestel honderd(.)**
210 **neuhvijftig- hondernegenenvijftig blijft de-**
211 **blijft de lijn hangen .hh op het toestel**
212 **waarvan (0.8) waarvandaan gebeld is.>**
213 C Ja.
214 CT Ik zit hier een letter teveel te typen
215 #
216 .hhhh **<toestel hondernegenvijftig geeft nu geen signaal**
meer>
217 (2.1)
218 Dat was het zo'n beetje [denk ik.
219 C [Ja Ja.
220 CT Ja okay e:::h dan gaan we dit aanmelden bij de KPN?=
221 C =Okay.

....

182 CT ((2.2 seconds of typing))
183 <After calling extensio:n (.) hundred [(.)the line
184 [#
185 ((4.6 seconds of typing))
186 [the
187 [#
188 ((1.9 seconds of typing))
189 [Really I'm tinkering a bit here
190 [## #
191 C [mhm
192 [## #
193 CT [<line of the extension from which the call was made]
194 [((...............typing......................))]
195 (2.0)
196 Full st[op.
197 [##
198 [extension onefivenine has no signal anymore full stop>]
199 [((................typing..........................))]
200 **After calling extension onehundredandninetyfive**
201 **the line keeps hanging to the extension of which (.)**
202 **wh[i-**
203 C → [>after calling extension< hundredandninetyfive?
204 (.)

205 → hundredand fifty-nine
206 (1.4)
207 → [onefivenine
208 CT [Yea::: extension hundredandfiftynine indeed.
209 .hhhh **<after calling extension one hundred (.)**
210 **ninehfive- a hundredandfiftynine the keeps-**
211 **the line keeps hanging .hh to the extension**
212 **from where- (0.8) from which is called.>**
213 C Yes.
214 CT I'm typing a letter too many here
215 #
216 .hhhh **<extension hundredandfiftynine does not give any sign anymore now>**
217 (2.1)
218 That was it[I think.
219 C [Yes Yes
220 CT Yes okay. U:::h well we'll report it to K.P.N.?=
221 C =Okay.

In the next and final fragment presented in this paper, the CT's reading back of her problem description makes clear to caller that only half of the reported problem has been recorded.

(18) call 1582

10 C Ik krijg elke keer lijntjes binnen en dan versta ik ze niet.
11 (1.0)
12 Dan bellen mensen teru:g .hhh en het is soms to- tot
13 drie keer aan toe en dan moeten ze bellen voordat
14 ik het hoo:r.
15 CT Ja u kraakt nu ook
16 C .hh Ik kraak. [Hahaha
17 CT [Hahaha
18 Ik ben al oud hoor dus dat klopt
19 C [Hahaha
20 CT [hahaha
21 CT U zegt het. ik heb het niet eens gedacht.
22 [Hahaha
23 C [hahaha
24 CT E:::h even kijken wat ik ga doen is even een melding
25 aan[maken
26 C [ja want ik heb er nog eentje h↑oor,
27 tzij- ik heb twee dingen eigen[lijk
28 CT [u heeft twee dingen=
29 C =ja [i-
30 CT [nou dat mag=
31 C =Ik heb ook nog een- dan bellen colleg↑a's (.)

32 en nou moet ik heel even denken hoor. Ik ben een
33 paar dagen weg geweest (.) er bellen collega's mij ↑op,
34 oh ja en dan valt de eh lijn
35 gewoon helemaal we[g
36 CT [mhm
37 (1.0)
38 C Dus dan hoor ik ook helemaal niks dus ut-
39 ja het is iets anders maar het komt bijna op hetzelfde
neer.
40 CT °Ja° e:::::::hm welk kantoornummer is het?
.
.
.
84 C Hallo daar ben ik wee:r.
85 CT Hallo. Ja.
86 (1.0)
87 .hh Ik heb hier genoteerd eh **de bedienpost (.) kraakt.**
88 (2.0)
89 C → ja maar hij ik hoor ook niks hé. Ik hoor niks.=
90 CT =mhm
91 C als mensen bellen
92 (1.0)
93 CT Ok↓ay::[**<°telefoniste (.) hoort (.) niets°>**]
94 [((............typen ....))]
95 okay
96 **Bedienpost kraakt, lijnen vallen soms weg, en (.)**
97 **telefoniste hoort n[iets**
98 C [ja
99 CT okay staat genoteerd.
100 Ik ga ernaar kijk↑en,
101 C Ja↑ ::
102 CT En op moment dat een en ander opgelost is,
103 dan ontvangt u vanzelf een fax van ons
104 C Dankuwe:l.

....

10 C All the time I receive these lines that I can't hear.
11 (1.0)
12 Next people call ba:ck .hhh and sometimes it happens
13 three times no less and than they should call before
14 I hear it.
15 CT Yes now you crack too
16 C .hh I crack. [Hahaha
17 CT [Hahaha
18 I'm old you see so that's right
19 C [Hahaha
20 CT [hahaha
21 CT You tell me. The thought didn't even occur to me.

22 [Hahaha
23 C [hahaha
24 CT U::h just see what I will do is just make a ticket
25 he[re
26 C [Yes since I have another one you ↑see,
27 It's- I have two things in [fact
28 CT [You've two things=
29 C =yes [i-
30 CT [that's alright=
31 C =I have also an- then colleagues c↑all (.)
32 and now let me think a minute. I've been away for
33 a few days (.) colleagues call ↑me,
34 oh right and next the line falls away
35 just like th[at
36 CT [mhm
37 (1.0)
38 C Then I don't hear anything at all so ut- yes it is
39 a bit different but it almost comes down to the same
thing.
40 CT °Yes° u:::::::hm which office number is it?
.
.
.
84 C Hello I'm back agai:n.
85 CT Hello. Yes.
86 (1.0)
87 .hh I have made a note here uh **the switchboard (.) cracks.**
88 (2.0)
89 C → yes but I don't hear anything either you see.
I don't hear anything.=
90 CT =mhm
91 C if people call
92 (1.0)
93 CT Ok↓ay::[<°**operator (.) hears (.) nothing**°>]
94 [((......typing.................))]
95 okay
96 **Switchboard cracks, lines fall away sometimes, and (.)**
97 **operator hears n[othing**
98 C [yes
99 CT okay noted down.
100 I'll have a look at ↑it,
101 C Ye↑ ::s
102 CT and the moment that matters are solved,
103 you'll automatically receive a fax from us
104 C Thanks very much.

Caller presents a hearing problem with incoming calls (lines 10–13). When CT responds "*Yes now you crack too*" (line 15), she incorrectly suggests that caller has described the troublesome incoming calls as "cracking." After some joking remarks in lines 16–22, CT announces that she is going to make a report (line 24). In response to this, caller states that he has not yet finished his problem presentation, as he has "*two things in fact*" (line 27). While his first problem is a hearing problem with incoming calls, his second problem is that when colleagues phone him, the line disappears altogether.

CT then asks for various names and numbers, which she enters into her computer. Caller then answers another call, and when he returns, CT reports what she has entered in the computer: "*I have made a note here uh the switchboard* (.) *cracks*" (line 87). Apart from the fact that caller did not talk about the line "cracking," this is but the first of the two problems caller reported. This may explain why caller does not immediately respond to CT. When CT does not continue her problem description after 2 seconds of silence (line 88), caller reminds her of the second problem: "*yes but I don't hear anything* (...) *if people call.*" CT acknowledges this correction by caller ("*Ok↓ay:*"), and continues her problem description "<°*operator* (.) *hears* (.) *nothing*°>" while typing. After marking the end of her recording the problem description ("*Okay*"), she reads aloud what she has just entered: "*switchboard cracks, lines fall away sometimes, and* (.) *operator hears nothing.*" After caller's acknowledgment ("*yes*"), CT states that the reporting of the problem has now been completed ("*Okay noted down*"), and she announces her future action: "*I'll have a look at it.*" This then initiates the closing of the call.

5. Conclusion

By working aloud when creating an electronic ticket, and/or by reading back what has been recorded, CTs enable callers to assist in ensuring that the problem is entered correctly and completely. Working aloud and reading back allows for the production of a collaborative, interactionally constructed problem description by CT and caller together. This, of course, does not guarantee that at the end of the day the problem will be solved. It does guarantee, however, that the information on the ticket that will be relayed to the problem solving organization matches the information caller provides. As pointed out in the beginning of this chapter, this correlation is important for these help desk calls, and it is the primary reason for introducing a script that reminds the help desk agents to report back what they have recorded as the problem description.

It turns out that CTs not only read back what they enter, but they also work aloud when recording names, numbers and the problem description. Our analysis shows that working aloud is very effective for allowing callers to monitor what is being done with their problem reports and to correct CTs, if needed. The CTs' working aloud may also explain the fact that callers design their talk with an orientation to their information being written down. Once CTs have indicated that they are going to make an electronic ticket, callers tend to present their information in a recordable format. That is, callers tend to produce "talk for forms" (Longman & Mercer 1993). They use telegraphic speech, they spell their family names, and they provide fax and telephone numbers digit by digit. In short, they speak the language of the written form.

Note

* We thank Paul ten Have, Kate Henning and Jürgen Streeck for their helpful comments on an earlier version of this paper.

References

Baker, C., Emmison, M., & Firth, A. (2002). 'Discovering order in opening sequences in calls to a software helpline'. In A. McHoul & M. Rapley (Eds.), *How to Analyse Talk in Institutional Settings* (pp. 41–57). London: Continuum Press.

Bergmann, J. (1993). 'Alarmiertes Verstehen; Kommunikation in Feuerwehrnotrufen'. In T. Jung & S. Müller-Doohm (Eds.), *"Wirklichkeit" im Deutungsprozess. Verstehen und Methoden in den Kultur- und Sozialwissenschaften* (pp. 283–328). Frankfurt am Main: Suhrkamp.

Chafe, W. L. (1982). 'Integration and involvement in speaking, writing, and oral literaure'. In D. Tannen (Ed.), *Spoken and Written Language: Exploring Orality and Literacy* (pp. 35–54). Norwood, NJ: Ablex.

Frankel, R. M. (1989). '"I wz wondering – uhm could *Raid* uhm effect the brain permanently d'y know?": Some observations on the intersection of speaking and writing in calls to a poison control center'. *Western Journal of Speech Communication, 53*, 195–226.

Heritage, J. & Sorjonen, M. L. (1994). 'Constituting and maintaining activities across sequences: *and*-prefacing as a feature of questioning design'. *Language in Society, 23*, 1–29.

Houtkoop-Steenstra, H. (1990). 'Normative and analytical perspectives on interviewing techniques'. In H. Pinkster & I. Genée (Eds.), *Unity in Diversity; Papers Presented to Simon C. Dik on his 50th birthday* (pp. 232–251). Dordrecht/Providence: Foris.

Houtkoop-Steenstra, H. (2000). *Interaction and the Standardized Survey Interview. The Living Questionnaire.* Cambridge: Cambridge University Press.

Jefferson, G. (1989). 'Preliminary notes on a possible metric which provides for a "standard maximum" silence of approximately one second in conversation'. In D. Roger & P. Bull (Eds.), *Conversation: An Interdisciplinary Perspective* (pp. 156–197). Clevedon: Multilingual Matters.

Lerner, G. (1995). 'Turn design and the organization of participation in instructional activities'. *Discourse Processes, 19*, 111–131.

Lerner, G. (1996). 'On the 'semi-permeable' character of grammatical units in conversation: Conditional entry into the turn space of another speaker'. In E. Ochs, E. A. Schegloff, & S. A. Thompson (Eds.), *Interaction and Grammar* (pp. 238–277). Cambridge: Cambridge University Press.

Longman, J. & Mercer, N. (1993). 'Forms for talk and talk for forms: Oral and literate dimensions of language use in employment training interviews'. *Text, 13*(1), 91–116.

Meehan, A. J. (1989). 'Assessing the "police-worthiness" of citizen's complaints to the police: Accountability and the negotiation of "facts"'. In D. T. Helm, W. T. Anderson, A. J. Meehan, & A. W. Rawls (Eds.), *The Interactional Order: New Directions in the Study of Social Order* (pp. 116–140). New York: Irvington.

Schegloff, E. A. (1986). 'The routine as achievement'. *Human Studies, 9*, 111–151.

Schegloff, E. A. (1992). 'Repair after next turn: the last structurally provided defense of intersubjectivity in conversation'. *American Journal of Sociology, 98*, 1295–1345.

Schegloff, E. A. (1996). 'Turn organization: one intersection of grammar and interaction'. In E. Ochs, E. A. Schegloff, & S. A. Thompson (Eds.), *Interaction and Grammar* (pp. 52–134). Cambridge: Cambridge University Press.

Schegloff, E. A. & Sacks, H. (1973). 'Opening up closings'. *Semiotica, 8*, 289–327.

Schegloff, E. A., Jefferson, G., & Sacks, H. (1977). 'The preference for self-correction in the organization of repair in conversation'. *Language, 53*(2), 361–382.

Schegloff, E. A., Ochs, E., & Thompson, S. A. (1996). 'Introduction'. In E. Ochs, E. A. Schegloff, & S. A. Thompson (Eds.), *Interaction and Grammar* (pp. 1–52). Cambridge: Cambridge University Press.

Silverman, D. (1997). *Discourses of Counselling. HIV Counselling as Social Interaction.* London/Thousand Oaks/New Delhi: Sage.

Whalen, J. (1995). 'A technology of order production: Computer-aided dispatch in public safety communication'. In P. ten Have & G. Psathas (Eds.), *Situated Order. Studies in the Social Organization of Talk and Embodied Activities* (pp. 187–231). Washington, DC: University Press of America.

Whalen, J., Zimmerman, D. H., & Whalen, M. R. (1988). 'When words fail: A single case analysis'. *Social Problems, 35*(4), 172–185.

Whalen, M. R. & Zimmerman, D. H. (1990). 'Describing trouble: Practical epistemology in citizen calls to the police'. *Language in Society, 19*(4), 465–492.

Zimmerman, D. H. (1984). 'Talk and its occasion: The case of calling the police'. In D. Schiffrin (Ed.), *Meaning, Form, and Use in Context: Linguistic Applications* (pp. 210–228). Georgetown University Round Table on Language and Linguistics. Washington, DC: Georgetown University Press.

Zimmerman, D. H. (1992). 'The interactional organization of calls for emergency assistance'. In J. C. Heritage & P. Drew (Eds.), *Talk at Work* (pp. 418–470). Cambridge: Cambridge University Press.

Chapter 4

The metaphoric use of space in expert-lay interaction about computing systems

Wilbert Kraan

1. Introduction

Over any given stretch of talk, participants have a number of different devices at their disposal that will regulate the discourse, structure thought about the topic of the talk and allow the strategic pursuit of their interactional goals. In the helpline interaction data analysed here, three metaphoric conceptual models appear to play a role in all of these functions. The *agent-trajectory* model appears to be the most dominant of the three. It involves the experiencer (the person who experienced the action narrated in a particular stretch of discourse), perceiving and conceiving herself in terms of a person moving along trajectories through a space and manipulating mostly container-like objects. Common phrases that exemplify this model include "starting up a program to get at the data", "putting files on a network drive" and "logging into a server via VPN". In this model, the computing system is conceived and talked about as a space, and the various files, programs, servers and other components of a computing system as containers and objects in that space.

The *direct interface* conceptual model, by contrast, is limited to actions by experiencers on directly perceived elements of the computer's interface: both the hardware (keyboard, mouse) and the software (icons, menus and windows). This model is characterised by more elaborate phrases such "clicking on the 'start' button, then 'applications' and then 'wordpad'" or "then I hit 'return' to get rid of the dialog box". It is less metaphoric than the agent-trajectory model, but it is still less direct or literal than the manipulation of 'real' objects that it resembles (Hutchins et al. 1986; Turner & Fauconnier 1995).

The *personification* conceptual model, finally, puts the computer in an agentive, human-like role and computer users in a patient role. In this model, the actions of the users are conceived to be subject to the independent will of the machine. Typical examples include "the computer won't let me in", "it threw an error" or "it says that the file is corrupt". Note that other machines in similar situations do not tend to be talked about as sentient, communicative beings. One "can't get into" a building or car, but not that they "won't let me in", unless access is regulated by some computer system.

While these conceptual models appear to have a primary role in the way users conceive the operation of computers, the alternation between them also has consequences for the negotiation of different stages or *frames* of the helpline call. As will be demonstrated, the opening stage of a call is often, but not exclusively, characterized by a notable shift from an agent-trajectory model to a personification model. These shifts are strategically motivated in the sense that callers appear to shift between the conceptual models in ways that will help them reach their interactional goals (ultimately, having their problem solved by the helpdesk agent) as efficiently as possible.

In the following sections, the appropriateness of the agent-trajectory conceptual model to the task of (talking about) operating a computer will be examined first. This will be followed by a consideration of the metaphoric nature of the computer interfaces at various levels, and how this aspect relates to domain specific knowledge and the nature of particular computer operation tasks. The second half of the chapter will consist of a detailed examination of the role of the different conceptual models in the strategic negotiation of different frames in the early stages of a call.

2. Spatial scalability in concepts

Given the graphical nature of modern computer operating system user interfaces like Apple Macintosh and Microsoft Windows, it might seem quite straightforward that talk that deals with computer operation is mainly spatially structured. The representations on the screen and the nature of input devices as keyboards and mice may appear to lead fairly directly to the spatial descriptions evident when people discuss operating a computer in the helpdesk calls investigated here. Generally, however, only users who appear to have just started to familiarise themselves with modern PCs have a tendency to talk about operating the machine in terms of the direct interface model (cf. Maglio & Matlock 1998). Even in such cases, a spatial description is rarely the exclusive conceptualisation of a computer system over stretches of discourse longer than a few utterances. It is therefore important to distinguish talk that is structured by directly perceived elements of the interface in this manner from equally spatially structured, but rather different, metaphoric conceptualisations. This is illustrated in example (I).

Example (I)
(caller C has just completed a call with A about the same topic)

```
1   A: hello, computing
2   C: yeah, I just talked to you about the hotmail
3   A: yeah
4   C: yeah, so here I am with my double you, double you, double
       you ["w.w.w."], hotmail dot com
5   A: hm, hm
6   C: I don't see return (0.5) I mean
7   A: no, I mean, is that got you into the page?
8   C: yeah, well, what do I do? is it search or netscape or what
       do I hit?
9   A: er, are you seeing the hotmail home page?
10  C: er, no, not yet, I'm still on the Bangor page actually
11  A: okay, you have typed that in, just press the return key
12  C: oh, the return key, wherever it is,
13  A: the enter key
14  C: oh, the enter, okay
```

Where the agent A in turn 7 adopts the pervasive and metaphoric agent-trajectory conceptual model that functions as a common ground between more experienced users and helpdesk staff, caller C's problem lies at the level of the direct interface. Consequently, he largely, if not exclusively, talks about operating his computer in those terms. Notice that A shifts to a similar conceptualisation in turn 9 and later after C's less than positive reaction in 8. In comparison to the direct interface model, two things are notable about the agent-trajectory conceptualisation and will be further developed here: The fact that it is scalable in complexity and the fact that the alternation of this with any other conceptualisation appears to be mainly governed by considerations of perspective and *framing* (in the sense of Goffman 1974, 1979).

It is scalability that sets the agent-trajectory conceptual model apart from the direct interface model. Rather than describe actions in terms which correspond to directly perceived elements in the experiencer's environment, it abstracts from that experience and provides a spatial structure that is entirely its own. The main characteristic of the agent-trajectory conceptual model is that the trajectories that experiencers describe can easily scale in the level of detail they comprise. Complex trajectories through, alongside or inside a variety of landmarks in the computer space can be quickly collapsed by merely mentioning the superordinate landmark of the computing system that is being discussed.

Alternatively, and much more frequently in helpline interaction, a more schematic trajectory can be decomposed in a much more detailed description by skilful helpdesk agents. The direct interface scheme, with its reliance on directly perceived elements, cannot be so fortuitously scaled and is therefore less economical in use with more experienced computer system users. In that sense, it appears that the more skilled a speaker considers herself and her interlocutor to be in a particular computer operation, the more schematic the trajectories are likely to be and vice versa. In this, the agent-trajectory model appears to leverage affordances that are commonly used in tasks like giving someone directions around an unfamiliar town. (II) illustrates:

Example (II)

```
23  A3: yeah, that maybe a facility there, the only other way I can
        think of doing it of the top of my head, (1.) is, if you log
        onto publix, (1.2)
        [d'you have ]
24  C4: [publix?, ye ]
25  A3: have you? you go through teemtalk or terraterm
26  C4: what? internet?
27  A3: no no no, erm, you go log onto windows ninety-five and
        one of the options under network apps is (0.7) erm (0.8)
        teemtalk
28  C4: right
29  A3: yeah,
30  C4: yeah
31  A3: start that one up
32  C4: yeah
33  A3: and the same, the default option in the box is, er, is
        publix
34  C4: yeah
35  A3: you log onto that using the same username and password
```

When A3 uses a comparatively schematic trajectory to describe how to obtain an email list in turns 23 and 25, C4 clearly doesn't understand, after which A3 breaks down the trajectory into much more detailed, constituent parts. This ability of helpdesk staff and other advanced computer users to quickly and flexibly scale what can be quite complex spatial arrays according to communicative need, may go some way to explain why knowledge transfer in helpdesk calls is often comparatively successful.

Conceptual models can also play a role in miscommunication in helpdesk calls, however. One important property of them is that the same model can be *profiled* (Langacker 1987) against several closely related as well as very different domains of experience. That is, constructions and referents that are familiar to callers may actually be used from a perspective that may only be familiar to the helpdesk agent or other experts. Call 1a6 (from which examples (III), (IV) and (VI) are drawn) illustrates the problem. This call is part of a series that revolves around a problem with local access to a centrally stored database. Early on in the first call of the series, the following happens:

Example (III)

24 C5: 'cause once he realised we didn't have an in-house helper he'd decided to come up
25 A4: oh yeah=
26 C5: =and could help. now that he whizzed through it in a very professional manner
27 A4: yes
28 C5: and he er said that oh I hadn't logged on properly
29 A4 yes
30 C5: but if I hadn't logged on properly I get a little error message saying you have not logged on properly redo, or whatever it is the message is
31 A4: yeah
32 C5: so I feel I would have known if that had been the case, anyway, he said oh there we are, you have got your files now, there you are, and off he went

The key phrase here is line 28 and further: "hadn't logged on properly", which recurs in a later call. C5 clearly took this is as a severely face threatening challenge to her (professional) competence in the interface domain and therefore seeks to clear the matter with some urgency. From the stated circumstances of the case as well as the responses of A4, however, it is clear that A4's colleague profiled his comment within the domain of network hardware: C5's personal computer was physically relatively far removed from the servers she was trying to log into, which led A4's colleague to hypothesise that the low-level process of establishing a reliable network connection had not been executed successfully by the system. The properties of 'logging in' – both the concept and its verbalisation – can, therefore, be quite different depending on the domain in which it is profiled. Alternatively, what 'logging in' means in context can depend on the perspective you take, or are able to take, on that particular action.

Conversely, computer systems or the people who operate them also need to be able to represent essentially the same entities in markedly different conceptual models as well as domains in order to facilitate task specific control. A webpage, for example, needs to be represented as just pictures and text for a user seeking information about a topic, but as textual HTML code for the editor of the page and possibly as binary code packets for an application– or network engineer.

This underlines both the pervasiveness of metaphor in talk about computer operation, and the fact that the use of a literal versus figurative dichotomy does not seem very helpful in its analysis. Specifically, to contend that one of the low level conceptualisations (or the output representations that facilitate them) is more 'true' or 'real' in an even subjectively perceived sense seems, at the very least, not as important as its appropriateness for a particular task. It is the inaccessibility of the network hardware domain to C5, and hence the construction's inappropriateness for the task of explaining the problem, that led to the misunderstanding in (III), not the notion that the network hardware conceptualisation would be more true or factive. Another illustration of the inadequacy of a binary classification of conceptualisations is provided by the usage of a third widespread metaphor in helpdesk interaction: the personification conceptual model mentioned previously. In this representation, users present not themselves, but the system in an agentive role. Line 49 in (IV) illustrates.

Example (IV)

45 C3: so I have gone into the help thing to see, well, you know to see if I could solve it
46 A4: yeah
47 C3: too complicated for me
48 A4: [chuckles sympathetically]
49 C3: it said things like set your net control file directory

In these few lines, C3 smoothly shifts from thinking about the computing system she discusses in terms of a spatial array she traverses to thinking about it in terms of a rather adversarial, independent agent. Neither is objective or literal in a conventional sense, but, as we shall see in the next section, is perfectly suited for the particular task C3 is engaged in. To sum up, binary notions of verity, whether derived from perceptual, logical or socio-cultural factors, appear to be just one of the many dimensions that impinge on talk. In the helpline data used here, it is frequently task or activity type (Levinson 1992) specific considerations that appear to govern conceptual-semantic structures.

3. Conceptual models and interactional structure

So far, I have mainly discussed some conceptual-semantic properties of helpdesk talk. Yet as the discussion of the perspective mismatch in (III) already shows, there is a potentially considerable strategic-interactional dimension to conceptualisation choice as well. In fact, the conclusion of the previous section – that the choice of conceptualisations appears to be mainly governed by task specific considerations – leads fairly directly to questions about the nature of the relation between the conceptual-semantic and the strategic-interactional levels of discourse. That is, if conceptualisation choice is task specific, which task determines what conceptualisation? Either the task described in one of the helpdesk phone calls could licence any of the three major conceptual models described above, or the task of the helpdesk call itself can licence them. To be more precise, what triggers switches between different conceptualisations could be either a matter of perception or conception at the time of the computer operation task described, or strategic projection at the time of the helpdesk call. After dealing with that question, I will outline in more detail an important way in which different conceptual models can be exploited strategically at the interactional level.

An answer to the question about which level governs conceptualisation choice is suggested by the fact that a considerable number of the subjects switch between conceptualisations at interactionally expedient points. Most notably, a prototypical helpdesk call will start (after conventional openers) with a caller's exposition of the problem that she wants help with. This stage is most likely to be verbalised in the agent-trajectory structure evident in line 7 of (I), most of A3's utterances in (II) and line 45 in (IV). The latter example also makes clear that switches to a different structure – what I have called the *personification* model – occur at the point at which the caller finishes her exposition and hands the floor to the agent. Line 2 in (V) is a very short, but representative example.

Example (V)

```
1   A1: hello, advisory
2   C1: hiya, I'm trying to get onto my email, I've been given your
        number, erm, it doesn't seem to be letting me in at all=
3   A1: =okay, what's your number?
```

Since the point at which the caller loses control over the computer ("it doesn't seem to be letting me in at all") in experienced time and the specific point at which help is requested of the helpdesk employee at the time of the call is cotemporal in the data, it could be the computer operating experience or strategic expediency within the call which triggers such conceptual switches. Evidence from repairs – in which callers restate an action begun in one conceptualisation to another – does suggest a primacy of interactional strategy, though. (VI) illustrates:

Example (VI)

```
104 C5: (0.9) right, now then, I have got paradox seven, (1.6) main
        thing in front of me, working directory or basically and it
        says in the little, er, (0.6) well showing you which file
        I'm on, it's en, colon, slash, halls, slash, fridd,
        slash, database=
105 A4: =yeah, so that's there?
106 C5: yes, and I can see underneath that a list of all files that
        I would normally just click on and bring up
107 A4: yes
108 C5: and have access to
109 A4: yeah, and when you do that it says [you do should bring it
        back]
110 C5: [and when I do ] that now
111 A4: yeah
112 C5: I click on one of my files and it says, an erro, error box
        comes up, unable to open file
113 A4: yeah
114 C5: so I go okay
```

Clearly, when subject C5 either comes across a problem that she understands (i.e. when she feels she still has the system under control) or when she doesn't want to project a conceptualisation switch yet, the agent-trajectory structure prevails. Since these repairs are phenomena local to the call – as opposed to some sort of perception repair during experienced time – it seems that local interactional strategic reasons prompt her to the repair and defer a conceptualisation switch from a agent-trajectory to a personification structure.

4. Interactional strategy at the beginning of a helpdesk call

Since switches from a agent-trajectory to a personification conceptual model are not the only strategic devices used in helpdesk calls, it is time to broaden the scope to other interactional phenomena that occur at a specific stage in these interactions. Specifically, I have used the word 'strategic' to describe the interactional dimension of conceptualisation switches. This implies that a speaker has a choice between different strategies to attain their interactional goal(s). Such strategies were indeed apparent in an investigation of the opening moves in an arbitrary subset of twenty helpdesk calls of the total data set of this study. Overall, four different strategies of managing the transition between the problem exposition frame and the resolution frame where identified in this way. The conceptualisation switch was the most frequent strategy in the subset (11 out of 20).

To understand why conceptualisation switches are such frequent frame switching cues, it is important to, first, understand the competing subgoals of the activity type at this stage of the call and, second, to outline strategic interactional advantages of each of the frame switching cue strategies. All callers in the twenty instances have called the agents for broadly the same interactional goal: to resolve a computer related task that is beyond their own capabilities. The agents' goal has been to help the callers as quickly and as efficiently as possible. It seems that in the negotiation of these two goals, the structure of the activity type generally and the switch between the two particular frames focused on here starts to make sense. To start at the socio-institutional level; these interactional goals have been formally identified. The organisation of which both agent and callers are a part has recognised the right of ordinary computer users to be helped with queries: "...any user with a computing problem can come for advice; from the trivial: 'How do I send my output to the laser printer?', to obtaining help with writing Fortran or C programs" (ASG, 1997). Conversely, the agents are obliged in principle to provide that help. As the occurrence of a variety of frame switching strategies indicates, though, the institutionally codified rights and obligations are not the only relevant ones.

First, in spite of the assertion that trivial questions are welcome, it is likely to be clear to both participants that the helpline is a finite resource. The 'Advisory' helpline investigated is managed by one person at a time, who has to deal with all queries from all channels (face to face, phone and, to a lesser extent, email). There is, therefore, only a finite amount of time and attention available for one caller. There is some indication in the data for this in the sense that, as noted earlier, the majority of calls are made by users who got stuck while trying to solve a problem on their own, in spite of the fact that the facility is there to provide advice. Personal experience within the organisation as well as personal communication with advisory staff indicates that quite a few computer related problems are referred to more or less knowledgeable direct colleagues, presumably for the same reason.

Second, information technology skills tend to be highly valued in the developed world, considerations of the domain's complexity, constant state of flux, recentness of the phenomenon and general lack of formal training notwithstanding . This also works conversely: people who have not been able to develop skills in the domain are in danger of being referred to with the pejorative 'computer illiterate'. Requesting help with a computer related skill, therefore, inherently carries a threat to the speaker's own positive face (Brown & Levinson 1987). Within the speech activity studied here, this is likely to be exacerbated considerably by the fact that computer skills in general, and often the specific task callers request help on, can easily be viewed as an integral part of a user's core professional competence. The indignity C5 expresses on being told that "*you have not logged on properly*" in line 30 of (III), and the lengths she goes to in the rest of the series of calls to show A4 that she is competent in the user action of logging on, certainly underlines that fact.

Third, a request for help, even if the requester has an institutional right to be helped, is always an imposition on the requestee. Effort is requested on the part of the agent, which is combined with a dependency of the caller on the agent's compliance. The entrenched relatedness of these two factors in the activity type can be seen clearly in (VII), where C3 needs to spend some effort in cancelling the request for help A3 infers by indicating that he is not dependent on A3's resolution of his problem by stating "*that's okay*" and "*no great loss*".

Example (VII)

1 A3: hello
2 C3: hello, is that [A3's first name]?
3 A3: speaking
4 C3: hi, its [C3's first name] here, er [full name]
5 A3: oh, hello
6 C3: hi, thanks I got my emails back=
7 A3: =oh, right=
8 C3: =and my addressbook seems to be working=
9 A3: =oh, good=
10 C3: =the only thing is, on on, the web pages, er, I have lost my bookmarks, but, that's okay=
11 A3: ermm, that's, errr (2) erh hhhhh
12 C3: no great loss
13 A3: (1) are you sure?
14 C3: yeah, yeah

Given these institutional and interpersonal constraints, and most of all the understood goal of the activity, some of the regularities across the different calls can be motivated. The frame structure of the interaction at the stage focused on here fairly directly follows from that overall goal: after an obligatory conversation opener, it makes sense for both participants to let the caller outline what the problem is, and, if at all possible, what exactly she wants help on.

The agent can then check her understanding of the caller's situation by asking further questions or, based on the evidence just received, offer some advice.

For the caller, that sequencing of the process means that the subgoals at this stage are to

1. convince the agent that her request is worthy of attention (minimising imposition)
2. minimise positive face threat to self
3. provide sufficient information, and no more, for the agent to provide a speedy and satisfactory solution to the request.

Many dimensions of subgoal (3) are call specific, contextual givens, but one could hazard that in most cases it tends to be a function of the degree of complexity of the problem and the state of knowledge the caller has or wishes to project. What is a strategic choice that reappears in all instantiations of the activity type, is the caller's judgment regarding the optimum balance between providing too much information – thus wasting time and effort – and providing sufficient information for an efficient response. Reconciling subgoals (1) and (2) is a potentially delicate balancing act.

The question, then, is how the participants are to negotiate the early interaction frames. In some (emergency) cases, the strategy is almost determined by the circumstances of the caller's situation or the circumstances override any conscious choice. For most other cases it appears that the caller has a basic strategic choice between some five points on a continuum of directness. 'Directness' in the sense of overtly stating one's intent. Callers can either go for a maximally indirect frame switch cue or a maximally direct frame switch cue or one of the points inbetween. The most indirect strategy appears to be a 'cue-less' frame switch like the one used by C7 in (VIII).

Example (VIII)

```
1  A8: hello there
2  C7: hello, hiya, I think I've exceeded my quota, so I've had a
       message
3  A8: (0.8) right, er (8.0) hold the line a minute=
4  C7: =yeah
5  A8: (0.9) you're not gone, yeah? (2.0) righto (1.7)username
       please?
6  C7: [username]
        (17.5)
```

In this case, the problem is so routine (as indicated by the fact that users receive automatically generated emails when it occurs), that a simple statement of the issue appears to be enough for A8 to both determine that C7 wants to switch from an exposition to a response frame almost immediately, as well as to determine what course of action to take as his response. Taking such an indirect strategy means relying on the basic structure of the activity type frames for the interpretation of the utterance by the agent. The advantage is that it allows the caller to avoid making a threat to her own positive face (subgoal (2)). In some circumstances, like those in (VIII), it may also be very effective in terms of subgoal (3) in that it facilitates a minimally extensive exposition frame. It does very little for subgoal (1), however. There is little in the way of minimising the imposition of the caller on the agent by making the worthiness of the caller's cause clear. This is further exacerbated by off-loading the interpretation of the caller's intention onto the agent, which is a risky strategy.

A slightly less indirect strategy is to use contrasts in tense and/or aspect to cue the transition from the exposition to the initial advice frame. Such purely temporal frame switch cues are comparable to a 'cue-less' strategy with regard to the off-loading of the burden of intention interpretation from the caller to the agent; even if one of the two examples in the data set, (IX), has a temporally foregrounded clause that is in effect a statement of incompetence: "*I have just not been able to open it*". This is sure to further subgoal (1), even if it does rather compromise subgoal (2).

Example (IX)

1 A2: [Welsh:] dau pedwar dau dau [English:] two four two two, can I help you?=
2 C4: =hello, erm, .. someone sent me a message (0.5) and the document is a rowr es ar oh ["rorle s.r.o."] (1) I have just not been able to open it
3 A2: what's it called again?
4 C4: it is, erm, well, in the message this is, it says that the file=
5 A2: =yeah
6 C4: is called ror, rorle, ar oh ar el ee dot es ar oh ["rorle.sro"]
7 A2: (1) dot es ar oh (1) I don't know that one
8 C4: right, er, it says that, er, (1) it is, er (1) if you're using pegasus mail, it says, then you can use the browser's extract function to lift the original content

The third frame transition strategy are reported speech cues. These are quite closely related to the personification conceptual model mentioned above; mainly because the speech being reported is that of dialog boxes that the computer interface puts out. Still, since speech reported from the interface does not put the machine in an agentive role, reported speech cues are analysed as a different category here. In terms of directness, reported speech cues such as "*login failed*" in turn 4 of (X) are less indirect than temporal cues and therefore offer some advantages in the way of subgoal (1) because they further the conveyed worthiness of the caller's request by implying that an unnamed third party scuppered the caller's best attempt to remain in control over the situation. Unlike the more indirect strategies, this also requires less utterance interpretation effort of the agent, thus furthering subgoal (1) The implied agent also neatly side-steps the worst of the inherent threat to the caller's own positive face of a request for help: it is the system's unhelpful or baffling behaviour that brought them into this predicament, not their own lack of computer skills. This strategy also allows one to exploit the enriched implicature for economy of time and effort with regard to the provision of sufficient information: the exposition of the problem can double up as an argument for quite how unreasonable the machine is.

Example (X)

```
1  A4: hello, advisory
2  C6: oh, hello, its [full name] speaking here, from modern lan-
       guages
3  A4: hi
4  C6: I've been trying to get into my email today, and although
       I'm certain I'm putting in the right password, I'm getting
       'login failed', all the time=
5  A4: =what are you using?
6  C6: erm, (0.2) netscape (1.1) mail
```

The fourth strategy callers use to manage the transition from an expository to an initial advice frame are agentive switch cues. These rely on the sudden switch from a agent-trajectory to a personification conceptual model. Agentivity switches take all of the advantages of reported speech cues one step further: The responsible agent is named and the variety of actions that can be attributed to it can be even more convincing of the worthiness of the request and the threat to the caller's face can be further attenuated by off-loading the responsibility. There may be a potential pay off with regard to subgoal (3), however. Relating the action of the interface that stopped one's trajectory through the system is not the most direct and unambiguous way of indicating what exactly the caller requires help on. Call (XI) points to that possibility by combining the agentivity strategy with the fifth frame transition device; the question cue.

Example (XI)

```
1  OA1: hello [full name]
2  A4:  oh, hello, [OA1's first name], ehm, I wonder if you can give
        me some advice here, a user has a laptop from sweden that
        talks windows ninety eight in swedish=
3  OA1: =yeah
4  A4:  (0.5) in order to get it to speak english, does he have to
        reinstall windows?
5  OA1: (1.2 ) buy a copy of english windows 98 and then install
        that
```

OA1's response in turn 5 indicates, however, that the question cue's advantage of lesser ambiguity with regard to what exactly help is required on may not be effective in all instances. A4's question conventionally requires a yes/no response, but OA1 chooses to respond to the first, personified part of turn 4. In most other instances, though, a caller's choice of the question cued strategy is effective in terms of subgoal (3), certainly with regard to a speedy response. Put differently, if the caller thinks he has a good idea what the problem is or what the solution is likely to be, a question strategy is the most specific and therefore potentially the most effective in terms of the overall activity type goal. As such it could potentially even minimise threat to the caller's positive face since projecting a potential answer to the request shows competence. Provided the agent agrees with the implied solution, that is – which makes this strategy rather risky. With regard to subgoal (1), though, the question cue strategy is inherently risky; the adjacency pair turn structure that a question carries leaves the agent very little option but to reply. In this sense, the strategy could almost bypass subgoal (1) altogether in favour of subgoal (3) It should be noted, however, that the question cue strategy is most frequently employed by callers who have chosen to contact the helpline before attempting to solve the question. This choice means that the goal of showing the worthiness of your request is achieved by different means. Where reported speech and agentivity switched cues rely on a 'helplessness through loss of control' implicature, callers using question cues rely on a 'deference to authority' implicature: it is the positive face of the agent that is appealed to. That is, subgoal (2) is heavily traded off in favour of subgoal (1).

In all, there are some good reasons why spatially structured conceptual switches are the most frequent frame switch cues at the early stages of helpdesk interactions. Apart from structural characteristics such as the simple given that it juxtaposes two of the most widely accepted computer system conceptualisations, it allows the helpdesk client to state her problem fairly directly, while saving her face and establishing the worthiness of her cause. In this sense, such interactional considerations appear to be more important (in this activity type, at least) than questions of perception at the time of experience.

5. Conclusion

From the data analysed so far, it is clear that the conceptual metaphoric use of conventional spatial structure can play an important mediating role (Lakoff & Johnson 1980; Lakoff 1992). It facilitates both the presentation and the understanding of complex and abstract computer systems as well as lay person – expert interaction about such systems. It also functions as a resource for negotiating interactional goals structurally and strategically. Given this interconnection, it is clear that the spatial structure actually used in interaction is not a simple function of the model that the interface presents of the system to the user. Equally, that same structure appears not so much an immutable, conventional given, but a flexible, contextually sensitive resource.

References

Application Support Group (1997). *Application Support.* Web page: http://www.bangor.ac.uk/is/asg/asgfront.html [31-01-2000]

Brown, P. & Levinson, S. (1987). *Politeness; Some Universals in Language Usage.* Cambridge: Cambridge University Press.

Turner, M. & Fauconnier, G. (1995). 'Conceptual Integration and Formal Expression. *Journal of Metaphor and Symbolic Activity, 10*(3).

Goffman, E. (1974). *Frame Analysis.* New York: Harper and Row.

Goffman, E. (1979) 'Footing'. *Semiotica, 25*(1/2).

Hutchins, Edwin L., Hollan, James D., & Norman, Donald A. (1986). 'Direct Manipulation Interfaces'. In Donald A. Norman & Stephen W. Draper (Eds.), *User Centred System Design.* Hillsdale, NJ: Lawrence Erlbaum.

Lakoff, G. & Johnson, M. (1980). *Metaphors We Live By.* Chicago: University of Chicago Press.

Lakoff, G. (1992). 'The Contemporary theory of metaphor'. In Andrew Ortony (Ed.), *Metaphor and Thought* (2nd edition). Cambridge University Press.

Langacker, R. (1987). *Foundations of Cognitive Grammar; Volume 1 Theoretical Prerequisites.* Stanford: Stanford University Press.

Levinson, S. C. (1992). 'Activity types and language'. In P. Drew & J. Heritage (Eds.), *Talk at Work; Interaction in Institutional Settings.* Cambridge: Cambridge University Press.

Maglio, P. P. & Matlock, T. (1998). 'Metaphors we surf the web by.' In *Workshop on Personal and Social Navigation in Information Space.* Stockholm, Sweden. http://www2.ucsc.edu/~tmatlock/pubs.html [20/7/1999]

Part II

Emotional support

Chapter 5

The mitigation of advice

Interactional dilemmas of peers on a telephone support service

Christopher Pudlinski

1. Introduction

The most recent types of telephone support service to arise in the United States have been warm lines. Similar to crisis hotlines, which can be traced back to the establishment of the Los Angeles Suicide Prevention Center and the Samaritans in England in the late 1950s (Farberow & Shneidman 1961; Lester & Brockopp 1973; Fox 1968), consumer-run warm lines have emerged within the past decade to compensate for shortcomings of existing professional programs. Both hotlines and warm lines mediate between a person's need for help and their reluctance to turn to the bureaucratic entanglements of existing health and social services. Both operate "after hours" when therapists, counselors, and other mental health support staff are not available. In contrast to crisis hotlines, consumer-run warm lines have three defining characteristics. They are a peer-run, pre-crisis service, designed for providing social support.

As a peer-run telephone support service, warm lines are staffed by working consumers. Working consumers is the term I chose to represent clients of the community mental health system who are employed on the warm line. They are typically clients who show some interest in working on the line. They are trained by professional staff, who discuss issues of confidentiality, setting boundaries, and being respectful of callers; share information on community resources; and conduct role plays of typical calls and crisis calls.

As a pre-crisis service, warm lines let clients discuss issues before they become serious. Working consumers are instructed to forward or refer the occasional crisis call to an associated crisis hotline service. Warm lines thus differ from crisis hotlines as "hot" calls, including suicidal issues and other urgent problems, are typically not dealt with by the warm line. Nonetheless, warm lines play a beneficial role in crisis prevention (Klein, Cnaan, & Whitecraft 1998); they permit hotlines to deal more exclusively with urgent problems and have led to a reduction of crisis calls from the community mental health clientele.

As a third characteristic, warm lines differ from help lines – information and referral services – because their main objective is listening and supporting, not referring or advising. This particular characteristic is called into question in my subsequent analysis, for giving advice is one of the activities working consumers, like other social support providers, perform.

2. The practice of giving advice

Studies using conversation analysis have started to map out the situated practice of advice-giving in a variety of contexts: medical settings (Heritage & Lindstrom 1998; Heritage & Sefi 1992; Leppänen 1998; Ragan, Beck, & White 1995); HIV counseling (Kinnell & Maynard 1996; Silverman 1997); career counseling (Vehvilainen 2001); service encounters (Jefferson & Lee 1992); and a call-in radio advice program (Hutchby 1995). These studies can be categorized into two types: (a) in situations where the advice-giver is clearly the authority, more overt advice tends to be given; (b) in situations where the advice-giver also wishes to be seen as a friend or equal, more mitigated methods are typically used.

In medical encounters, the practitioner's need to address client problems typically leads to advice-giving that is straightforward or overt. For instance, Heritage and Sefi (1992) examine advice giving by British health visitors (HVs) to first-time mothers. HVs give advice by using strongly prescriptive forms: overt recommendations ("my advice to you is that"); imperatives ("always be very very quiet at night"); modal verbs of obligation ("should"; "ought"); and factual generalizations. Leppänen (1998), in a study of Swedish district nurses, also found frequent use of imperatives and modal verbs of obligation. Sometimes, however, they used slightly more mitigated forms: presenting advice as an allowable alternative ("if you then go to the doctor … can you ask him to…") or by describing patients' future actions. These slightly mitigated forms were used when patients are not seemingly aware of the relevant issues and nurses are concerned about the risk of a possible misunderstanding.

Mitigated or delicate methods, used in response to particular institutional dilemmas, have also been found in other professional settings, including the practice of therapy (e.g., Ratliff & Morris 1995) and HIV counseling. For example, professional counselors, obligated to give advice on safe sex practices, do so in the format of information (Kinnell & Maynard 1996; Silverman 1997). "The advice-as-information device heightens ambiguity between talk as advice for the client or as information, and provides more latitude for the client to hear the talk as generally rather than personally relevant" (Kinnell & Maynard p. 422). This strategy allows the counselors to help clients solve problems by putting forth information as advice, while allowing the clients some autonomy: leaving it up to the client to determine whether the information was relevant to their specific circumstances.

Across these situations, advice giving has a natural asymmetry, often implying that the advice recipient is not as knowledgeable on that issue. This asymmetry often creates dilemmas for both advice-giver and advice-recipient (e.g., Goldsmith 1992; Goldsmith & Fitch 1997; Heritage & Sefi 1992). As a foundational work on troubles telling and advice giving, Jefferson and Lee (1992) explain this by describing how responding to troubles, especially in service encounters, is inherently dilemmatic: a convergence of two distinct environments.

> Jefferson and Lee (1992) state that recipients of troubles-telling in ordinary conversation are frequently expected to express great concern for troubles-teller's feelings and perceptions, and often find their advice ignored – if not openly rejected – by troubles-tellers reluctant to accept the lesser interactional status of advice recipient. In institutional settings, on the other hand, service providers often express indifference to troubles-tellers' feelings while insisting that others acquiesce in their preferred solutions.
>
> (quoted in Silverman 1997: 188)

The person expressing concern/empathy is playing a role that involves focusing on the other person and not on the information being shared. The role of the person expressing indifference, however, plays an agency-supported role of information giver. Jefferson and Lee argue that advice giving is not a fitted pair to troubles telling. A troubles recipient, concerned about the trouble teller's feelings, and an advice-giver, orienting to the other as an advice-seeker, are seemingly separate and contradictory roles. In this way, providing empathy/concern and advice giving can be viewed as aspects of two distinct environments.

Yet, these contrary roles can merge in certain contexts. Putting forth options as information – a strategy that actively involves the clients in developing instructions – is a way to deal with the dilemmas inherent in balancing the role of the expert with the needs for rapport and patient involvement and education. In the opening chapter of his book, Silverman (1997) notes that HIV counselors are pulled in two different directions: health-promotion (information-giving) and non-direct counseling. This is a contrast between telling the clients what they should do (giving information, such as regarding the use of condoms and other safe sex practices), and facilitating clients' own efforts to arrive at their own decisions.

These studies provide two relevant insights into the practice of advice giving. First, the relationship between parties is a significant factor. Professionals, asserting their status as knowledgeable experts, typically give advice in straightforward and/or direct ways.[1] On the other hand, counselors tend toward more indirect and/or mitigated advice, since expressing concern and/or befriending clients also is part of that agenda. However, little is know about what occurs in peer-to-peer advice giving. While it would seem likely that mitigated advice would occur when there is symmetry between parties, it seems unlikely that power asymmetry plays such a clearly defined role as to affect advice giving in terms of mitigation across all circumstances.

Second, one cannot speak to advice giving (or any practice per se) without a clear description (and understanding) of relevant contextual and sequential elements (see Goldsmith & Fitch 1997), including the relationship between the advice-giver and advice-recipient. "The study of advice should both carefully explicate the details of the production of advice and show how these details are systematic products of the interactants' orientations to specific features of the institutions" (Leppänen 1998: 210). To do this, this study focuses on a particular circumstance in which advice is given and examines instances of the most direct forms of advice given on a peer support line in response to a client's troubles telling.

3. Dilemmas of advice-giving on consumer-run warm lines

Three different consumer-run warm lines located in the northeastern United States were studied. At each of the sites, as a prelude to my recording of selected calls, I attended at least two months' worth of staff meetings, read training manuals, and consulted with supervisors and working consumers. A previous article (Pudlinski 2001) concluded that warm lines are typified by ideological dilemmas (see Billig et al. 1988), consisting of three contrary themes: connectedness, problem solving, and nondirectiveness. These themes evidence themselves in the practices of these peer workers. Connectedness involves building a peer support network and establishing client relationships with both the warm

Table 1. Connecting themes to training injunctions

Theme	Training injunction
1. Connectedness	Establish and seek commonalities
	Build peer friendships
	Develop a peer support network
2. Nondirectiveness	Let the client talk
	Maintain appropriate boundaries
	Respect clients as decision-makers:
	Let clients take responsibility for their own solutions
3. Problem solving	Help clients to solve problems
	Be knowledgeable care-givers
	Refer clients to other community resources
	Ensure client safety:
	Support medical professionals' recommendations

line and individual workers. Problem solving involves ensuring client safety and putting forth solutions to client problems. Nondirectiveness, which values respecting client autonomy, involves active listening and maintaining appropriate worker-client boundaries. These three themes are further summarized in Table 1.

These two latter themes directly conflict when solutions are put forth in response to a client's report. Indeed, these two themes of nondirectiveness and problem solving are contrary for the participants as follows. Respecting client autonomy involves letting clients take responsibility for making their own decisions. In contrast, helping clients to solve problems involves working consumers taking some responsibility for clients' problems and solutions. These themes can be seen as counter-values, with each value taught in training as something that supports the warm line's interests, and with each value having a counter-value that seemingly instructs workers in an opposing direction.

Supporting this indigenously derived characterization of this particular dilemma is the fact that professional helpers (e.g., therapists and counselors) experience a similar dilemma of responsibility (e.g., Dryden 1987; Kinnell & Maynard 1996; Silverman 1997). "In a way it is the question of how much one stands back, feeling that the person's freedom is very important even if he [sic] is going to do something that appears to you self-destructive" (Dryden, p. 98). This specific dilemma is tied to two contradictory societal beliefs: the belief that individuals are responsible for themselves and their actions, and yet the recognition that some people (e.g., suicidal clients; young adults suffering from eating disorders) require guidance while dealing with their significant problem(s).

4. Methodology: Procedures for data collection and analysis

Given that working consumers reported aspects of calls as problematic and that these problems often stemmed from wrestling with contrary themes, I turned to the actual phone calls to further explicate these contrary themes and to explore how working consumers dealt with them. From February 1996 to January 1997, I periodically observed three different warm lines and recorded select calls. These three lines, founded in 1993 or 1994, varied in terms of structure. At site 1, 20 volunteers staffed the lines, with 2 or 3 of them together working 4 hour shifts in an office setting. At site 2, a beeper was employed, allowing the 2 paid working consumers to take calls in their homes. At site 3, 5 consumers worked the warm line. Calls were forwarded into the home of the consumer working that evening. While hours varied between sites, all lines were open for at least 4 hours in the early evening, 7 days a week, and 365 days of the year (except at site 1).

At each site, I monitored two weeks of calls. Only at site 2 was call volume atypical during this study period. At site 1, 37 calls from 15 different callers were received over 12 days and 67 hours of operation during May and June 1996. At site 2, in December 1996 and January 1997, 34 calls from 8 different callers were received. On the average, this line – open 6 hours per night – usually receives 12 calls per week. At site 3, over a period of 14 consecutive days in December 1996, 19 calls from 7 different callers were received during its 56 hours of operation. All in all, informed consent was received on 44 calls.[2] I then transcribed all 44 calls: 3 ½ hours of calls (18 calls) at site 1, over 5 ½ hours of calls (12 calls) at site 2, and 3 hours from 14 calls at site 3.

When working consumers responded to clients' reports of troubles, they made choices that attempted to manage these contrary themes in some way. I thus used the ethnographic information I learned about the organization, summarized above, to connect my findings derived from analysis of talk-in-interaction back to the agency and warm line itself.[3]

Given the fact that responding to client problems seemed to be the type of situation in which working consumers had the most difficulty, I searched for various types of responses tied to clients' reports. After having found nine general categories of responses, I focused my analysis on the category with the largest number of instances: offering recommendations and solutions. I reviewed all 129 instances (within 89 different episodes) of offering recommendations and solutions to find out what different methods were used. These instances were found in three sequential environments: a client described a problem and the working consumer responded by putting forth an option (49 episodes); a working consumer got concerned over a client's non-problematic report (24 episodes); the client described a solution and the working consumer put forth an alternate option (16 episodes). This paper examines the first two of these three sequential environments.

5. Encouraging clients to adopt a solution

A previous article (Pudlinski 1998) described the three most common methods used by working consumers to put forth options in response to client reports: incorporating a solution within a query ("did you see the doctor"); sharing one's own problem and solution; and merely giving information about a solution ("I know that bereavement class is ... on the twenty-sixth"). These methods fit the form [speaker appears to be doing X while also doing Y], where [doing Y] is putting forth an option for the client to adopt. Working consumers primarily used these implicit methods to put forth options in response to client reports because of the contrary themes operating in the warm line setting.

However, the strength of such analysis often hinges upon deviant cases (Heritage 1988); this chapter focuses on those instances where working consumers seemingly put forth options in a more direct and straightforward manner in response to a client's report of a problem. They do this by presenting that option as necessary or desirable. With this strategy, working consumers actively encouraged clients to adopt a solution. Encouraging adoption of a solution involved formulating a course of action and either positively evaluating it ("it's very important") or using a form of obligation ("you should"; "why don't ya"). These forms of obligation and/or positive evaluations of an option were also found in warm line calls in other circumstances, including when working consumers disagreed with current client plans, put forth options on issues mutually viewed as not problematic, or gave a client a recommendation to be passed on to a third party. In those other circumstances, however, these characteristics served a much different function.

Encouraging clients to adopt a solution occurred in two particular circumstances: (a) in response to an urgent medical or psychological problem (after and/or in the course of the client telling about a problem); (b) when a client implicitly sought a workable plan.

Client tells of an urgent problem

Crisis calls were especially troubling for working consumers. For example, a working consumer at site 3 saw crisis calls as the biggest challenge of working on the warm line: "every time the phone rings I do a 'flip in my stomach' that this could be the one: a crisis. I am always wondering if it is going to be a crisis call." Indeed, stories of the worst calls other working consumers have ever received invariably consisted of crisis calls. Similarly, another working consumer, at site 2, reported to me his own feelings of inadequacy at the end of call #26 – a particularly difficult call:[4]

> I don't think I'm gonna sleep that well tonight. I'm worried that she might be suicidal... I blew it big time! It was my worst hour on the warm line... She felt I was doing well, but I thought I wasn't.

Crisis calls were the most challenging for three reasons. First, the determination of whether a client was in crisis was often difficult to make. Uncertainty as to what to do or say was common. Rare was the call that presented itself as a clear-cut crisis. Thus, it was difficult for working consumers to know what to do. Although supervisors instructed working consumers to err on the side of caution, referring a client to a crisis service when a client was not in crisis, or would have preferred to talk to some one on the warm line, could have negative ramifications. For example, a working consumer at site 3 called Crisis on a friend who disclosed suicidal thoughts to her over the warm line; the friend, who typically expresses such thoughts in a semi-serious manner, was very upset that Crisis was subsequently involved. Second, working consumers were, by default, responsible for the client's welfare up until the time they referred the call to Crisis. Yet, I observed instances where crisis services were not able to mollify the client and clients sought out the warm line again for further assistance. Lastly, working consumers often felt powerless in a situation in which clients needed help the most. This feeling stemmed from contrary instructions, for all three sites trained working consumers how to deal with a client in crisis and yet all three sites also urged these peers that the best option was to forward these calls to an appropriate expert: an affiliated crisis service.

The first example is the only recorded one of a client expressing suicidal thoughts:

EXAMPLE 1: CALL #26 (site 2; DH = working consumer)

```
1   C:   Well you know there are times when I get so depressed
2   DH:  Mm hmm
```

C: That I would like to take all of my phenobarbital at one
night with a bottle of bourbon and say (0.9) that's all she
wrote
(0.6)
DH: Do you have a doctor by the way
C: Oh yes
DH: Do you tell him or her
C: No I don't
DH: Why don't ya- (0.2) you don't- you're afraid to tell her
C: Um hmm
DH: Uhm: (3.2) wh- (0.6) well it's very important that you
speak to a your doctor about this
(0.6)
C: I mean 'n that- that's how depressed I get
DH: Mm hmm
C: .Hhh I do take accenden which is a anti-depressant
when I first took it dahm, (0.2) twelve years- uhh
fifteen years ago[, I was=
DH: [mhm
C: =taking four one hundred milligrams a night .hh and
now I only take ah two milligram- two hundred
milligrams a night (0.3) sote you know I hav- I have come
down

It was noticeable that the working consumer did not immediately encourage the client to speak to her doctor once the suicidal thoughts were expressed. Instead, he elicited information relevant to helping the client solve her problem. His first question (line 7) established a condition necessary for follow-up inquiries, functioning like a pre-sequence. First, establishing that the client had a doctor allowed a follow-up inquiry as to whether the client discussed the suicidal thoughts with the doctor (line 9). Had she reported that she told her doctor, the working consumer might have asked the client to report what the doctor recommended. Had this occurred, it would have fit the warm line's interest in supporting medical professionals' recommendations, as an aspect of the theme of problem solving.

However, in response to the client's assertion that she had not discussed the matter with the doctor (line 10), the working consumer inquired about her motives (line 11) and then strongly encouraged the client to speak to her doctor (lines 13–14). The working consumer portrayed the course of action as quite desirable for the client to do. He claimed it was "very important."

The working consumer's encouraging adoption of this solution was consistent with the warm line's interest in solving client problems. However, it was contrary to the warm line's interest in having working consumers not be responsible for advising clients, as an aspect of the contrary theme of nondirectiveness. When the client asserted that she had not discussed her suicidal thoughts with her doctor and confirmed that it was because of fear, the working consumer encouraged this course of action, though with some initial hesitation in the turn (line 13).

Encouraging the adoption of a solution implicates a dilemma of responsibility. For therapists and counsellors, dilemmas of responsibility consist of a choice between taking responsibility for client welfare (and to what degree), and respecting the clients' own innate abilities to make informed decisions about their lives (Dryden 1987). Even if the recommended action turned out to be successful, this may heighten the potential for the client to become reliant on the warm line service to such an extent that they would have been less able to help themselves; without self-help, a client's progress might have been short-lived. Indeed, clients might "experience a reduced sense of achievement because he or she needed assistance" (Goldsmith 1992: 270). It might then not be consistent with the warm line's interests in empowering clients as decision-makers. For these and other reasons, giving advice was not recommended at sites 1 or 2, who instructed working consumers to "suggest, not advise" and "offer options; don't give advice" respectively.

Yet situations arose where clients seemingly were in danger of harming themselves (e.g., consuming a bottle of pills and a bottle of alcohol in a suicide attempt). In these situations, a pro-active approach was recommended across sites. The difficulty of handling this type of situation would be characterized as follows: In urgent circumstances, if working consumers were not pro-active, clients might have needlessly suffered. Yet, if working consumers were pro-active, they risked putting forth options that might have been incorrect for clients or have backfired, where the working consumers got blamed for their recommendations. However, alternate strategies, such as more indirect ones, lacking in persuasive force, could have been viewed by clients as not pertinent to them and/or rejected easily by them by treating them as irrelevant (Silverman 1997). Thus, a more explicitly persuasive approach, albeit hedged, in offering solutions was treated as appropriate in urgent circumstances.

In this second example, the client displayed an urgent medical problem. Here, the working consumer encouraged the adoption of a solution within the following circumstances: It was not the initial discussion of this problem and a serious problem availed itself to the working consumer. The coughing fit provided an opportunity for the working consumer to reopen discussion of a problem previously discussed.[5]

EXAMPLE 2: CALL #41 (Site 3; T = working consumer)

T: =[Mm hm
C: =[((twenty coughs, with some loud inhales))=oh brother
T: That sounds pretty bad. who- who told ya ta go to the
hospital and get this checked ou°t°
C: My landlord's wife
T: Yeah (0.5) ahh by the sound of that cough Nell I don't
know maybe- maybe that's pretty good advice maybe you
should get it checked out=
C: =I bouldn't be surprised if I got walking nahmoanya
T: You should getit- you should ya know, you should probably
what do you usually do go to the emergency room when your
sick
C: Yeah
T: Why don't ya go there and jus tell em: tell em: your
[not feelin well
C: [((coughs))=how am I gonna get a ride there
T: Hah
C: How am I gonna get a [ride there
T: [Call Rides do you have the
number
(0.5)
C: Yeah but they're closed
T: Well [you don't have ta do it- you don't hafta do it=
C: [.Hh they close at eight
T: =today call like on Monday and ask em ta set up an
appointment
C: Okay I'll do that

The 15 second long coughing fit showed the seriousness of the client's problem. The working consumer did not have to rely on client reports of medical problems; the physical symptoms were readily apparent over the phone. A working consumer could know – without report – if a client was not feeling well by their tone of voice, coughing, etc. Since the physical symptoms were available to the working consumer, the client was in no position to deny that the problem existed.

Right at the end of the client's own assessment of her coughing fit, the working consumer inquired about an option authored by a third party. Within the inquiry, the working consumer brought up the option of going to the hospital and getting her condition checked out (lines 3–4). She followed the inquiry with her own endorsement of this option (lines 6–8). And in her next turn (lines 10–12) the working consumer strengthened her endorsement by using a form of obligation ("should") and incorporating a similar solution within a query. After the client's acknowledgment ("yeah"), the working consumer provided a script for the client to use at the emergency room (lines 14–15). By portraying these actions as simple to do, the working consumer was encouraging the client to go to the emergency room.

While the working consumer encouraged the client to go to the emergency room, she did a number of things that hedged her taking over responsibility for the client's solution. Each move she made was mitigated in some way. In lines 6–8, by using the same terms for the option as she had previously used to describe the third party's "advice," the working consumer clearly could have been seen to be reiterating the original advice: "get this checked out." Also, she again prefaced the option with "maybe." In lines 10–11, the working consumer started to reiterate her prior utterance, but changed to reaffirming it by asking a double question. She moved away from a potentially unmitigated reiteration ("you should get it (checked out)") and ended up going with a safer alternative by inquiring about a more specific course of action. This was a safer alternative as the working consumer elicited the client's perspective and carefully worded the specific option. The emergency room, being a specific part of a hospital, carried with it the fact that the stay at the hospital would have been temporary, yet reiterated the urgent nature of the problem. Had the working consumer completed her attempts at encouraging a specific activity, she would have assumed more responsibility for the client's action. By eliciting the client's perspective, the working consumer allowed the client to participate in the decision about this matter.

In addition to mitigating her endorsement and eliciting the client's perspective, another way to diminish or share responsibility was to put forth an option that deferred responsibility to a particularly relevant expert: a "doctor." In these instances of encouraging adoption of a solution, the working consumer encouraged the client to consult with health professionals and/or use community resources. The working consumers encouraged clients to talk to a psychiatrist or counselor in examples 1 and 3; to visit an emergency room (example 2); and to call to confirm an already made appointment with a local transportation service (example 4). In three of the instances, the working consumer deferred the diagnosis and solution to an appropriate health professional. Going to an expert in response to a serious problem was a fairly uncontroversial course of action; it would be consistent with the warm line's interests in supporting medical authorities and community services. In this way, the working consumers diminished their responsibility for the client's future course of action. Referrals to a professional – one that had the expertise to solve long-term crises and other client problems – was also a typical and valued function of other telephone support services, such as crisis hotlines.

In a post-call interview, the working consumer felt the client's illness was the "main focus of call," as indicated on the post-call tracking sheet: "I advised her to definitely go to emergency room & get checked out for bronchial cough she said she's had it quite some time. She was already advised to get it checked out. Told her to call Rides or Hot Spot for transportation (I made sure she had Rides #)." The working consumer used such strong language as "advise" and "definitely" to show her taking some responsibility for the client's welfare. Yet, she was also careful to note that her advice was secondary to – and consistent with – another person who "already advised." She diminished her own responsibility for the client's solution by seconding a third party's recommendation. Thus, the working consumer's written summary of the call also reflected the delicate middle ground between taking too much and taking too little responsibility for a client's subsequent actions.

Although urgent situations were recognized by working consumers and supervisors across sites as circumstances in which it might be in the warm line's interests to look after clients' welfare and help them solve their problems, even in these situations, working consumers mitigated their endorsement of a solution and put forth safe options. By encouraging use of medical authorities and community services, the working consumers were both addressing the clients' problems and limiting their liability for the clients' future well-being.

Client implicitly seeks a solution

Encouraging adoption of a solution was also used as an initial strategy in response to a problem in circumstances when clients implicitly sought a workable plan. When clients implicitly sought a recommendation, working consumers encouraged the client to adopt a noncontroversial, safe option:

EXAMPLE 3: CALL #36 (Site 3; N = working consumer)

```
C:  Yeah I'm jus- I'm really scared because like, (0.4)
    they cut my social security
N:  Mm hmm
C:  Down because I was working [and like] I don't know if=
N:                             [ Mm hmm ]
C:  =social security I know after like three months
    they'll bring it back up [but ] 'hat I'm scared about=
N:                           [Yeah]
C:  =is that um (0.3) they're gonna ah not give me the
    social security for either quitting or gettin fired
    but ya know I don't know if this is a good enough
    reason to quit or not because of my anxieties
```

```
N:  Yeah, (0.3) note umm why don't ya talk to your
    psychiatrist about it
C:  Yeah well [I'm gonna talk to her tomorrow]
N:            [I mean if he can say he thinks] ya want-
    He thinks you should quit (0.2) or he or she
C:  Yeah and she can give [me a (paper)
N:                        [If ya get a note
C:  Yah
N:  Ats all ya need
C:  Yeah I'm gonna talk to umm actually not my my
    psychiatrist tomorra cause she's gonna be out
    [but one] of the therapists at ah=
N:  [Mm hmm]
C:  =Southern Rutherford
N:  Mm hmm
C:  (And its-) I'm gonna talk to her tomorrow and=
N:  =Yeah sall ya need is a note from a- even a therapist
    or something that ya know
    (1.2)
C:  That I'm not ready yet (0.6) and ah I'm also worried
    about losin my state if I [(do get it      )
N:                            [yeah if ya keep working you
    will
    .
    .
    .
N:  Yeah s'd that's why ya need to talk to like your
    psychiatrist or your therapist somebody get- somebody
    ta write ya a note
C:  Yeah [(that say something)
N:       [Make copies make copies of ah ya know get one
    letter make a cop- make hom- like y'know five copies
C:  Yeah give one to ta state give one to [social=
N:                                         [social=
C:  =security and one to the employer]
N:  = security one to your job        ]
    (1.2)
C:  In fact I'm gonna write that done I'm gonna write that
    down on a piece of paper (0.4) I'm gonna [write that=
N:                                           [ okay
C:  =down right no[w uhm]=
N:                [okay ]
C:  =actually that's some pretty good advice I thank you,
    for that
N:  But you said you didn't want any advice I'm
    sor[ry about that
C:     [Oh I just I didn't I didn't expect it [before
```

N: [I kn(h)ow

w(h)ell (h)okay

EXAMPLE 4: CALL #39 (site 3: Z = working consumer)

C: I'm gonna go the MHA Christmas party

Z: Oh alright

(1.2)

Z: That's something to look forward to

.

.

.

C: I just hope (0.5) taxi shows up (0.7) just my luck it won't

Z: Who's this

C: The taxi

Z: Oh the taxi

C: .Hhhhh °yeh°

(0.9)

Z: Why don't you call right now and confirm it

C: °Prescut° (0.7) .hhh I can't cause they're not in

(2.0)

C: Its ah through the state

(0.4)

Z: Okay

C: I can call in the morning and confirm it

(0.2)

Z: Good

(0.3)

C: Good idea thank you

Z: Okay

(3.0)

In examples 3 and 4, the clients implicitly asked for working consumers' recommendations by claiming uncertainty about how to deal with this problem and/or disclosing their anxiety over some future action. Implicitly asking was in contrast to explicitly seeking and not seeking. Explicitly seeking likely would have involved an information-seeking query (e.g., "what can/should I do about this?"). In contrast, not seeking would have involved presenting a report of a current problem that did not involve an information-seeking query about possible solutions or involve uncertainty/worries about future actions; examples 1 and 2 did not seek solutions in this way.

In example 3, the client expressed uncertainty about quitting his job (lines 10–12) and portrayed himself as scared about ramifications for social security entitlement. In example 4, the client expressed worries over the reliability of transportation to a social event. In both instances, the client implicitly sought advice on how to handle their situation.

The ways that the working consumers encouraged adoption of a solution and the ways that the clients responded suggest that contrary themes were operating. Working consumers, by encouraging safe options, addressed injunctions within a theme of problem solving. While putting forth recommendations by encouraging adoption of a solution was consistent with the warm line's interests in helping clients to solve problems, the putting forth of a safe option was consistent with the warm line's interests in supporting medical authorities and/or community resources. The clients' active involvement shows that they may be orienting to the contrary theme of nondirectiveness and its emphasis on autonomy.

In these examples, the working consumers encouraged the adoption of a solution. It was appropriate for the working consumers to help clients solve their problems when the clients sought solutions from them. Withholding advice, when sought, can be problematic. "It would be foolish, if not unethical, to withhold crucial information or advice when the client is extremely vulnerable or confused, or when she is not likely to know certain things, or arrive at important conclusions, without assistance" (Feltham 1995: 18). As such, both clients sought certainty and assistance. By calming the clients' current feelings of anxiety, the working consumers could have met site 3's main objective: make sure the client was safe for the night.

In these situations in which clients implicitly sought a solution, the working consumers recommended safe courses of action. In example 3, a recommendation tied directly to the issue which the client sought advice would have involved encouraging quitting and risking loss of social security or encouraging continuing to work despite his high anxieties. Either recommendation would involve the working consumer's being liable for the client's future action on this matter, and potentially liable for creating more client anxiety, if the option that the working consumer put forth did not work out as planned and/or the client lost out on additional income. Instead, the working consumer solved the problem by encouraging the client to talk to her psychiatrist about it. This was consistent with the warm line's interest in supporting medical authorities.

One indication of contrary themes was how the working consumer encouraged a safe option. Another indication that contrary themes were operating for clients was their displays of knowledge after working consumers' encouraged adoption of a solution. Although the focus of this analysis is on working consumer strategies, the clients' implicitly seeking a solution seemed to be followed by an eagerness to describe how they would implement the solution that the working consumers put forth.[6] This was a significant response to the working consumer's strategy. Once a solution was put forth, clients were quick to get involved in formulating the details of a workable plan, demonstrating their capabilities to solve a problem, once such guidance was provided. This was important as a subsequent move for the clients; it balanced seeking help with a demonstration of their own abilities to make the solution work for them. The clients in examples 3 and 4 declared their intent to implement the recommendation: "*I'm gonna talk to her tomorrow*"; "*I can call in the morning and confirm it.*" This indicated acceptance of the solution put forth, and yet also added sufficient details, such as when they would do the intended action. Such involvement allowed the client/patient to take some responsibility for their solutions (Ragan et al. 1995). Such responses also are consistent with how clients responded to advice in certain medical settings through assertions of their own knowledge or competence (e.g., Heritage & Sefi 1992; Leppänen 1998).

Client involvement was consistent with the warm line's interests in not having working consumers be responsible for client actions. As such, if clients would typically take responsibility for detailed solutions they themselves had sought, encouraging adoption of a solution would likely support warm line interests two fold: helping clients to solve problems and yet avoiding working consumer responsibility for these solutions.

The working consumers' encouraging adoption of a safe, uncontroversial course of action was appropriate when a client implicitly sought a solution or advice regarding a troubling circumstance. Both clients displayed their appreciation for the working consumers' recommendation: "*good idea thank you*" (line 24; example 4); and "*actually that's some pretty good advice I thank you, for that*" (lines 55–56; example 3). This appreciation even occurred in example 3, when the working consumer encouraged a safe course of action that did not specifically address the client's worry, and the client earlier had indicated that he did not want advice (as recalled in lines 57–58 by the working consumer). Finding similar tokens of appreciation in observations of peer advice giving, Goldsmith and Fitch (1997) note that "these closing formulas explicitly convey *respect* for the quality of the information provided and the expertise of the provider and *gratitude* for the effort and *caring* that are demonstrated in the provision" (p. 469). As such, offering safe, uncontroversial advice, in these circumstances, elicited client gratitude and enhanced the helping relationship between working consumers and clients. In circumstances where a solution was implicitly sought and a recommended course of action was uncontroversial, it was especially appropriate for warm line purposes.

6. Discussion

Warm lines, as peer support services for people with mental disabilities, foster client decision-making through a web of contrary themes: connectedness, nondirectiveness, and problem solving. As part and parcel of this dilemmatic ideology that characterizes consumer-run warm lines, circumstances arise in which working consumers must put forth advice – typical of a third theme of solving client problems. Putting forth safe, uncontroversial options and mitigating these options were ways to be consistent with contrary themes. It was in the warm line's interests for working consumers not to take responsibility for client solutions yet it was also in the warm line's interests to help clients solve problems. These features arose when working consumers encouraged adoption of a solution.

Encouraging adoption of a solution involved formulating a course of action and either positively evaluating it or using a form of obligation. It seems to occur in more urgent circumstances and seems particularly fitted to such circumstances. Where a client's problem is especially serious or urgent, working consumers seemingly assume more responsibility for the client. Contrary themes of problem solving and nondirectiveness are addressed when this strategy is used. Working consumers tend to deal with these dilemmas by seeking that elusive middle ground between taking responsibility for client welfare and respecting clients as decision-makers. They diminish the responsibility they take in certain ways: options (a) tend to be safe ones (e.g., talk to an expert), (b) put forth with mitigation, and sometimes (c) put forth after other, more indirect methods are tried first. This limits working consumers' liability and minimizes resistance (to some extent).

These mitigated aspects also occur when clients implicitly seek a solution, for working consumers are careful not to take too much responsibility for the option. As a feature of working consumers' encouraging adoption of a solution, clients' putting forth details of the solution facilitates their involvement. Clients might want to be seen as competent, especially if they sought out help initially (e.g., Heritage & Sefi 1992). So, clients quickly get involved in describing how to implement the plan and share the responsibility for adoption.

This study has implications for the practice of advice giving, combining ethnography and conversation analysis, and calling for help. Recognizing that options can be put forth in many ways, this study presents a different conception of advice giving than previous ones. While encouraging adoption of an option is a common strategy for giving advice, its mitigation is less common. Even when urgent problems present themselves (e.g., suicidal thoughts) or callers implicitly seek solutions, options are put forth with some mitigation. This is likely a characteristic of advice giving being a dispreferred action on peer-run warm lines, while advice is often a necessary and preferred action in other contexts. In professional contexts and others where asymmetrical relationships are typical, environments marked by urgency or clients seeking advice are typically characterized by advice given in an unmitigated and straightforward way. Mitigation, if it occurs in such environments, occurs in more uncertain circumstances, where clients disagree with or misunderstand the professional's portrayal of the problem (e.g., Leppänen 1998). To further understand these differences between peers giving advice and other more asymmetrical relationships, additional studies should examine other environments in which peer advice is given and in which both peers and professionals give advice, so as to compare these in a similar contextual environment.

As a second implication, I used ethnography to not only collect data that helped me to explicate concepts (e.g., themes and contrary themes) at the level of the organization itself but also to explicate concepts relevant to the lived experiences of the working consumers. The ethnographic findings were then helpful to connect claims about features of strategies to these organizational concepts. As previous research on advice giving clearly indicated, the contextual environment plays a significant role in the ways in which advice is given. Recognizing this, one could make an assertion about organizational-level (contextual) factors that are not demonstrated in one instance if they are clearly evident in another similar instance in another similar circumstance. The information I presented prior to example 1 about a common working consumer orientation to crisis calls exemplifies this. By describing how working consumers commonly orient to crisis calls, this one example can inform other practitioners about general issues that extend beyond the peculiarities of this one excerpt.

So, having a sense of what working consumers were taught in training and thought they should do during calls allowed me to make connections between the warm line in general and the practices of working consumers. By connecting to institutional features, I connect in similar ways as Kinnell and Maynard (1996) have to goals for pre-HIV test counselors. I recognize that people might not always be aware of these goals, but awareness can display itself in certain moments in the interaction. While conversation analysis is beginning to speak to multiple functions and goals (e.g., Pomerantz, Fehr, & Ende 1997), it is ethnographic methods that help to explicate the source and complexity of these goals and possible tension between goals at the level of the organization. As such, claims can be made about institutional aspects by looking at the interaction and doing some sort of correlation. Do features of strategies used by working consumers in interaction (e.g., keeping the client as teller) reflect warm line interests (e.g., let the client talk)? When they do, correlations can be tentatively asserted. Do features of strategies used by working consumers (e.g., respecting client autonomy while also putting forth a solution) correlate with contrary themes as observed (e.g., be nondirective yet also help solve problems)? When they do, correlations can be tentatively asserted. Thus, what might seem like a coherent analysis is really the result of careful interplay between my conversation analytic findings and relevant ethnographic data. Refinement of the analysis was multi-directional; it involved interplay between the interaction itself and the observational data.

Additional studies which combine conversation analysis with ethnography in a delicate back-and-forth interplay might help researchers to better understand these connections between practices and organizational culture/ideology. After all, responding to calls for help often occurs in an institutional context of some sort. Understanding the relevant themes/objectives to which help providers orient within that context has the potential for creating a richer, better informed analysis of telephone-mediated help.

Notes

1. Even a study of mother-daughter conversations notes clear asymmetry in the advice-giving practices of the mothers (Randall 1997).

2. Informed consent was primarily performed by working consumers at the beginning of the phone call. They read a brief script, assuring clients of anonymity.

3. For a fuller treatment of the debate on integrating ethnography and conversation analysis, consult colloquiums on this issue in the 1990/1991 issue of *Research on Language and Social Interaction* (e.g., Hopper), a special 1994 issue of *Journal of Contemporary Ethnography* (e.g., Nelson), and a special 1998 issue of *Research on Language and Social Interaction* (Tracy).
My view is that both conversation analysis and ethnography can get at contrary themes, albeit in different ways. Combining these two methods has been an increasingly valued enterprise within a particular intellectual tradition that studies the lived experiences of interactants (e.g., Atkinson & Drew 1979; Maynard 1984; Moerman 1988; Sharrock & Anderson 1982). While ethnomethodology/conversation analysis traditionally found "ethnographic data unacceptable for use in … studying the organization of practical activities in and of interpersonal interaction" (Nelson 1994: 308), recent conversation analytic studies of talk, especially within institutional contexts, have used ethnographic data to varying extent so as to support claims and extend one's analysis (e.g., Goodwin 1990; Heath 1997; Kinnell & Maynard 1996; Komter 1998; Ragan et al. 1995; Silverman 1997). Ethnographic data is especially helpful when specialized training and prior social experience are necessary for a

fuller understanding of a particular institutional context (Cicourel 1992). Indeed, my attendance at staff meetings and my discussions concerning training practices and phone calls with working consumers and supervisors give me insights that made it easier to understand some of the peculiarities of warm line interaction.

4. Examples 1 and 4 are from the only 2 calls in my corpus of 44 recorded calls which were considered crisis-like calls by the working consumer.

5. This prior sequence occurs at the opening of this phone call, and is analyzed in Pudlinski (1998):

```
1.  C:  Just wonderful [I feel great I got the flu but I'm=
2.  T:                 [Yeah
3.  C:  =feeling good
4.  T:  Uhoh: you got the flu
5.  C:  Yeah pains in my chest 'n my
6.  T:  Yeah
7.  C:  An my stomach feels queasy in the mornin an=
8.  T:  =Mmhm
9.  C:  And I've got a- a headache and I've got a sore throat
10.     and I've got wr- achy joints
11. T:  Did you- are you doin anything for it did you see the
12.     doctor er
13. C:  No I didn't
14. T:  You didn't
15. C:  No I my- my landlord's wife suggested that I go to
16.     the hospital
17. T:  Yeah
18. C:  But I dunwant to I'll be okay
19. T:  You think you'll be alright
20. C:  Oh yeah I'm drinkin- I just had a grapefruit for
21.     dinner [pleny a hi- vitamin c w'n when the Can-teen=
22. T:         [Did ya
23. C:  =gets here I'm wanna ask em ta go get me a gallon of
24.     orange juice 'n a gallon a milk
25. T:  Mm hmm
```

6. See Pudlinski (2002) for further discussion of the tactics used by clients/callers to accept and reject advice on these warm lines.

References

Atkinson, J. M. & Drew, P. (1979). *Order in Court: The Organisation of Verbal Interaction in Judicial Settings.* London: Macmillan Press.

Billig, M., Condor, S., Edwards, D., Gane, M., Middleton, D., & Radley, A. (1988). *Ideological Dilemmas: A Social Psychology of Everyday Thinking.* London: Sage.

Cicourel, A. V. (1992). 'The interpenetration of communicative contexts: Examples from medical encounters'. In A. Duranti & C. Goodwin (Eds.), *Rethinking Context* (pp. 291–310). Cambridge, UK: Cambridge University Press.

Dryden, W. (Ed.). (1987). *Therapists' Dilemmas.* Cambridge, UK: Hemisphere.

Farberow, N. L. & Shneidman, E. S. (1961). 'A Survey of agencies for the prevention of suicide.' In N. L. Farberow & E. S. Shneidman (Eds.), *The Cry For Help* (pp. 136–149). New York: McGraw-Hill.

Feltham, C. (1995). *What is Counselling?* London: Sage.

Fox, R. (1968). 'The Samaritans.' In H. L. P. Resnik (Ed.), *Suicidal Behaviors: Diagnosis and Management* (pp. 405–417). Boston: Little Brown.

Goldsmith, D. J. (1992). 'Managing conflicting goals in supportive interaction: An integrative theoretical framework.' *Communication Research, 19,* 264–286.

Goldsmith, D. J. & Fitch, K. (1997). 'The normative context of advice as social support.' *Human Communication Research, 23,* 454–476.

Goodwin, M. H. (1990). *He Said She Said: Talk as Social Organization Among Black Children.* Bloomington, IN: Indiana University Press.

Heath, C. (1997). 'The analysis of activities in face to face interaction using video.' In D. Silverman (Ed.), *Qualitative research: Theory, Method and Practice* (pp. 183–200). London: Sage.

Heritage, J. (1988). 'Explanations as accounts: A conversation analytic perspective.' In C. Antaki (Ed.), *Analyzing Lay Explanation: A Case Book of Methods* (pp. 127–144). London: Sage.

Heritage, J. & Lindstrom, A. (1998). 'Motherhood, medicine, and morality: Scenes from a medical encounter.' *Research on Language and Social Interaction, 31,* 397–438.

Heritage, J. & Sefi, S. (1992). 'Dilemmas of advice: Aspects of the delivery and reception of advice in interactions between health visitors and first-time mothers.' In P. Drew & J. Heritage (Eds.), *Talk at Work: Interaction in Institutional Settings* (pp. 359–417). Cambridge, UK: Cambridge University Press.

Hopper, R. (Ed.). (1990/1991). 'Ethnography and conversation analysis after *Talking Culture* [Special Issue].' *Research on Language and Social Interaction, 24.*

Hutchby, I. (1995). 'Aspects of recipient design in expert advice-giving on call-in radio.' *Discourse Processes, 19,* 219–238.

Jefferson, G. & Lee, J. R. E. (1992). 'The rejection of advice: Managing the problematic convergence of a "troubles-telling" and a "service encounter."' In P. Drew & J. Heritage (Eds.), *Talk at Work: Interaction in Institutional Settings* (pp. 521–548). Cambridge, UK: Cambridge University Press.

Kinnell, A. M. & Maynard, D. W. (1996). 'The delivery and receipt of safer sex advice in pre-test counseling sessions for HIV and AIDS.' *Journal of Contemporary Ethnography, 24,* 405–437.

Klein, A. R., Cnaan, R. A., & Whitecraft, J. (1998). 'Significance of peer social support with dually diagnosed clients: Findings from a pilot study.' *Research on Social Work Practice, 8,* 529–551.

Komter, M. L. (1998). *Dilemmas in the Courtroom: A Study of Trials of Violent Crime in the Netherlands.* Mahwah, NJ: Erlbaum.

Leppänen, V. (1998). 'The straightforwardness of advice: Advice-giving in interactions between Swedish district nurses and patients.' *Research on Language and Social Interaction, 31,* 209–239.

Lester, D. & Brockopp, G. W. (Eds.). (1973). *Crisis Intervention and Counseling by Telephone.* Springfield, IL: Charles C. Thomas.

Maynard, D. W. (1984). *Inside Plea Bargaining: The Language of Negotiation.* New York: Plenum.

Moerman, M. (1988). *Talking Culture: Ethnography and Conversation Analysis.* Philadelphia: University of Pennsylvania Press.

Nelson, C. K. (1994). 'Ethnomethodological positions on the use of ethnographic data in conversation analytic research.' *Journal of Contemporary Ethnography, 23,* 307–329.

Pomerantz, A., Fehr, B. J., & Ende, J. (1997). 'When supervising physicians see patients: Strategies used in difficult situations.' *Human Communication Research, 23,* 589–615.

Pudlinski, C. (1998). 'Giving advice on a consumer-run warm line: Implicit and dilemmatic practices.' *Communication Studies, 49,* 322–341.

Pudlinski, C. (2001). 'Contrary themes on three consumer-run warm lines.' *Psychiatric Rehabilitation Journal, 24,* 397–400.

Pudlinski, C. (2002). 'Accepting and rejecting advice as competent peers: Caller dilemmas on a warm line.' *Discourse Studies, 4,* 481–499.

Ragan, S. L., Beck, C. S., & White, M. D. (1995). 'Educating the patient: Interactive learning in an OB-GYN context.' In G. H. Morris & R. J. Chenail (Eds.), *The Talk of the Clinic: Explorations in the Analysis of Medical and Therapeutic Discourse* (pp. 185–207). Newark, NJ: Erlbaum.

Randall, D. M. (1997, May). 'Advice giving sequences in everyday mother-daughter conversations.' Paper presented at the meeting of the *International Communication Association*, Montreal, Quebec.

Ratliff, D. A. & Morris, G. H. (1995). 'Telling how to say it: A way of giving suggestions in family therapy supervision.' In G. H. Morris & R. J. Chenail (Eds.), *The Talk of the Clinic: Explorations in the Analysis of Medical and Therapeutic Discourse* (pp. 131–147). Newark, NJ: Erlbaum.

Sharrock, W. W. & Anderson, R. J. (1982). 'On the demise of the native: Some observations on and a proposal for ethnography.' *Human Studies, 5*(2), 119–135.

Silverman, D. (1997). *Discourse of Counselling: HIV Counselling as Social Interaction.* London: Sage.

Tracy, K. (Ed.) (1998). 'Analyzing context [Special issue].' *Research on Language and Social Interaction, 31.*

Vehvilainen, S. (2001). 'Evaluative advice in educational counseling: The use of disagreement in the "stepwise entry" to advice.' *Research on Language and Social Interaction, 34,* 371–398.

Chapter 6

Four observations on openings in calls to *Kids Help Line*

Susan Danby, Carolyn Baker and Michael Emmison

1. Introduction

This chapter presents an initial investigation of communication between children and young people with telephone counsellors at *Kids Help Line*, a national Australian children's helpline. It focuses on the openings of the calls to show the ways in which these young callers disclose and describe their troubles and delicate or sensitive matters to adult counsellors, and how these counsellors, in turn, display an awareness of the interactional sensitivity of these descriptions.

The chapter makes four observations concerning what appear to be regularities in the talk. First, an investigation of the greetings draws attention to what Sacks (1992 Vol. 1:4) refers to as a 'procedural rule' which operates during the work of identification and recognition in the opening sequences (see also Schegloff 1979). The greeting term used in the counsellor's first turn in which only an organisational identification is provided, serves to propose – although not ensure – that the caller, should they choose, can remain anonymous. Second, we observe how the format of the counsellor's first turn gives the initiative to the caller to competently design their entry into the talk. Third, we observe the counsellors using a form of recipient design. We note that the second and subsequent turns by the counsellor (after the greeting/hello) typically follows an interactional pattern closely resembling the callers, in both greeting and pace and tone of the call. And fourth, our analysis shows that the counsellor searches for the 'reason' for the call, and not for an identification of a 'problem' within the call. There is an important distinction to be made between the 'reason' and the 'problem.' Once the reason has been identified, the counsellor no longer searches for a problem. Each of these four observations will be discussed in turn. In later sections of the paper, we indicate how these observed regularities reflect significant elements of the philosophy of the help line.

2. *Kids Help Line*

Kids Help Line is the only Australian national help line that deals specifically and entirely with a population of callers aged five to eighteen years. The helpine receives in excess of one million calls per year but currently has only the capacity to answer 400,000 calls, indicating the importance of this resource for young people needing personal support. For this study, fifty telephone calls chosen randomly over a six-month period were taken from a corpus of calls already audio-recorded for quality assurance. The calls were transcribed using the conventions of conversation analysis (Psathas 1995), with names and other identifying information changed to protect the anonymity of the caller and counsellor. This paper focuses on the openings of a subset of these calls, up to the point where the counsellors do more than minimal uptakes and make their first substantive insertions into the talk.

Observation 1

First, the ways that the counsellor respond to the caller's summons (the ringing phone) are important for the subsequent interactional work. Some examples of the first few turns are:

```
(rings)
CT: hi there Kids help↑line
    (0.8)
C:  um hello .hh um
    (0.2)

(rings)
CT: hi Kids helpline
C:  hi yeah um I've got a bit of a problem um I get teased at
    school a lot
    (0.4)

(rings)
CT: hi Kids help↑line
    (0.2)
C:  .hh hi um gidday how are you
CT: hi good thanks
```

The counsellor's first utterance on answering the phone is multifunctional. It serves, first, as an 'answer' to the phone summons (Schegloff 1968, 1986); second it serves to identify the organization; and third it can be oriented to as a greeting by the caller. As we see in the examples above the 'hi Kids help line' opening provides for a range of possible responses. In the Baker, Emmison and Firth (2001) computer helpline data, the call taker's greeting is longer. It typically takes the form:

> "thank you for calling technical support,
> my name is Penny,
> how can I help you?"

By contrast, the *Kids Help Line* counsellors' initial turn is arguably less constraining in its 'sequential implicativeness' (Schegloff & Sacks 1973). For example within their initial turn, the counsellors do not give their names, thus avoiding giving callers an implied obligation to provide their names. Such an obligation was noted by Sacks in one his earliest lectures in referring to a procedural rule whereby 'a person who speaks first in a telephone conversation can chose their form of address, and ... thereby chooses the form of address the other uses' (Sacks 1992 Vol. 1:4). Conceivably, if the caller was not a young person but an adult representing an organization, such as a school or probation service, then, according to Sacks, the procedural rule would dictate that the caller should offer their own organizational membership in the second slot. The counsellors do name 'kids help line' to confirm that the caller has reached the right place. They do say 'hi', – rather than a more formal 'hello' – thus inviting a return 'hi' or equivalent in reply, which secures a form of connection. Significantly they do not propose that the caller needs help (cf. "how can I help you?" in the computer helpline).

The design of this initial turn is important. We have observed that it does *not* include a proposed exchange of names, possibly preserving the notion of anonymity, and it does *not* presuppose that the caller wants help, which might seem curious since this is *Kids Help Line*. The contrasting form 'how can I help you' in the computer helpline effectively proposes that the caller begin by describing how they can be helped. The *Kids Help Line* design avoids this.

At the same time, the counsellors' initial greeting suggests that any form of first turn by the caller is permissible. This puts a greater onus on the caller to announce what comes next, but it also provides the caller with more choice about how to enter into the talk. And as seen in the examples above, callers design their first turn in quite different ways. This seems to reflect a philosophy that the caller be given the space to design their entry to the talk as they wish.

Observation 2

The second observation is concerned with the interactional management of getting into the calls. In these opening sequences, pauses are evident in many of the initial turns by the callers, as are placeholders such as *um*. These could be heard as the caller being unsure of exactly how to describe the problem. Extract 1 is an example where we see the caller describing his "*problems at school*" not in one continuous description but in segments, interspersed with various hesitation markers.

Extract 1

1_1_2

```
1   CT: hi there Kids help↑line
2        (0.8)
3   C:  um hello .hh um
4        (0.2)
5   C:  ah-uh like I wanted to talk to er to someone about like I
6       have some problems at school↑
7        (0.4)
8   CT: °sure°
9   C:  yeah
10  C:  .hh um ha I kind of have this trouble↑ (.) you see I don't
11      know why it happened?
12       (0.2)
13  C:  basically um .hh
14       (0.4)
15  C:  I was sick fo:or a couple of days↑ .hh
16  CT: yeah
17  C:  and yeah
18       (0.4)
19  C:  I sort of u:um
20       (1.0)
21  C:  um um what=would=you ah like went to the doctor's↑ .h and
22      like he he didn't give us he gave us some prescriptions so I
23      like yeah I stayed away from school for a couple of days↑
24  CT: mm
```

The caller's pauses make interactional spaces for the counsellors to offer acknowledgment tokens such as *yeah*, *okay* and *mm* to show that the counsellor is listening. These pauses also offer interactional space for the counsellors to begin speaking substantively, that is, offering something more than listening tokens, but this is not what typically occurs in these opening sequences.

Extract 2

2_1_11

CT: hi Kids help↑line
(0.2)
C: .hh hi um gidday how are you
CT: hi good thanks
C: um look (.) I'm just a bit worried right now (.).hh
CT: mm hm?
C: I'm in a stage of my life
(0.2)
C: where I'm um (.) developing
(0.4)
C: I don't know (.) different like to the rest of the boys in my
class?
CT: mm
C: a:and I (.) it's sort of becoming like
(0.4)
C: I wait for them (.) to get to me
(0.2)
C: and tease me↑
CT: °okay°
C: it's become really irritating no:w
(0.2)
CT: °right°
(1.0)

The caller in this case (Extract 2) says that he is "*just a bit worried right now*" (line 5) and goes onto present his concern, again leaving open some interactional spaces. We observe that after an initial *mm hm*? in line 6, the counsellor lets the caller go on with the description, even though the caller presented many interactional spaces where the counsellor could have inserted questions and made substantive comments. The next 'continuer' from the counsellor, *mm* in line 13, follows interrogative intonation by the caller in line 12, which is hearable as an invitation to respond (see intonation-response also in line 18–19).

Extract 3

1_1_3

CT: hi there kids helpline
(0.6)
C: hello um
(0.4)
C: my friend just got kicked out of home and she's got like
nowhere to sta(hh hh):ay
CT: mm
C: and um

9 (0.6)
10 C: she doesn't and she wants to make a few phone calls but she's
11 got no money on her pho:ne
12 (1.0)
13 CT: right
14 C: and we don't know what to do
15 (1.0)
16 CT: okay
17 (0.8)

The counsellors in these initial responses uses *mm* often. For instance, in Extract 3 above, in line 7 the counsellor uses *mm*. Gardner (1997) shows that *mm* is oriented to three uses: "as a weak acknowledger, as a continuer, and as a weak assessment token" (p. 133). The primary purpose of this particular receipt token is to indicate that the listener has no problem of comprehension or of hearing the talk (Gardner 1998). Also, the use of this particular token suggests that the counsellor has nothing significant to add to the topic of the talk. Unlike *yeah*, which can have an affirming response, *mm* "is just about as empty in semantic terms as a systematic vocalization can be" (Gardner 1997:132). Here, the use of *mm* is used to a neutral response, which neither affirms nor agrees with the caller's topic of talk. It suggests that there is "nothing further to say on the topic" (Gardner 1997:133). Here, the use of *mm* has the consequence of acting as a continuer, and thus there is a need for further talk from the caller.

In Extract 4 below, we find an exception to the noncommittal *mm* or equivalent. In this case the counselor issues a token of newsworthiness: "*ohh*?" (cf. Heritage 1984a). Soon after the caller expresses confusion about his trouble, "*I don't know why*" (line 16), the counselor recaptures the sense of the earlier formulation 'all of a sudden my friends started to hate me' by commenting: "*well it's strange isn't it*". Then the counsellor enters into some diagnostic questioning. This appears to us to be a relatively early intervention by the counsellor in comparison to other calls.

Extract 4

2_1_12

1 CT: hi kids helpline
2 (2.2)
3 C: HELLO?
4 (.)
5 CT: hello
6 (0.2)
7 C: HI .hh um

(0.4)
C: it's about my friends they um .hh (.) like all of a sudden
they started to hate me
(0.2)
CT: ohh?
(0.6)
C: yeah
(2.0)
C: like I don't (.) I don't know why?
CT: mm
(0.4)
CT: well it's strange isn't it
C: yeah=
CT: =so so everything was going okay
(0.2)
C: hmm

Extract 5 below is more in line with the general pattern of issuing tokens of having heard, and 'continuers', such as *mm hm* or *right*.

Extract 5
2_1_10

CT: hi Kids helpline
C: hi yeah um I've got a bit of a problem um I get teased at
school a lot
(0.4)
CT: [mm hm
C: ['cause 'cause of my hair it's like really thin and I'm
really small as well
CT: right
C: and I don't know how to like stand up to the bullies and that
CT: [mm hm
C: [um do you have any like
(0.6)
C: any things to suggest like
CT: mm?
C: who I should speak to and that
CT: oh okay

In the use of *mm hm* here in lines 5 and 10, the counsellor is passing up the opportunity to speak, handing the floor back to the speaker (Gardner 1997). The counsellor is giving the caller permission to design their own entry into the talk. The initiative is with the caller. The counsellor is creating the interactional space for the caller to competently decide how they will tell their own troubles and how they want the counsellor to listen to them. This is a different interactional strategy to that used typically by adults with children and young people, as adults typically use a framework of questions, advice and so on. In Extract 5, for example, the counsellor leaves it with the caller to come to the point of asking for advice.

Extract 6, part of which was shown earlier in Extract 1, is an example of how long it can take for the caller to get to the point and for the counsellor to grasp what the dimensions of the troubles are.

Extract 6

```
4   C:  um hello .hh um
5       (0.2)
6   C:  ah-uh like I wanted to talk to er to someone about like I
7       have some problems at school↑
8       (0.4)
9   CT: °sure°
10  C:  yeah
11  C:  .hh um ha I kind of have this trouble↑ (.) you see I don't
12      know why it happened?
13      (0.2)
14  C:  basically um .hh
15      (0.4)
16  C:  I was sick fo:or a couple of days↑ .hh
17  CT: yeah
18  C:  and yeah
19      (0.4)
20  C:  I sort of u:um
21      (1.0)
22  C:  um um what=would=you ah like went to the doctor's↑ .h and
23      like he he didn't give us he gave us some prescriptions so I
24      like yeah I stayed away from school for a couple of days↑
25  CT: mm
```

The first six lines are preliminary to the troubles talk, as are lines 11–12. The 'basic' issue is begun in line 14. However, as we will show later, these preliminaries are crucial for sending important messages to the counsellor. In lines 22–24, the caller alludes to the problem, which then leads to the narrowing of the problem. By line 46 (see Extract 7), the counsellor formulates, but does not fully state, what the caller has done; that is, he assists in the production of the story. After confirmation in 48 from the caller (see continuation of the extract below), the counsellor finishes the statement he began in 46, which is a next event in the caller's story: 'you told them it was for this time'. The caller then adds the consequence 'and he caught me'.

Extract 7

CT: mm
C: um and then like a:ahm I had to go back to school like today↑
(0.4)
C: um and like I was sort of like
(0.4)
C: I don't know why but like I basically photocopied the medical
certificate (.) from my last time↑ and like (.) °yeah°
(0.4)
CT: sorry
(0.2)
CT: can you run that by me again you
C: ah I I made a photocopy of the medical certificate
from um
(0.6)
C: ah from previous time I was sick
CT: yeah
C: and like I I gave it to my teacher and told them like °yeah°
CT: you told them it was for
(0.4)
C: yeah [yeah
CT: [this this time
C: and he caught me on it (0.2) doing that
CT: oh okay (0.6) so what happened then?
C: o:oh I just like um

This caller's description of the problem can be looked at another way. It takes the form of an extended story that includes an abstract ('I don't know why it happened'), an orientation ('I was sick, went to the doctor, stayed home'), a complicating action ('I went back to school and photocopied my certificate from last time') and result ('I got caught'). All that is missing from the Labov and Waletzky (1967) narrative model is the coda, which we could show to be worked out with this counsellor through the rest of this call. This took the form of the caller asking how she can evaluate this course of action, how she can make sense of it. We observe that this story has some potentially confessional elements in that it is a story partly about deceit, but the caller and the counsellor avoid this implication very delicately. The counsellor, having heard that the caller was caught out in her fraud, simply asks what happened next. On the caller's part, as the call continues, rather than seeking absolution for a sin, asks how she could have done this. She had signalled this interest at the beginning of her call in her statement "*you see I don't know why it happened?*"

Observation 3

This leads us to the third observation, how the counsellors listen for how the callers want to be heard. This is a crucial skill and is evidenced in the pattern of non-intervention we have described above. However it also encompasses the many ways in which counsellors design their subsequent turns so as to 'receive' the caller's messages about how they want to be heard. Extract 8 below is an example of this. We note that the second and subsequent turns by the counsellor (after the greeting/hello) typically follow an interactional pattern closely resembling the callers, in both greeting and pace and tone of the call.

Extract 8

1_2_7

```
CT: Kids helpline
C:  hi
CT: hi:i
C:  um
    (0.6)
C:  oh I'm just
    (0.2)
C:  I I just=needed to talk to somebody I'm so stressed out ha
    [ha
CT: [yeah?
    (0.6)
CT: what's going on↑
C:  um
    (0.2)
C:  oh everything I've just moved up the North Coast on my own ah
    ha
    (0.6)
CT: moved up the North Coast from whe:ere
C:  fro:om Durrino about an hour and a half away hh
CT: Durrino (.) that's near Farrina? isn't it?
C:  yeah=
CT: =oh yeah
    (0.4)
CT: how old are you
C:  fifteen
```

In this call, the caller signals clearly that she "*just=needed to talk to somebody*" and we see the counsellor following this invitation through. The caller next announces that she is "*so stressed out*" which the counsellor replies to first with a continuer in the form of "*yeah?*" and following the pause, with a second question that matches the allusion to things going on but lightens her reply through a rising inflection: "*what's going on?*"

```
C:  I I just=needed to talk to somebody I'm so stressed out ha
    [ha
CT: [yeah?
    (0.6)
CT: what's going on↑
```

In reply to "*what's going on*", the caller states "*oh everything*", an extreme case formulation (Pomerantz 1984). She begins a story about her recent move to the North Coast, but again there is a pause, which the counsellor fills with a repeat of the caller's words and the question "from where", assisting in the telling of the story by asking a possibly casual question.

```
CT: moved up the North Coast from whe:ere
C:  fro:om Durrino about an hour and a half away hh
CT: Durrino (.) that's near Farrina? isn't it?
C:  yeah=
```

This casualness is then repeated through the counsellor's guess about where Durrino is. Here the counsellor builds a small conversational sequence that puts on hold the *problem* of being "*so stressed out*", and "*oh everything*" going on, apparently in favour of the first statement of the caller that she "*just=needed to talk to somebody*" (line 8), which is her *reason for the call.* We attend to this distinction between problem and reason in the next section of the paper.

However we note here also the different ways that callers indicate the magnitude of their problem or distress, ranging from "oh everything" to "just a bit worried". The counsellors can hear these formulations of how serious things are, but do not engage directly in agreeing or disagreeing with such assessments (cf. Pomerantz 1984) as might be found in casual conversation or everyday troubles-telling, for example "it can't be that bad" or "don't be worried, it will be fine". Their approach seems to be to let stand the caller's initial assessment as a resource for their subsequent work with the caller. This will be pursued in more detail in a later paper.

Extract 9 offers further evidence of this phenomenon of recipient design. In order for counsellors to design their turns at talk to match the caller's implication of how they want to be heard, the counsellors have to listen first for this implication. In Extract 9, the caller provides a preferred interactional tone – "*if you could just give me the name of a service*" – and the counsellor attends to and reflects this tone in his own voice.

Extract 9

1_2_6

CT: hi there kids helpline
(0.6)
C: um hi I was actually wondering if you could just give me a
name of a service
CT: [[mm
C: [[in
(0.2)
C: just in the Dawnsville area for um
(1.0)
C: sexual=abuse=counselling
(1.0)
CT: sure can
(0.4)
CT: just give me a sec
(0.4)
CT: can I just get your first name
C: um Karen
(0.6)
CT: Karen
C: yeah
(3.0)
CT: and how old are you Karen
(1.0)
C: not quite a kid anymore I'm twenty-one
CT: okay
(1.0)
CT: just hold on a sec and I'll see what I can find for you
(10.6)
CT: I'm just trying to think of the postcode it's forty-eight ten
I think isn't it
(0.6)
C: oh yeah

We note that the caller takes a number of turns interspersed with hesitancy markers to name the kind of service she is after – suggesting the delicacy of the matter perhaps (Silverman & Perakyla 1990). The counsellor's reply to this naming of a possibly delicate matter is very upbeat: "*sure can*", "*just give me a sec*". We observe that the counsellor attends to and reflects the specific words used by the caller. For instance, "*just*" is used by the caller in line 3, and the counsellor uses "*just*" in lines 14, 27 and 30. We note that this caller is several years older than the intended age range of the service; she (the caller) also comments on this in line 24 but it does not seem to change the interactional tenor as the counsellor resumes his original tenor at 27.

The counsellors use a form of recipient design that acknowledges that the callers have rights and grounds for supposing that they (or their reasons for calling) can be so recognised (Schegloff 1979). Counsellors display their recognition by doing sequentially appropriate second parts for the type of sequence initiated by the caller. This is achieved through the normalisation of the problem and the design of the tone of interaction. Extract 10 shown earlier as Extract 3, is also an example of this aspect of recipient design.

Extract 10

1_1_3

```
CT: hi there kids helpline
    (0.6)
C:  hello um
    (0.4)
C:  my friend just got kicked out of home and she's got like
    nowhere to sta(hh hh):ay
CT: mm
C:  and um
    (0.6)
C:  she doesn't and she wants to make a few phone calls but she's
    got no money on her pho:ne
    (1.0)
CT: right
C:  and we don't know what to do
    (1.0)
CT: okay
    (0.8)
```

The caller has announced news that might elsewhere be treated with alarm or some such evaluation: a young person has been kicked out of home and she has nowhere to stay. She has announced this with a laughing tone at the end. The counsellor does not laugh with her and instead uses the "empty" *mm* to acknowledge the news. This use of *mm* makes this news perfectly normal or even routine, just as the caller seems to do. The announcement that the friend also has no money on her phone is replied to with "*right*", acknowledgement of information. Finally, the statement "*we don't know what to do*" receives the reply "*okay*", suggesting a somewhat stronger uptake and possibly a matter that the counsellor could help with.

Extract 11 shows another dimension of recipient design. In this call Valerie has asked to speak with a counsellor named Rose, but has been told that Rose is not there. She resolves this by stating that she will talk instead with the current counsellor, indicating that he will do as a substitute.

Extract 11

```
15  C:   oh well I'll just talk to you then ha ha
16  CT:  okay
17  C:   .hh um
18        (0.2)
19  C:   well my name's Valerie anyway?
20  CT:  Valerie how are you
21  C:   um
22        (0.4)
23  CT:  I usually talk to Rose um
24  CT:  mm hm
25  C:   .hhh hh.(.)I um
26        (0.4)
```

In this case the counsellor does not appear to be following the caller's pattern of interaction in one puzzling respect. Up until this point the call has been about finding whether the counsellor named Rose is available. The caller then offers her name to the counsellor, which the counsellor acknowledges by repeating it, but he does not give his own name. There is some similarity here with Sacks' well-known example of the caller to a psychiatric clinic who does not provide his name in return to the name offered by the clinical worker who answered the phone. Sacks (1992:325) notes that people not wanting to give their names in return to a name offered by the other, "use some more or less elaborate means which served to never have the place where their name belongs occur, such that they never gave their names but their names were never absent, i.e., that the slot never occurred". In our example above, by repeating the name 'Valerie' in the slot where he could have given his own name (line 20), the counsellor is adroitly indicating that a 'name' is 'officially due' but by this means he neatly avoids having to give a return name.

The caller then gives him a *second* chance to provide a name through the indirect "I usually talk to Rose", but again the counsellor passes up the invitation to offer his name. This counsellor's interactions would be described by Heritage (1984b) and Silverman (1985, 1994) as a departure from the norm in this non-reciprocality in relation to first names. However we consider that there could be good reasons for not providing his name, possibly in deference to the special relationship that Valerie has with Rose. And we show below that this does not deter the caller, Valerie, from proceeding to talk as she intended to.

2.1 Observation 4

The fourth observation is the counsellor's search for the reason for the call. This observation arose in relation to looking for the caller's problem or trouble that led to the call. We started by looking for what the caller's problem was, and how it was presented to the counsellor. We soon found that this way of listening to the calls did not properly capture how the counsellors were listening to the calls, and therefore was missing an important philosophical and procedural aspect of the counsellors' work. This work is: "we listen, we care" rather than "we can solve your problems".

Our first step was to separate analytically the callers's statements of problems from their statements or allusions to what they were seeking from the counsellor: how they wished to be listened to, as discussed above. We studied the Extracts for what the callers presented first: the problem (for example, being kicked out of home) or the reason for the call (for example, not knowing what to do, needing some advice), or whether these were presented together. What we found that was most significant was that the counsellors search for the reason for the call; the clues from the callers about how they wished to be heard and responded to. The counsellors are listening, not for the problem, but for the reason for the call. When they find this, they stop searching.

For instance, in the following extract, (Extract 12 part of which was shown earlier as Extract 5) the *problem* is named and described in lines 2 and 3 (I get teased at school) but the *reason for the call* starts in line 9 and continues in lines 12, 14 and 16 (do you have any suggestions?). The counsellor's use of *oh* in line 17 is used in response to being informed about the reason for the call. *Oh* is produced when this crucial information from the caller seems possibly complete (Heritage 1984a).

Extract 12

2_1_10

```
CT: hi Kids helpline
C:  hi yeah um I've got a bit of a problem um I get teased at
    school a lot
    (0.4)
CT: [mm hm
C:  ['cause 'cause of my hair it's like really thin and I'm
    really small as well
CT: right
C:  and I don't know how to like stand up to the bullies and that
CT: [mm hm
C:  [um do you have any like
    (0.6)
```

```
C:  any things to suggest like
CT: mm?
C:  who I should speak to and that
CT: oh okay=
C:  =I've spoken to my teachers and that and they seem like
     (0.4)
C:  I think they've spoken to the bully but you know
CT: right
C:  it keeps happenin'
CT: right
     (0.2)
CT: is it (.) there's there's one one person in particular who's
C:  oh it's actually a group of kids
CT: right
```

After the counsellor has heard why the caller is calling, what he wants from the service, the counsellor proceeds with talk about the problem. There are cases in the materials, however, where what the caller wants is not what the counsellor can provide. This occurs when the caller wants advice, in which case the counsellors work to avoid giving advice directly. They will provide information, as in the call for a telephone number. They will listen as "just somebody to talk with", as seen above. In another case, the case of the girl with nowhere to stay and no money for her phone, it turned out that the caller wanted the counsellor to provide money. A careful study of how counsellors handle requests that are in effect not practically or philosophically possible to fulfill will be the topic of another paper. The reason for the call, then, needs to be (or to be turned into) a request for listening and caring, not solving problems. The *Kids Help Line* works on the principle that young people have resources for solving their own problems, and that the telephone counseling services is such a resource, but not a repository of advice.

In this respect, the *Kids Help Line* counselling service seems to offer just the opposite of a conventional adult response to children's announcements of problems, which is to give advice and solve the problems for them. Instead of a format of 'problem identification – advice giving', what we see in these calls is a format 'talking about problems – listening to that talk' or even just 'talking – listening'.

Extract 13, part of which was shown in Extract 11, is an example of a caller who has used the Kids Help Line before. She first attempts to locate a counsellor she has spoken with previously, and on finding out that this counsellor is not available, states "oh well I'll just talk to you then". The reason for her call is evident: "just to talk".

Extract 13

2_1_13

CT: hi Kids helpline
C: hello is Rose there please
CT: Rose
C: yeah
CT: okay I'll just check for you?
(10.0)
CT: a:ah not tonight?
C: not tonight?
CT: no she'll be
(0.2)
CT: Sunday afternoon
(0.2)
C: ohh will she?
CT: yep
C: oh well I'll just talk to you hen ha ha
CT: okay
C: .hh um
(0.2)
C: well my name's Valerie anyway?
CT: Valerie how are you
C: um
(0.4)
CT: I usually talk to Rose um
CT: mm hm
C: .hhh hh.(.)I um
(0.4)
C: my mum's just had a massive brain operation
CT: mm
C: u:um it took nine and a half
(0.4)
C: hours?

Valerie proceeds to talk to the counsellor about dramatic situations in her life, not asking for advice but using the service exactly as intended. By listening, the counsellor can show he cares.

3. Conclusion

This paper has investigated the openings of calls to *Kids Help Line*. In the course of describing our four observations, we have shown ways in which the counsellors provided space for the callers to define the terms and the tenor of the talk, and we have observed the competent ways in which the children and young people drew on their resources of talk to display their troubles, and particularly their reason for calling.

This paper has attended to four initial observations. In the very first turns of the telephone conversation, the counsellor does not provide their name. This closes off the obligation for the caller to give their name and also places greater onus on the caller to announce what comes next. Any opening is possible. In the interactive management of getting into the calls, the pauses in the talk provided opportunities for the counsellor to speak. Instead, for the most part, the counsellors used receipt tokens such as *mm* to acknowledge that they were listening but had nothing significant to add at that moment. This is a further, powerful way of indicating to the callers that the initiative is with the callers. We noted that the callers were competent in designing their own entry into the talk. We then described the work of the counsellors in the form of recipient design. This involves normalising the problem as well as aligning with the caller's tone and pacing of the interaction. Our analysis of the openings show that the interest of the counsellor is to search for and find the reason for the call, which may or may not include the identification of the problem.

In this initial paper we have located in the data itself a clear sense of the *Kids Help Line* philosophy-in-action. By looking at very fine details of how turns are designed, at how next turns are related to prior turns, at how spaces in the talk are filled or not filled, we find the conversational machinery (and the conversational art) through which the listening and caring is done. It is not possible to locate such crucial detail through more broad-brush approaches to studying counselling calls, such as checklists. The connection between caller and counsellor has to be built up and maintained turn by turn, as the conversation unfolds. We have been able to show some dimensions of the interactional sensitivity and skill of both callers and counsellors. In later papers we will be attending to many further aspects of the calls. This has been an initial investigation that has shown us some details that we could not have imagined were there, in the important opening moments of the calls.

References

Baker, C., Emmison, M., & Firth, A. (2001). 'Discovering order in opening sequences in calls to a software helpline'. In A. McHoul & M. Rapley (Eds.), *How to Analyse Talk in Institutional Settings* (pp. 41–56). London: Continuum Press.

Gardner, R. (1997). 'The conversation object *Mm*: A weak and variable acknowledging token'. *Research on Language and Social Interaction, 30*(2), 131–156.

Gardner, R. (1998). 'Between speaking and listening: The vocalisation of understandings'. *Applied Linguistics, 19*(2), 204–224.

Heritage, J. (1984a). 'A change-of-state token and aspects of its sequential placement'. In J. M. Atkinson & J. Heritage (Eds.), *Structures of Social Action: Studies in Conversation Analysis* (pp. 299–345). Cambridge: Cambridge University Press.

Heritage, J. (1984b). *Garfinkel and Ethnomethodology*. Oxford: Polity Press.

Labov, W. & Waletzky, J. (1967). 'Narrative analysis: oral versions of personal experience'. In J. Helm (Ed.), *Essays on the Verbal and Visual Arts* (pp. 12–44). Seattle: University of Washington Press.

Pomerantz, A. (1984). 'Agreeing and disagreeing with assessments: Some features of preferred/dispreferred turn shapes'. In J. M. Atkinson & J. Heritage (Eds.), *Structures of Social Action: Studies in Conversation Analysis* (pp. 299–345). Cambridge: Cambridge University Press.

Pomerantz, A. (1986). 'Extreme case formulations: A way of legitimizing claims'. *Human Studies, 9*, 219–229.

Psathas, G. (1995). *Conversation Analysis: The Study of Talk-in-Interaction*. Thousand Oaks, CA: Sage.

Sacks, H. (1992). *Lectures on Conversation Vol. 1 and 2*. Edited by Gail Jefferson. Cambridge, MA: Blackwell.

Schegloff, E. A. (1968). 'Sequencing in conversational openings'. *American Anthropologist, 70*, 1075–1095.

Schegloff, E. A. (1979). 'Identification and recognition in telephone conversation openings'. In G. Psathas (Ed.), *Everyday Language: Studies in Ethnomethodology* (pp. 23–78). New York: Irvington Publishers.

Schegloff, E. A. (1986). 'The routine as achievement'. *Human Studies, 9*, 111–112.

Schegloff, E. A. & Sacks, H. (1973). 'Opening up closings'. *Semiotica, 8*, 289–327.

Silverman, D. (1985). *Qualitative Methodology and Sociology*. Aldershot, Hants: Gower.

Silverman, D. (1994). 'Analysing naturally-occurring data on AIDS counselling: Some methodological and practical issues'. In M. Boulton (Ed.), *Challenge and Innovation: Methodological Advances in Social Research on HIV/AIDS* (pp. 69–93). London: Taylor & Francis.

Silverman, D. & Perakyla, A. (1990). 'AIDS counselling: The interactional organization of talk about 'delicate' issues'. *Sociology of Health and Illness, 12*(3), 293–318.

Chapter 7

'I just want to hear somebody right now'

Managing identities on a telephone helpline

Hedwig te Molder

1. Introduction

In this chapter I shall explore some specific ways in which callers and call-takers may categorize themselves and each other, and the kind of interactionally sensitive business which is being performed through these identity ascriptions. In drawing upon a single case, I examine how one of the participants undermines the institutional nature of the interaction by constructing a 'deviant' identity. It is shown how the caller turns the interaction into everyday talk by portraying herself as an 'ordinary person' in various ways. At the same time, however, we shall see how this identity and its matching set of possible inferential implications are far from being undisputed. On the whole, the analysis will provide insight into some of the social and rhetorical activities which are performed through the subtle management of identities and counter-identities in helpline talk.

As such, this paper combines an interest in the detailed analysis of institutional talk (for example Drew & Heritage 1992), with a second strand of recent work in psychology which aims to respecify cognition from an analyst's category into a participant's resource. This discursive psychological approach starts from the idea that cognitive notions such as attitudes, scripts, motives and identities, are part and parcel of participants' interactional work rather than simply reflecting their inner worlds (Edwards 1997; Edwards & Potter 1992; te Molder 1999; te Molder & Potter 2005; Potter 1998). From this perspective, identities are no longer private mental states which can be 'switched on' in relevant situations, but social phenomena which are locally produced and managed (see also Antaki & Widdicombe 1998).

2. Analytic material and context

My focus will be on one specific piece of data, namely an extract from a conversation at a telephone helpline in the Netherlands. This national helpline receives more than 40,000 requests per year, all with respect to health problems in the widest sense of the word.

The extract is taken from a sample of more than sixty recorded and transcribed interactions between callers and call-takers. The sample contains a heterogeneous set of topics and covers more or less the range of issues that the helpline most frequently deals with. The analysis is work-in-progress, and at this stage I do not want to make bold generalist claims about the materials. Here my aim is to make a more theoretical point by treating the analysis of this extract as an indication of a research phenomenon that deserves further attention, and as such it will ultimately inform further analysis over a larger data corpus (cf. Schegloff 1987).

The case itself has been selected because of its potential richness in terms of interactional aspects: the extract shows a notable change in initiative (i.e. in this particular organisational context), which is followed by an intriguing and unusual exchange about star signs. While it concerns the only conversation in the corpus in which the caller regularly controls the initiation and shaping of topics (for example by questioning the counsellor), the analysis indicates that the interactional business performed with it, i.e. a particular kind of identity work, is a recurrent phenomenon in the data.

The general objective of the project is to contribute to the improvement of helpline expertise as well as to the extension of interaction-analytic theory, in both cases by providing insight into the most important interactional skills of the people involved in helpline interactions.

3. Analysis

The extract[1] is taken from a conversation between a caller and a counsellor (the typical professional background and title of call-takers in this organisation) and lasts about fourteen minutes. In the first half of the conversation, the caller has described herself as suffering from agoraphobia and depression. The fragment starts right after she has explained her situation:

[26A – c2 – 14]

Cal: I am seeing someone from the RIAGG,[2] she comes to visit me here
Cou: Oh yes
Cal: You know,once in a while, but it doesn't he:lp
Cou: It doesn't help
Cal: °Ahh of cou:rse not°
Cou: Of course not?
Cal: I know, yes, you have to do it on your o:wn
(0.5)
Cou: Ye:s, but you're also calling right now hoping it will help
somewhat(.)I suppose
(0.9)
Cal: No, I just want to hea:r somebody right now=
Cou: =[Yes
Cal: [Just to know that I am not all [alo:ne on this planet
Cou: [°yes°
Cal: ((sob))=
Cou: =Yeah I see but that may he:lp right?
Cal: .hhhhhhh
(2.1)
Cou: For instance(.)and another time an a:rm around you may
help, another time an advi:ce would help, of course you
have to do it on your own, bu:t (1.3) we are here with so
many people on earth for a reason, I'd say that it would be
really nice if you could support one another a bit (1.1)
rather than doing it all all on your own
Cal: Tell me something about yourself, why do you do this work?
(0.7)
Cou: .hh Why I do this work?=
Cal: =Mm hm
Cou: We:ll sim- because this simply is my job, this is eh, my
heart's in it
Cal: Oh!>that's what I just wanted to say, if it's simply is
your job or whether your heart's in it, that's a bi:g
difference<
Cou: Ye:s=
Cal: =Now then!
Cou: Well that's what I do with my work, here's where my heart
lies, that's my work
Cal: .hhhh °what star sign are you?°
Cou: What did you say?
Cal: .hh What star sign are you?
(1.0)
Cal: >°ah it doesn't mean anything of course°<
Cou: I- I can't hear you, now what did you say?
Cal: What star sign are you?
Cou: O::h, what star sign am I? ah! you like astrology?

Cal: [Yes, I
Cou: [You're familiar with that
Cal: No::, I believe that no:thing exists for no:thing (.)in the end
Cou: I believe that too
Cal: Well?
Cou: Yes, I am Taurus
Cal: A:hh
Cou: And you?
Cal: I'm Leo
Cou: Leo
Cal: With Sagittarius for ascendant
Cou: A↑ha, a↑ha
Cal: What aha ha
Cou: Leos with Sagittarius for ascendant don't sit at home
(0.8)
Cal: No eh?=
Cou: =Having agoraphobia
Cal: °No eh?°
Cou: No, I don't think so
Cal: No
Cou: They hunt (.) they go outside
Cal: How nice of you that you ( )
Cou: Isn't that right?
Cal: Ye::ah [it's true
Cou: [( )but they're very strong animals=
Cal: =Don't say that!
Cou: Aren't they?
Cal: Ehh: I ( )
Cou: [The lion is
Cal: [This is this is ehh psychological peeling and I really
think that you do see that through ( ) I find you ( )
Cou: The lion is the king of the jungle you can't
[alter that]
Cal: [°Ye:s no↑°]yes
Cou: And the Sagittarius is a ehh hunter
Cal: ( )
Cou: So that's all I'm- I'm saying
(1.1)
Cal: .hh Yes=
Cou: =That doesn't fit the image of agoraphobia and being depressed
Cal: No! that( )just doesn't work
Cou: No
Cal: Because I am (.) I also have a very sensitive side from
Cancer probably
[((sob))

92 Cou: [Mm yes (0.9) but say let's get back to (.) you say from
93 the RIAGG, someone comes to your home but that doesn't, of
94 course help you say

What happens in this fragment? For the purpose of analysis, I shall deal with it in three parts (lines 1–25, lines 26–57 and lines 58–94).

Seeking help? (segment 1)

1 Cal: I am seeing someone from the RIAGG, she comes to visit me here
2 Cou: Oh yes
3 Cal: You know,once in a while, but it doesn't he:lp
4 Cou: It doesn't help
5 Cal: °Ahh of cou:rse not°
6 Cou: Of course not?
7 Cal: I know, yes, you have to do it on your o:wn
8 (0.5)
9 Cou: Ye:s, but you're also calling right now hoping it will help
10 somewhat(.)I suppose
11 (0.9)
12 Cal: No, I just want to hea:r somebody right now=
13 Cou: =[Yes
14 Cal: [Just to know that I am not all [alo:ne on this planet
15 Cou: [°yes°
16 Cal: ((sob))=
17 Cou: =Yeah I see but that may he:lp right?
18 Cal: .hhhhhhh
19 (2.1)
20 Cou: For instance(.)and another time an a:rm around you may
21 help, another time an advi:ce would help, of course you
22 have to do it on your own, bu:t (1.3) we are here with so
23 many people on earth for a reason, I'd say that it would be
24 really nice if you could support one another a bit (1.1)
25 rather than doing it all all on your own

Let us first focus on a specific part of segment one, in which the caller explicitly formulates her reason(s)-for-calling: "*I just want to hea:r somebody right now*" (line 12), "*Just to know that I am not all alo:ne on this planet*" (line 14). Interestingly enough, the counsellor treats these utterances not as a neutral description of the caller's need at that very moment, but as an indirect request for help. In line 17, we see how the counsellor reformulates 'hearing somebody' as a possible form of help, whereas the caller defines 'hoping to receive help' and 'wanting to hear somebody' in terms of a contrast (lines 9–12). The counsellor's redefinition is underlined in lines 20–25, in which she characterises the assumed request for help as a matter of course ("*we are here with so many peo-*

ple on earth for a reason"), with which you might even do others a favour ("*it would be really nice if you could support one another a bit*").

By thus lowering the threshold for a help request, the counsellor suggests that the caller is having difficulties posing such a question in explicit terms: her 'real' need does not concern hearing a voice but receiving help.

A first observation is that the caller's apparently clear-cut reason-for-calling is not treated as something pre-given and fixed. On the contrary: it is interactionally at issue and approached as a contentious phenomenon (cf. Edwards 1997). As I will point out, the analytic issue here is not to determine which of the participants is giving the right picture. What is essential, is how the participants themselves practically manage these issues, and, more specifically, what they are *doing* interactionally by formulating their own and others' 'needs'. As we shall see, by couching the caller's need in specific terms, both the caller and the counsellor suggest their membership of particular categories.

A competent participant

First, we shall examine how the caller deals with the characterisation of her reason-for-calling as an indirect request for help. If we look in more detail at the sequential position of the utterance "*No, I just want to hea:r somebody right now*", we see that it is a response to the counsellor's utterance in lines 9–10 ("*Ye:s, but you're also calling right now hoping it will help somewhat* (.) *I suppose*"). By denying that she is calling because she wants to be helped, the caller portrays herself as someone with a minimal request: she does not expect help but merely wants to hear a voice. It is interesting to see how the caller also turns her request in something 'small' and mundane by using the word "just" ("*I **just** want to hea:r somebody right now*" (line 12), "***Just** to know that I am not all alo:ne on this planet*" (line 14)). With retrospective effect, this marks seeking help as a less mundane and more 'ambitious' undertaking. Note how this characterisation receives more emphasis in a context in which seeking and receiving help is considered a normal activity, that is, in terms of institutional expectations. By not letting herself be categorised as someone who is seeking help, and instead presenting herself as a person with 'ordinary', not too sweeping demands, the caller displays a watchful care with respect to her own competence as a 'normal' and thus equal interlocutor.

Earlier in the conversation, we find similar indirect references to the issue of competence. Here, the caller makes clear that helping does not help (line 3), because: "*I know, yes, you have to do on your o:wn*" (line 7). By noting, in a pre-emptive way, that she has to solve it herself, the caller presents herself as an 'informed' person. She does not need that particular advice anymore. Notice how she uses the term *know* to underline her competence in this respect: her remark indicates not so much a guess but something she knows or learnt by experience.

Heritage and Sefi (1992) show that the display of knowledgeability and competence is a recurrent theme in service encounters. In their study of interactions between health visitors and first-time mothers, they describe how the mothers indirectly resist advice by claiming that its content is already known, and/or acted upon. The fact that there is an official obligation for health visitors to perform these visits might explain the mothers' resistance and their ways of emphasising their own knowledge. In that respect, it is striking that also in this situation, in which the caller herself has initiated the conversation, competence is such an important issue. Apparently, calling a helpline does not necessarily involve presenting yourself as someone requiring help. This issue will return prominently in the second part of the conversation.

However, the utterance "*I know, yes, you have to do on your o:wn*" (line 7) seems to do more interactional work than only displaying knowledgeability. It is also hearable as a disappointment: in the end, other people's help gets you nowhere; you have to do it on your own anyway. Notice first how the caller, in depicting help as something which is doomed to failure (see also lines 3 and 5), undermines the basic idea of a helpline. Helplines provide help, in one way or another, or they will not have a right to exist. A help-does-not-help scenario does not fit the particular institutional nature of a helpline.

Imagine that the counsellor would confirm the inefficacy of help. This would not only make her contribution redundant but also contest the conventional obligations for a helpline. Clients may expect help and even have a right to be helped – this can be considered part of the conventional set of cultural resources by which we come to understand helpline activities (see also below). So, by defining help as something useless in this particular context, the caller also 'invites' the counsellor to correct her. In lines 9–10, we can indeed observe how the counsellor confirms her identity of a 'help giver', namely by describing the caller's reason for contacting the helpline as a quiet hope to be helped anyway.

Needs and identities

We may conclude that in the first segment of the conversation (lines 1–25), a subtle negotiation takes place about the nature of the caller's request. Notice how this exchange about 'needs' performs a lot of work in relation to the identity of both the caller and the counsellor. By qualifying the conversation as potential help, the counsellor defines the caller as someone who is seeking help. In this way, she identifies the caller as belonging to a certain social category, which is common-sensically associated with particular kinds of activities. Sacks (1992 Vol. 1; see also 1972, 1979) describes these as category-bound activities (CBAs). The category 'help seeker' *could* be associated with, for example, being supported by others, fear, embarrassment, crying or other forms of possible dependent behaviour. Conversely, by describing someone's behaviour, we can also provide for what his or her social identity might be. It is crucial to add, however, that the connection between identity and activity is not linear, but part of participants' 'negotiation package' (Antaki & Widdicombe 1998; Edwards 1998). This is also what happens in the case of the caller: she does not allow herself to be categorised as a help seeker without a 'battle' but carefully displays her autonomy, by presenting herself as a competent participant with (no more than) ordinary needs. This includes some subtle room for disagreement, such as in the case of her remark that you have to do it on your own – also notice her quiet sobbing.

What is your star sign? (segment 2)

26 Cal: Tell me something about yourself, why do you do this work?
27 (0.7)
28 Cou: .hh Why I do this work?=
29 Cal: =Mm hm
30 Cou: We:ll sim- because this simply is my job, this is eh, my
31 heart's in it
32 Cal: Oh!>that's what I just wanted to say, if it's simply is
33 your job or whether your heart's in it, that's a bi:g
34 difference<
35 Cou: Ye:s=
36 Cal: =Now then!
37 Cou: Well that's what I do with my work, here's where my heart
38 lies, that's my work
39 Cal: .hhhh °what star sign are you?°
40 Cou: What did you say?
41 Cal: .hh What star sign are you?
42 (1.0)
43 Cal: >°ah it doesn't mean anything of course°<
44 Cou: I- I can't hear you, now what did you say?
45 Cal: What star sign are you?
46 Cou: O::h, what star sign am I? ah! you like astrology?
47 Cal: [Yes, I
48 Cou: [You're familiar with that

Cal: No::, I believe that no:thing exists for no:thing (.) in the end
Cou: I believe that too
Cal: Well?
Cou: Yes, I am Taurus
Cal: A:hh
Cou: And you?
Cal: I'm Leo
Cou: Leo
Cal: With Sagittarius for ascendant

Line 26 ("*Tell me something about yourself, why do you do this work?*") marks a sudden transition, that is followed by an intriguing exchange. The caller takes the initiative and starts to 'interrogate' the counsellor. The abrupt character of this transition is, first and foremost, caused by the fact that the caller does not meet the normative expectation that she will respond to the counsellor's first assessment (Pomerantz 1984) in lines 20–25. (The term expectation does not refer to individual wants or desires but to normative preferences in the sense of conversational structures, see Schegloff & Sacks 1973). Instead of agreeing or disagreeing with it, the caller poses a question without delay, apparently not related to the previous utterance, and for which she gives no account. A completely new topic is being introduced, seemingly without the previous one being closed. The caller's initiative to change the topic is remarkable in an organisational context such as this one, which tends to show the professional's control over the initiation and shaping of topics (Drew & Heritage 1992).

Also in other respects, the caller's question establishes an important caesura. Note how she invites the counsellor to talk about things which affect the counsellor, rather than maintaining the counsellor's focus on *her* affairs. This runs contrary to the institutional routine that the person who contacts the helpline also forms the topic of conversation. Moreover, by changing the topic, the caller redefines the exchange as a more or less informal conversation. The talk does not concern help or advice anymore, but an ostensibly non-committal topic as star signs.

However, by 'interrogating' the counsellor about her personal motives to do this job and, after that, asking her in an indirect sense about her personality, the caller defines the counsellor's position in a way which is similar to the situation in which she has just been herself. Whereas, in institutional terms, the counsellor poses the questions and the caller provides the answers, the caller now reverses these conversational roles. The counsellor has to give answers, and these answers also concern 'private' matters. Lines 32–34 show how the caller precisely selects this personal element ("*whether your heart's in it*") from the counsellor's response, thereby defining it as informative. By subsequently focusing on her star sign, the counsellor as a person is further put in the forefront.

These activities ultimately put the caller on an 'equal footing' with the counsellor. We see this in line 50, in which the counsellor presents her answer in terms of a personal opinion that she wants to share rather than as a professional point of view. The caller also treats this agreement as an incentive to pose her question again – a question which still has not been answered (I shall return to this aspect presently).

From this perspective, the transition in line 26 is not as abrupt as it first seems: again we see how the caller underlines her competence, now by positioning herself explicitly as an equal participant in the conversation. Note how, similar to what happens in the first segment of the extract ("*I just want to hea:r somebody right now*"), the caller portrays herself as a 'normal' partner in an ordinary conversation, thereby excluding herself from the standardized relational pair (Sacks 1972) of help giver: help receiver.

With her initiative to select and develop a topic, the caller claims, so to speak, her identity as an *ordinary person* (Sacks 1984), and with that, her right to participate to the interaction as a 'normal', competent member of society. Categories may not only be associated with particular activities, they may also entitle participants to certain experiences or certain knowledge (Whalen & Zimmerman 1990; Potter 1996). Whereas the caller as a help seeker could lose her 'right' to a symmetrical approach, as an 'ordinary' conversational partner she will be entitled to this form of equality.

In this respect, Jefferson and Lee (1992) make an interesting distinction between two categories which participants may draw upon when they describe some sort of trouble, namely, advice seekers and troubles tellers. Troubles tellers are treated as self-responsible persons, who, despite their problem, still have to meet the requirements of the community. The advice seeker, on the other hand, is predominantly approached as a (sufferer of a) problem. In other words, by defining yourself as a help or advice seeker, you may loose the right to be treated as a 'person'.

It is this right that the caller seems to claim, first by depicting herself explicitly as someone who does not seek help but something more 'ordinary', and subsequently as a 'normal' and thus equal partner in conversation.

I can't hear you, now what did you say?

Let us focus in more detail on the counsellor's contributions. The caller's question in line 26 has created the (normative) expectation that the counsellor will produce the second part of the adjacency pair, that is, an answer to the question. Instead, we see a pause (line 27), a hearable in-breath and a repeated question (both line 28). The answer itself only follows in lines 30–31. We find similar reactions with respect to the caller's second question, with which she inquires after the counsellor's star sign. In lines 40 and 44, we see a double request for repetition ("*What did you say?*" and "*I- I can't hear you, now what did you say?*") and in line 46, the counsellor repeats the question herself ("*O::h, what star sign am I?*"). Finally, before she gives an answer ("*Yes, I am Taurus*") in line 52, she poses two return questions in lines 46 and 48 ("*You like astrology?*" "*You're familiar with that*").

What happens here? Let us first have a look at the caller's uptake. If we restrict ourselves to the first exchange about the counsellor's motives for doing the job, we see that the caller reacts very promptly to the counsellor's remarks (see the immediate link between lines 28–29 and lines 35–36). Moreover, she pointedly marks the 'information' that is given to her as 'old news', that is, insights which are already available to her ("*Oh!>that's what I just wanted to say*" in line 32 and "*Now then!*" in line 36). In doing so, she claims the initiative in the conversation in a more enduring manner.

In the exchange about star signs, the caller responds differently to the counsellor's contributions, that is, in the first instance. When there is no answer to her question ("*.hhhh °what star sign are you?°*", line 39), she starts by repeating it (line 41). When, however, this louder version does not produce a positive result, she mentions that she does not attach any specific value to star signs ("*>°ah it doesn't mean anything of course°<*", line 43). This formulation allows the caller to establish that she has no particular commitment to the topic and thus no specific stake in receiving an answer to her question. By displaying some pre-emptive indifference towards to the topic, the caller innoculates herself against the possibility of a sceptical response, or no response at all. In doing so, she treats the continued absence of the reply not only as a volume matter but also as an indication that the topic is perhaps not going to be dealt with for some other, more sensitive reason.

We can observe the same cautiousness in some of her other utterances with respect to star signs. Look, for example, at line 39, in which the caller poses her question in a rather low voice and only after a hearable in-breath. In doing so, the caller construes the question as a more or less delicate object (cf.

Silverman 1997). A related phenomenon can be found in line 49, in which the caller adds an explanation with respect to her interest in star signs, so as to give her question an accountable basis.

However, after having accomplished these more and less explicit forms of pre-emptive accounting, the caller again claims the initiative with emphasis. Line 51 makes clear that the caller treats the counsellor's return questions ("*You like astrology?*" "*You're familiar with that*") and the subsequent utterances as an insertion sequence: with her "*Well?*", she presents herself as someone who is still entitled to an answer.

The analytic task here is not to cast doubt on the authenticity of the volume problem or to decide what the counsellor's 'real' motives are. Much more relevant is the question what exactly the postponement of the answer is *doing* in the interaction. Interestingly, the counsellor's contributions are similar to the preludes of a dispreferred response (Pomerantz 1984), which can also be preceded by a hesitation, a pause or a request for repetition of the previous utterance. We can explain this similarity by analysing this part of the exchange nót as a 'simple' question-answer pattern, but as an initiative to 'interrogate' the counsellor, which can either be accepted (preferred response) or rejected (dispreferred response). From this point of view, the counsellor's 'answers' precede the rejection of the control over the agenda as claimed by the caller.

Moving on, I will argue that in line 58 ("*a↑ha, a↑ha*") the counsellor indeed (re)claims the initiative, namely by suggesting with retrospective effect that the initiative has been with her all the time.

We can conclude that the exchange about job motives and star signs is again and simultaneously a 'negotiation' about identities (between inverted commas, as I do not want to suggest that it concerns strategic behaviour. That is something for the participants to decide (see also Edwards 1997; cf. Heritage 1990/1991). Once more, the caller underlines her competence, namely by putting herself forward as an ordinary and thus equal partner, thereby undermining the institutional character of the interaction, whereas the counsellor, in postponing her answer, resists the caller's initiative. It is interesting to see how the caller's identity (the non-help-seeker, the ordinary partner in conversation) also implies a particular identity for the counsellor, who ceases to be a 'help giver' in that case. However, this identity might be contested. It is precisely what happens in the third segment of the talk, in which the counsellor reclaims the initiative, herewith (again) constructing her identity as someone who is providing help and the caller's 'matching' identity as someone who possibly needs help.

Aha aha: Outlining the client (segment 3)

58 Cou A↑ha, a↑ha
59 Cal What aha ha
60 Cou Leos with Sagittarius for ascendant don't sit at home
61 (0.8)
62 Cal No eh?=
63 Cou =Having agoraphobia
64 Cal °No eh?°
65 Cou No, I don't think so
66 Cal No
67 Cou They hunt (.) they go outside
68 Cal How nice of you that you ( )
69 Cou Isn't that right?
70 Cal Ye::ah [it's true
71 Cou [( )but they're very strong animals=
72 Cal =Don't say that!
73 Cou Aren't they?
74 Cal Ehh: I ( )
75 Cou [The lion is
76 Cal [This is this is ehh psychological peeling and I really
77 think that you do see that through( )I find you ( )
78 Cou The lion is the king of the jungle you can't
79 [alter that]
80 Cal [°Ye:s no↑°]yes
81 Cou And the Sagittarius is a ehh hunter
82 Cal ( )
83 Cou So that's all I'm- I'm saying
84 (1.1)
85 Cal .hh Yes=
86 Cou =That doesn't fit the image of agoraphobia and being depressed
87 Cal No! that( )just doesn't work
88 Cou No
89 Cal Because I am (.) I also have a very sensitive side from
90 Cancer probably
91 [((sob)
92 Cou [Mm yes (0.9) but say let's get back to (.) you say from
93 the RIAGG, someone comes to your home but that doesn't, of
94 course help you say

Line 58 ("*A↑ha, a↑ha*") again shows a remarkable turn in the conversation. The caller first displays some cautious resistance in response to the "*A↑ha, a↑ha*" by the counsellor. With some emphasis, she asks for an explanation ("*what aha ha*", line 59), thereby prompting the counsellor to give an answer.

What exactly is the counsellor's "*A↑ha, a↑ha*" doing? If we take a closer look, we see how this, at first sight perhaps insignificant, utterance allows the counsellor to do some crucial interactional business. By suggesting that it now 'becomes clear' to her, the counsellor defines the preceding exchange not as some informal and innocent talk but as part of a (professional) strategy which finally bears fruit. (Notice in this respect how the caller explicitly characterises the counsellor's observations as professional-strategic in line 76 ("*This is this is ehh psychological peeling* (...)". I shall elaborate on that in the next section). In doing so, the counsellor puts the change of initiative in the second segment of the conversation in a quite different perspective: by indirectly describing the recent remarks by the caller as informative in a more than non-committal respect, she retakes, so to speak, the initiative with retrospective effect.

In lines 62–70, however, the caller evaluates the counsellor's sudden and revealing 'discovery' in a more positive manner. After a pause, she now cautiously 'joins' the counsellor in her *aha erlebnis.* With her "*No eh?*" (line 62 and, in a lower voice, line 64), she presents herself as someone who would like to believe that she, adorned with the characteristics of the Leo, with Sagittarius for ascendant, is not susceptible of agoraphobia. (As mentioned before, the caller described herself earlier in the conversation as suffering from agoraphobia.) By expressing her sympathy in a personalised way ("*How nice of you that you* ( )"), she also defines the status of the counsellor as an equal partner rather than, say, a professional provider of help.

Having a very sensitive side (from Cancer probably)

However, the caller's display of alignment radically changes in line 72 ("*Don't say that!*"). Here she strongly and promptly resists the counsellor's description that lions are "*very strong animals*" (line 71). In using this description, the counsellor indirectly portrays the caller as an autonomous and powerful personality. In this sense, it connects with the counsellor's previous observations, which received a positive response from the caller. However, the category "*very strong*" people may also invoke other associations, such as particular expectations a 'very strong' person has to live up to.

In line 87 ("*No! that* ( ) *just doesn't work*") and lines 89–90 ("*Because I am* (.) *I also have a very sensitive side from Cancer probably*"), the caller is attending to the kinds of inferences which may be drawn about her character by virtue of her membership of this category. By putting forward that she "*also*" has "*a very sensitive side from Cancer probably*", she suggests first, that a very strong person might not be seen as sensitive and second, that this common assumption does not apply to her: she has a (very) strong and a (very) sensitive side. Note how this formulation allows the caller to claim sensitivity without having to deny her strength. Rather than constructing an other-than-very-strong identity, she defines both characteristics as two sides of the same coin (compare: 'I *am* very sensitive', which is a possible version of the self-repair in lines 89–90 ("*Because I am (.) I also have a very sensitive side from Cancer probably*") with: 'I also have a very sensitive *side*').

Through resisting this possible inferential implication of being 'very strong' and offering a "*very sensitive side from Cancer probably*" instead, the caller shows how it can plausibly be argued (i.e. within the astrological framework) that she suffers from agoraphobia and depression. It is her "*very sensitive side*" that makes her susceptible of these problems. In formulating this two-sided identity, the caller re-claims the problems that the counsellor has just 'denied' her (see lines 86–89: "*=That doesn't fit the image of agoraphobia and being depressed*" "*No! that ( ) just doesn't work*" "*No*" "*Because...*").

Note, however, that the caller does not describe her 'other side' as, say, weak or dependent, but as very sensitive. This "very sensitive side" does not affect the caller's competence in any negative sense: it suggests vulnerability but it also implies a capacity to perceive things that other people might not notice or realize. In other words, her sensitive side allows her to be 'only human' without damaging her overall strength. In drawing upon this self-description, the caller again carefully defines herself as a normal, capable member of society, now by presenting herself as someone who is strong and sensitive at the same time (but notice the sobbing and the way in which the counsellor treats the caller's self-description in lines 92–94 – I shall come back to that later on).

Psychological peeling

Let us return for a moment to lines 76–77, in which the caller evaluates the counsellor's remarks as a form of psychological peeling" ("*This is this is ehh psychological peeling and I really think that you do see that through*"). What happens here? By indirectly categorising the caller as a 'very strong' person, the counsellor attributes to her the kind of competent identity which the caller has claimed all along, but in a rather extreme version. Note how she also describes the caller as someone who does not sit at home having agoraphobia (lines 60 and 63). This obviously contradicts the caller's self-diagnosis earlier in the conversation, and contrasts with the counsellor's characterisation of the caller as a help seeker in the first part of the fragment. As we have seen, the caller also treats this image as one-sided and incomplete: she 'adds' a sensitive side so as to re-entitle her to the problems that she has claimed to suffer from.

However, by providing the caller with such an ample supply of competence, i.e. to a degree that there is no room for something other-than-being-very-strong, the counsellor makes the description of the caller's character 'suspicious' in terms of being part of a conversation between two equals. The caller orients to the counsellor's observations as belonging to a professional approach ("***psychological** peeling*") rather than an informal exchange of ideas. Moreover, "*psychological peeling*" suggests that the 'outer surface' is being removed in search of something deeper. In other words, the caller identifies the counsellor as someone who is constructing her not as a 'person' but as ('sufferer of a) problem' – an 'object' of analysis.

By adding "*and I really think that you do see that through*" (lines 76–77), the caller also treats the counsellor's remarks as strategic, that is, as observations which are wilfully made, as part of a plan. Interestingly, the caller does not say 'I see you through' but "*I really think that you do see that through*". Both utterances could be seen as 'unmasking' a strategy but whereas the first utterance attributes a special perceptivity to the speaker, the second utterance also ascribes a particular kind of 'competence' to the other participant. In putting forward that she thinks that the counsellor does see it through, the caller suggests that the counsellor is clever enough to know what she is doing. At the same time, she also confirms her own competence, namely by indicating that she in turn is able to observe all this. This formulation allows the caller, on the one hand, to refer to the violation of the informal nature of their talk (i.e. it is part of a strategy), and, on the other hand, to invite the counsellor to repair it (by carefully constructing her as a competent observer of her own behaviour).

In lines 78–86, however, the counsellor more or less confirms the institutional (help giver:help seeker) relationship by not denying the assumed psychologizing nature of her utterances: professional-strategic or not, my remarks are grounded in solid astrological facts (note: the terminology which the caller herself has introduced), against which there can be no argument.

In doing so, she defines 'psychological peeling' as a non-issue. In mentioning that the facts cannot be altered ("*The lion is the king of the jungle you can't alter that*", lines 78–79), she orients to the idea that independent of how calculated her actions are, the facts will remain the same. Moreover, whatever the status of her actions, the facts speak for themselves ("*And the Sagittarius is a ehh hunter*" "*So that's all I'm- I'm saying*", lines 81 and 83).

In suggesting that, according to the (facts of the) astrological map, the caller is not 'allowed' to suffer from agoraphobia and depression, the counsellor invites the caller to qualify herself for these problems. In lines 92–94, we see how the counsellor picks up the caller's 'qualification' ("*Because I am*

(.) *I also have a very sensitive side from Cancer probably*", lines 89–90) and the sobbing in line 91 in such a way so as to make the caller 'ready for intervention' (see Edwards 2000). That is, she treats it as an invitation to refocus on the caller's history of help, thereby reformulating the caller's identity as a (potential) help seeker rather than someone to chat to. With this, she explicitly (re-)establishes the conversation in institutional terms and (re-)confirms her lead in the conversation.

4. Conclusion

Having read the conversation between the counsellor and the caller, it is tempting to conclude that the caller's need can best be characterised as an implicit request for help. Probably just as tempting is the conclusion that the caller does not have such a request but merely wants to chat. Most tempting, perhaps, is the option that the truth is somewhere in the middle. However, the question as to what the caller 'really' wants is not relevant to an adequate understanding of this conversation, as I hope to have shown. What is relevant, is how the participants managed this issue in their talk.

We have seen that the formulation of the caller's reason-for-calling was part and parcel of the participants' interactional business. By constructing her reason-for-calling in a particular way, the caller worked up and implied a specific identity. In portraying herself as an ordinary person, the caller actively undermined the potentially negative inferential implications of the category help seeker (such as dependent behaviour) and also defined herself as entitled to the rights of an equal partner in conversation. The same identity allowed the caller to present herself as 'strong but sensitive', thus giving herself the right to be suffering from particular problems.

I also showed how the caller's identity had interactional implications for the counsellor's identity, and vice versa. The caller's 'ordinariness' deprived the counsellor, in the first instance, of her right to take the initiative in the introduction and shaping of the conversational topics. In a subtle way, however, she re-claimed the initiative, after which the caller could again be defined as a potential help seeker, that is, for the time being.

More broadly, this analysis shows that participants do not simply report their inner life when describing particular needs or problems (cf. Edwards & Potter 2005). On the contrary, we have seen how these descriptions are used interactionally and rhetorically, for example to built and refute certain identities. Interestingly, the caller's main concern in this respect was not to account for her help request, as we might expect from a helpline call, but to display her competence as a normal interlocutor.

Acknowledgements

I would like to thank the counsellors at the helpline for their kind cooperation and involvement. I hope that the results of this study are of use to them and to future callers. Finally, I am indebted to the editors of this volume, Alan Firth, Carolyn Baker and Michael Emmison, and to Jonathan Potter and Alexa Hepburn for their thoughtful comments on previous versions.

Notes

1. The fragment is translated from Dutch into English with the help of a professional translator. Like transcription, translation is already a form of analysis and, in that sense, the version provided must be considered as a free translation. The text of the original Dutch extract can be found at the end of this paper.

2. The RIAGG is a regional institute which provides for ambulatory mental assistence.

References

Antaki, C. & Widdicombe, S. (1998). 'Identity as an achievement and as a tool'. In C. Antaki & S. Widdicombe (Eds.), *Identities in Talk* (pp. 1–14). London: Sage.

Drew, P. & Heritage, J. (1992). 'Analyzing talk at work: An introduction.' In P. Drew & J. Heritage (Eds.), *Talk at Work: Interaction in Institutional Settings* (pp. 3–65). Cambridge: Cambridge University Press.

Edwards, D. (1997). *Discourse and Cognition*. London: Sage.

Edwards, D. (1998). 'The relevant thing about her: Social identity categories in use.' In C. Antaki & S. Widdicombe (Eds.), *Identities in Talk* (pp. 15–33). London: Sage.

Edwards, D. (2000). 'Couples, conflict and counselling'. Paper presented at the *7th International Pragmatics Conference*, Budapest, 9–14 July.

Edwards, D. & Potter, J. (2005). 'Discursive psychology, mental states and descriptions'. In H. te Molder & J. Potter (Eds.), *Conversation and Cognition* (pp. 241–259). Cambridge: Cambridge University Press.

Heritage, J. C. (1990–1991). 'Intention, meaning and strategy: Observations on constraints in interaction analysis.' *Research on Language and Social Interaction, 24*, 311–332.

Heritage, J. C & Sefi, S. (1992). 'Dilemmas of advice: Aspects of the delivery and reception of advice in interactions between Health Visitors and first-time mothers.' In P. Drew & J. Heritage (Eds.), *Talk at work: Interaction in Institutional Settings* (pp. 359–417). Cambridge: Cambridge University Press.

Jefferson, G. & Lee, J. R. E. (1992). 'The rejection of advice: managing the problematic convergence of a 'troubles telling' and a 'service encounter'.' In P. Drew & J. Heritage (Eds.), *Talk at Work: Interaction in Institutional settings* (pp. 521–548). Cambridge: Cambridge University Press.

Pomerantz, A. (1978). 'Compliment responses: Notes on the co-operation of multiple constraints.' In J. Schenkein (Ed.), *Studies in the Organization of Conversational Interaction* (pp. 79–112). New York: Academic Press.

Pomerantz, A. (1984). 'Agreeing and disagreeing with assessments: Some features of preferred/dispreferred turn shapes.' In J. M. Atkinson & J. Heritage (Eds.), *Structures of Social Action: Studies in Conversation Analysis* (pp. 57–102). Cambridge: Cambridge University Press.

Potter, J. (1996). *Representing Reality: Discourse, Rhetoric and Social Construction.* London: Sage.

Potter, J. (1998). 'Cognition as context (whose cognition?)'. *Research on Language and Social Interaction, 31,* 29–44.

Sacks, H. (1972). 'On the analyzability of stories by children'. In J. Gumperz & D. Hymes (Eds.), *Directions in Sociolinguistics* (pp. 325–345). New York: Holt Rinehart & Winston.

Sacks, H. (1979). 'Hotrodder: A revolutionary category.' In G. Psathas (Ed.), *Everyday Language: Studies in Ethnomethodology* (pp. 7–14). New York: Irvington.

Sacks, H. (1984). 'On doing 'being ordinary'.' Compiled by G. Jefferson in J. M. Atkinson & J. Heritage (Eds.), *Structures of Social Action: Studies in Conversation Analysis* (pp. 413–429). Cambridge: Cambridge University Press.

Sacks, H. (1992). *Lectures on Conversation* (Vols. I & II, edited by G. Jefferson). Oxford: Blackwell.

Schegloff, E. (1987). 'Analyzing single episodes of conversation: An exercise in conversation analysis.' *Social Psychology Quarterly, 50,* 101–114.

Schegloff, E. A. & Sacks, H. (1973). 'Opening up closings.' *Semiotica, 8,* 289–327.

Silverman, D. (1997). *Discourses of Counselling: HIV Counselling as Social Interaction.* London: Sage.

Te Molder, H. (1999). 'Discourse of dilemmas: An analysis of communication planners' accounts'. *British Journal of Social Psychology, 38,* 254–263.

Te Molder, H. & J. Potter (Eds.). (2005). *Conversation and Cognition.* Cambridge: Cambridge University Press.

Whalen, M. & Zimmerman, D. (1990). 'Describing trouble: Practical epistemology in citizen calls to the police.' *Language in Society, 19,* 465–492.

Original Dutch extract

B ik heb nu contact met een RIAGG-medewerkster, die komt hier
H oh ja
B weet je wel, een keer in de zoveel tijd, maar dat he:lpt niet
H dat helpt niet
B °ahh natuu:rlijk niet°
H tuurlijk niet?
B ik weet, ja, je moet toch ze:lf doen
(0.5)
H ja: maar je belt nu ook hier naar toe in de hoop dat het toch iets
helpt (.)neem ik aan
(0.9)
B nee, ik wil nu gewoon iemand ho:ren=
H =[ja
B [gewoon weten dat ik niet [allee:n ben op deze aardkloot zit
H [°ja°
B ((beller snikt zachtjes))=
H =ja maar dat kan toch wel he:lpen?
B .hhhhhhh
(2.1)
H bijvoorbeeld (.) en een andere keer kan een a:rm om je heen
helpen, een ander keer kan een advie:s helpen, natuurlijk moet
je het alleen doen maa:r (1.3) we zijn niet voor niks met
zoveel mensen hier op aarde, ik denk, dat is dan toch ook
prettig dat je elkaar een beetje kan steunen (1.1) en niet dat
je alles helemaal helemaal alleen moet doen
B vertel me iets van jezelf, waarom doe je dit werk?
(0.7)
H .hh waarom ik dit werk doe?=
B =mm hm
H tja: gew- omdat dit gewoon mijn werk is, dit is eh, hier ligt mijn
hart
B oh!>dat wou ik maar even zeggen, of het gewoon je
werk is of dat je hart erin ligt, dat is toch een hee:l
verschil<
H ja:=
B =nou dan!
H ja nou dat doe ik met mijn werk, hier ligt mijn hart, dat is mijn
werk
B .hhhh °wat voor sterrenbeeld ben je?°
H wat zeg je?
B .hh wat voor sterrenbeeld ben je?
(1.0)
B >°ah, het zegt niks natuurlijk°<
H ik- ik versta je niet, wat zeg je nou?
B wat voor sterrenbeeld ben je?
H o::h,wat voor sterrenbeeld ben ik?Ha!Je houdt wel van astrologie?
B [ja ik
H [ben je wel bekend mee
B nee::, ik geloof dat nie:ts voor nie:ts is (.) uiteindelijk
H dat geloof ik ook
B nou?
H ja, ik ben stier

B a:hh
H en jij?
B leeuw
H een leeuw
B met boogschutter ascendant
H a↑ha, a↑ha
B wat aha ha
H leeuwen met boogschutter als ascendant die zitten niet thuis
(0.8)
B nee he?=
H =straatvrees te hebben
B °nee: he?°
H nee, dat dacht ik niet
B nee
H die gaan op jacht (.) die gaan naar buiten
B wat leuk van je dat je dat zo ( )
H ja toch?
B jaa::h [is waar
H [( )maar ze zijn hele sterke dieren=
B =nouww:::!
H ja toch?
B ehh: ik ( )
H [de leeuw is
B [dit is dit is ehh psychologisch pellen hoor en ik denk heus wel dat
je dat doorhebt hoor( ) ik vind je ( )
H de leeuw is de koning van het woud daar kan je
[niets op afdoen]
B [°ja: nee↑°] ja
H en de boogschutter dat is een ehh jager
B ( )
H dus dat is alleen maar wat ik- ik zeg
(1.1)
B .hh ja=
H =dat past niet in het beeld van straatvrees en depressief zijn
B nee! die ( ) werkt gewoon niet
H nee
B want ik ben (.) ik heb ook een hele gevoelige kant van de kreeft
waarschijnlijk
[((snif))
H [mm ja (0.9) zeg maar even terug he (.) je zegt van het
RIAGG, die komt aan huis maar dat helpt natuurlijk niet
zeg je

Part III

Healthcare provision

Chapter 8

Callers' presentations of problems in telephone calls to Swedish primary care

Vesa Leppänen

1. Introduction

This chapter examines callers' help-seeking behaviour in telephone calls to Swedish primary care helplines. These calls are often the first steps that people in Sweden take when they experience medical problems in need of professional medical help. Usually nurses answer and listen to the callers' problem presentations, ask diagnostic questions, make decisions about what measures need to be taken, suggest their decisions to the callers in the form of advice, and finally sum up the new cases in medical files that often are sent to doctors or others who do the actual physical examinations of the patients.

The chapter analyses the first few seconds of these calls, the moments at which callers' subjectively experienced illnesses are formulated as possible medical problems. On the basis of analyses of audio-recordings of the calls, I describe the different talk-based 'formats' through which callers present their problems, and discuss the interactional consequences of the different formats.

1.1 The focus of this study

The overall aim of this research project is to describe and analyse a number of features of routine calls to Swedish primary care; how callers present their problems, how nurses perform diagnostic work, and how nurses respond to callers' problems. As mentioned above, this chapter only analyses the first part of these calls, the first few seconds when callers present their problems.

One possible limitation of the chapter is that it has no comparative aims. A number of studies have addressed how people talk about their troubles, concerns and problems in a variety of contexts. Jefferson (1980, 1984a, 1985,

1988) and Jefferson and Lee (1980, 1992) have analysed different features of troubles-talk in both institutional and non-institutional English-speaking settings. Troubles-tellings have also been studied by Coupland, Coupland and Robinson (1992) as they occur after the question "how are you?" in interviews with elderly Welsh persons. Coupland, Robinson and Coupland (1994) also analysed how the question "how are you?" elicited problem presentations in encounters between doctors and patients. Another study of problem presentations in routine doctor-patient interactions in primary care was conducted by Ruusuvuori (2000). Leppänen (1998) analysed the positions and forms that patients' presentations of concerns took in routine face-to-face encounters with Swedish district nurses. Whalen and Zimmerman (1987, 1990) and Zimmerman (1992) have analysed how problems are presented in telephone calls to the emergency number 911 in the US. Let me here mention that a careful comparison of the above findings, and ones presented in other contributions to this volume, would be very fruitful. It would show us, in more detail, which features of problem-talk are generic to "talk about problems" and which are related to the different contexts.

1.2 The database

Empirical data for this project were collected between February and November 1999 and consists of 276 recorded telephone calls to 13 different nurses working at 6 primary care centres. The database also consists of semi-structured interviews with 18 nurses working at 12 different centres, but these are not analysed here. Only 209 of the recordings were used in the present analysis. In some of the remaining calls the caller was not a potential patient, but a colleague at the centre or a caller from another institution. In other calls the nurses started the tape-recorders too late after the calls had begun, so that the first seconds were missing. The recordings have been transcribed using the conventions first defined by Jefferson (Sacks, Schegloff, & Jefferson 1974; Heritage 1984a). These conventions are especially suitable for sequential analyses of interaction – analyses that focus on the *interrelationships* between words, utterances and gestures.

Let me say a few words about the transcripts presented below. Each transcript has an identification code, for instance "5 VCTEL 10", indicating which nurse is talking (number 5) and from which recording the extract is taken (number 10). When possible, each utterance is translated word for word from Swedish to English. Therefore, the word order is sometimes a bit odd, but I hope it will be understandable for most readers. Although this study is about an activity that typically occurs in the beginnings of these calls, the presented transcripts do, in many cases, not include the first 15–20 seconds of talk. The reason is that the nurses usually informed the callers about this study and asked for consent to record the calls immediately after the opening summons/answer, identification and greeting sequences.

1.3 Working on the telephone and computer

Before we turn to the analysis, it will be illustrative to give a short description of the work setting of these nurses. The nurses working on the helpline could see the doctors' schedules on their computer screens. They could see which doctors worked at what times and if a particular time-slot was filled with a patient visit or administrative tasks. The lengths of the time-slots varied, but were usually 10, 15 or 20 minutes. At most centres, the nurses could choose between *different types of slots* for different types of patients. They had "acute slots" available for today's callers. They also had "time scheduled slots", where callers with less acute problems could be placed. The waiting time for these cases was usually between 3 weeks and 3 months. Some centres also had "semi-acute slots" reserved for callers who did not need a same-day appointment, but who could not wait for weeks (for instance, patients who needed renewed medical certificates to prolong their sick leaves).

When working with the computers, the nurses filled in a number of fields for each caller: the caller's address, personal identification number (birth date plus a 4-digit code), telephone number, and a short description of the caller's problem (for instance, "pain in the stomach", "heart medicine prescription" or "hypertension"). This means that the nurses' main task, similar to that of the alarm operators studied by Zimmerman (1992), was to construct "cases" to be handed over to the doctors or nurses to whom the callers were sent. These cases, typed into the computers, were the documents that the doctors used when they began their work with the patients.

2. Callers' presentations of problems

Callers' presentations of problems are regularly packaged in one of three formats: as requests to see a doctor, as requests to ask questions, or as narratives.

2.1 Requests to see a doctor

In 66 of the 209 analysed recordings, the callers simply requested to see a doctor. Callers tended to make their requests in slightly different ways, which will be presented below.[1]

(a) Requesting to see a doctor followed by a presentation of the problem
Callers would request to see a doctor, then briefly describe the problem (C=caller, N=nurse):

(1) 5 VCTEL 10

```
1  C:  ˙hh ja hejsan ja skulle vilja::
       ˙hh yes hello I would like to::
2      beställa en tid hh ja har en tvååring
       make an appointment hh I have a two year old
3      som har ont i halsen va(hh),
       that has a sore throat you see(hh),
4  N:  °ja ˙hh hur länge har han haft ont i halsen?
       °yes ˙hh how long has he had a sore throat?
```

(2) 5 VCTEL 21

```
1  C:  ˙h EEE ja skulle vilja komma in där för
       ˙h EEE I would like to come in there because
2      ja har gått o haft ont i huvudet o vatt
       I have gone round and had headaches and been
3      yr hela dan,
       dizzy all day,
4      (0.8)
5  N:  °ja:, e- men ja ha- ja har inte fler tider
       °yes:, e- but I hav- I don't have more time slots
6      nu i kväll
       now this evening
```

In these two extracts, the callers first utter *requests* that begin with "*I would like to*" which are followed by "*make an appointment*" in extract (1) and "*come in*" in extract (2). They then add short *presentations* of their problems. In extract (1) the caller says "*I have a two year old that has a sore throat you see*" and in extract (2) "*I have gone around and had headaches and been dizzy all day*". These presentations are *minimal versions of the callers' problems.* (Longer versions are given when callers use the narrative format, as will be shown below.) But observe that these presentations contain the following three important pieces of information: (a) *Who* has the problem ("*a two year old*" and "*I*", respectively); (b) A *description* of the problem ("*sore throat*" and "*headaches and dizzy*"). These descriptions are best viewed as "glosses" (Jefferson 1985), which are unpackaged in the following talk (This is what nurses' diagnostic work largely consists of; guiding callers unpackagings of the problems they initially presented.); (c) *How long* the problem has lasted ("*all day*" in extract (2)). This last piece of information is missing in the first extract, but observe that the nurse requests it in her first question (see also extract (6), below). Establishing the time of onset of the callers' problems seems to be of central concern to the nurses.

From the nurse's point of view, a sufficient initial problem presentation consists of these three pieces of information. We find evidence for this when we examine the following extract, where the caller only requests to see a doctor without presenting the problem. The nurse awaits the description by repeating part of the caller's request and then adds a "*yes*(*hh*)?":

(3) 12 VCTEL 34

```
1   C:  ja skulle gärna villa ha en tid hos
        I would like to make an appointment with
2       doktor bengtson?
        doctor bengtson?
3   N:  hos björn bengtsson  [ja(hh)?
        with björn bengtsson [yes(hh)?
4   C:                       [˙h h ja va där
                             [˙h h I was there
5       i förra veckan för ja har så ont i huvet hh,
        last week because I have such a headache hh,
```

This extract illustrates that the nurse expects a complete problem presentation to consist of more than a request to see a doctor. When this information is not provided, she asks for it. The "yes(hh)" (6) elicits the information that is missing. The following extract illustrates this pattern further: when the caller does not provide a problem presentation attached to the request, the nurse first waits for 1.5 seconds and then explicitly asks the caller why he wants to see the doctor:

(4) 10 VCTEL 5

```
1   C:  ˙hh ja ja skulle ha e (.) till e::
        ˙hh I I would have e (.) to e::
2       doktor lisa larsson,
        doctor lisa larsson,
3   N:  ja?
        yes?
4       (1.5)
5   C:  lisa larsson,
```

```
6   N:  jae va vill du me henne,
        yes what do you want with her,
```

When callers use the request format (including descriptions of their problems), the nurses treat the presentations as complete. From this point on, the nurses begin to perform diagnostic work. (In extract (1) the nurse starts to ask diagnostic questions and in extract (2) the nurse begins to close down the entire encounter by saying that there are no more time slots left for this particular evening.)

(b) Requesting to see a doctor followed by a diagnosis

In a number of instances, the callers first uttered a request and then added a *diagnosis*. That is, not only did the callers describe their problems, but they also expressed, using medical terminology, the nature of their problems:

(5) 5 VCTEL 14

```
1   C:  ja skulle vilja ha tid hos en doktor
        I would like to make an appointment with a doctor
2       där hh ˙hh ja tror ja haft halsfluss(hh),
        there hh ˙hh I think I've had tonsillitis(hh),
3   N:  ja::, hur länge har du haft de(hh)?
        yes::, how long have you had that(hh)?
```

(6) 5 VCTEL 3

```
1   C:  ˙hh ja ska vilja hh b- besöka en läkare
        ˙hh I would like to hh v- visit a doctor
2       ˙hh ja har (.) troligtvis magsår
        ˙hh I have (.) probably ulcers
3       ˙hh de känns så,
        ˙hh that's how it feels,
4   N:  ja::hh, har du pratat me din
        yes::hh, have you spoken to your
5       vårdcentral ida eller?
        primary care centre today or?
```

In extract (5), the caller first requests to see a doctor and then adds a diagnosis of the problem: "*I think I've had tonsillitis*". We see an identical pattern in extract (6), when the caller first requests to see a doctor and then adds a diagnosis of the problem: "*I have* (.) *probably ulcers ˙hh that's how it feels*". One important observation about these diagnostic utterances is that they are made very *cautiously*. In extract (5), the caller begins by saying "*I think*" before the diagnosis is uttered. In extract (6) the caller inserts "*probably*" into "*I have ulcers*". Then the caller continues by saying, "*that's how it feels*".

Callers' diagnoses are usually cautious. They achieve this cautiousness by marking their statements as matters of "belief" ("I think") or "subjectively felt symptoms" ("that's how it feels") rather than as objectively perceivable truths. By inserting these subjectivity markers, the callers move their diagnostic utterances from "objective truth" into the realm of subjectively felt symptoms.

Drew (1991) noted the existence of similar forms of cautiousness when he analysed doctor-patient interactions. He provides the following example:

(7) [Drew 1991:38]

```
Pt  B't this time I have a little problem.
    (0.9)
    I seem to have
    (0.8)
Dr  nYes[:s.
Pt      [what is it-contracted
    (0.4)
Dr  khn [Ye:s
Pt      [tendon:.
Dr  That's right. how long have you been
    in developing thi:s.
```

We see that the patient, when presenting the problem, says "*I seem to have*", thus lowering the epistemological claims she makes about her diagnosis. The patient also marks the diagnostic term as a question by inserting "*what is it*" and she inserts a silence between the first part of the diagnostic utterance, "*contracted*", and "*tendon*". In these cases, the callers are oriented to the nurses (and, in Drew's example, the doctor) as persons who have the *right* to the analytical terminology; they have the right to decide when these terms are properly used. This is treated, by callers, as a part of the nurses' professional domain. (This does not necessarily mean that there is a *cognitive* asymmetry of knowledge. Callers may very well know precisely what diagnosis will be used by the nurse or the doctor.) Thus the callers, by being cautious, construct the nurses as "more knowledgeable about medicine" than they are themselves. To understand in more detail why this interactional asymmetry is (re)created, we need to turn to the next set of cases:

(*c*) *Only diagnosis*

In some instances, the callers did not utter any explicit requests, but simply gave the diagnosis, as in the following extract. Observe that the caller inserts an "*I think*" before she utters the diagnosis, "*urinary infection*", thereby doing being cautious.

(8) 5 VCTEL 9

```
1   C:  ˙hh ja hej doris björn-bengtsson(hh)?
        ˙hh yes hello doris björn-bengtsson(hh)?
2   N:  ja  he[j?
        yes he[llo?
3   C:        [˙hh hej=de e så a=ja tror ja
              [˙hh hello=it's like this=I think I
4       har fått urinvägsinfektion?
        have had a urinary infection?
5   N:  ja::hh,
        yes::hh,
6       (.)
7       hur märker du de då?
        how do you notice that then?
```

This format, giving the diagnosis, functions *interactionally* as a request, although it is not explicitly formed as one. This is so for two reasons: (i) It is placed in a position where the caller is expected to present his or her problem, and (ii) "Diagnoses" are found later in the care-giving chain than "descriptions" of problems or "requests". Let me explicate this further. According to the ideal image of the history of medical cases, problems adhere to the following path: (a) First something problematic appears – a symptom. The caller feels ill or observes something on his body that may be a medical problem; (b) The caller then contacts the medical institution and the professional listens to the caller's problem; (c) Verbal and physical examinations are made; (d) A diagnosis is stated; and (e) Treatments are chosen and given. This is the view callers and nurses seem to have as to the order in which things should be performed. Now, when a caller provides a diagnosis, he moves directly to one of the later steps in this ideal chain of events. And, in doing so, the caller implies all the previous steps. If the caller utters a diagnosis, he also implicates that a request is being made. Therefore, a diagnostic utterance in this position implies a request.

Let us now turn to the nurse's response, "*how do you notice that then?*" (line 7). This utterance has some interesting properties, compared to extracts (1), (5) and (6): (i) In these extracts, the nurses, in the positions following the problem presentations, move on to related matters. For instance, in extract (1), when the caller tells the nurse that her son has a sore throat, the nurse wants to know how long he has had it. Similarly, in extract (6), when the caller has said that he probably has ulcers, the nurse moves on to ask the caller if he has contacted the primary care centre. But in this extract (8), the nurse does not move forward to talk about related matters, but orients to the previous utterance. Thus, whereas the nurses' treatments of the callers' problem presentations in extracts

(1, 2, 5 and 6) can be described as continuous or successive upon the previous presentations, the response in extract (8) is discontinuous. It is oriented towards the utterance that preceded it. (ii) But not only is the question oriented to the previous utterance, it is also a request for information about how the caller has concluded that this particular diagnostic label should be used. Thus, by asking this question, the nurse moves the call back, from "a stated diagnosis" (step (d) in the ideal image of the history of medical cases) to "symptoms" (step (a)).

From the nurse's point of view, this is a legitimate question. She, as a medically trained nurse working at this particular primary care center, is expected to decide what diagnostic label should be used when. If the patient has not earlier understood that she is talking to a medical expert, it becomes explicit now. Therefore, the utterance is also hearable as a questioning of the callers *right* to utter diagnoses. It can be viewed as a reprimand, a piece of socialization of callers about how to present problems to nurses on the telephone. As a client-socializing utterance it is well fitted into the environment of this call. It is an implicit way to correct the client's way to present problems to medical specialists. It is not as threatening to the relation to the patient as an explicit correction would be. The utterance can also easily be accounted for, if anyone should ask, as a medically relevant question and not as a correction.

(*d*) *Requesting to see a doctor followed by a specification of the outcome*

Now, if we follow the ideal image of how medical cases are processed, we can expect to find cases in which callers move on to the later steps of the chain by requesting to see a doctor and then specifying the outcome of the visit. Given the analysis above, we would expect that callers who tell nurses what needs to be done to them would be opposed even more harshly. But see the following example:

(9) 1 VCTEL 7

```
1   C:  jag skulle vilja att berit persson
        I would like berit persson to
2       skrev ut lite värktabletter till mig?
        prescribe some painkillers for me?
3   N:  ja- jag får bara säga att samtalen
        yes- I have just to say that the calls
4       som ja har idag bandas,
        that I have today are taped,
5       (.)
6   C:   ja[?
        yes[?
```

```
7   N:  [är de okej för dig?
        [is that okay for you?
8   C:  ja de är helt okej,
        yes it's all right,
9   N:  de låter bra det,
        that's good,
10  C:  ja,
        yes,
11  N:  då ska vi se när är du född¿
        now let's see when were you born¿
```

We see that the caller requests to see a doctor and specifies what she wants from her. But the nurse does not question the caller's right to do so. Why is this so? We can see that the caller asks for a specific doctor, which suggests that she has a working relationship with this doctor. The caller also asks for painkillers, which doctors prescribe in restricted amounts (since they can be abused). Therefore patients with chronic pain call regularly to get new prescriptions. This means that these patients are not, themselves, responsible for having to call regularly. The request, in this case, does not emanate from the patient, but from a third party (the doctor and the rules governing how much medicine should be prescribed). This seems to be common to all cases in which callers produced requests followed by specified outcomes. Thus, due to the nature of this particular problem, this caller is able to make her request for a specific outcome in a very straightforward way without being opposed by the nurse.

Let me finally say that, in a few cases, callers uttered *only the specified request without asking to see a doctor* ("I would like to have a new prescription for hh ˙hh I use tenormin and the prescription I have is too old they said at the pharmacy"). In these cases, as in (9), the responsibility for calling is placed upon a third party, not the caller. The nurses' responses reflect this fact in these cases as well.

2.2 Questions

In (20) of the (209) extracts, the callers began their problem presentations by *asking for permission to ask a question.* These problem presentations commonly take the following sequential shape: (i) The caller *asks for permission to ask a question* (arrows 1 in extracts (10) & (11), below); (ii) The nurse then produces an *acknowledgement/yes-answer* (arrow 2) conveying to the caller that she is prepared to listen to the question; (iii) But the caller does not go on and ask the question. Instead, she provides *background information* about the problem (arrow 3) before she finally (iv) states the *question* (arrow 4); (v) The nurse then begins to answer the question/inform the caller (arrow 5).

(10) 5 VCTEL 23

```
1   C: ja skulle bara vilja fråga om fästingar?        ← 1
       I would just like to ask about ticks?
2   N: jA,                                             ← 2
       yES,
3   C: ˙hh min dotter har fått en fästing i            ← 3
       ˙hh my daughter got a tick in
4      ljumsken igår hh ˙hh hon e                      ← 3
       her groin yesterday hh ˙hh she is
5      fyra år,                                        ← 3
       four years,
6   N: mhm?
7   C: ˙hh ja ska bara höra hu=huhh ja fick            ← 4
       ˙hh I'll just ask how=huhh I get
8      bort den men hh (1.6) så att de inte            ← 4
       it away but hh (1.6) so that it isn't
9      ä nåt farlet(hh)?                               ← 4
       anything dangerous(hh)?
10  N: nej de e ju inget FARLIT hh ee nu               ← 5
       no it isn't you know anything DANGEROUS hh ee now
```

(11) 5 VCTEL 26

```
1   C: ˙hhja hej ja har en enkel fråga                 ← 1
       ˙hh yes hello I have a simple question
2      angående ett öga(hh)?                           ← 1
       about an eye(hh)?
3   N: ja(hh)?                                         ← 2
       yes(hh)?
4   C: hh ja ha::r fått sån här blodsprängt öga hh     ← 3
       hh I ha::ve got one like this bloodshot eye hh
5   N: ja,
       yes,
6   C: eee ser ut som nån form utav bristning hh       ← 3
       eee looks like some kind of rupture hh
7   N: ja,
       yes,
8   C: ˙hh va bör man vidta för åtgärder(hh)?          ← 4
       ˙hh what measures should you take(hh)?
9   N: ja (.) om du har b- e: de bara rött             ← 5
       yes (.) if you have b- is: it just red
10     att ett blodkärl                                ← 5
       so that one blood vessel
```

Let me explicate this pattern further. I begin by offering two observations about the callers' first utterances (arrows 1): (a) In both extracts the callers begin by asking the nurses if they can ask a question. This is what Schegloff (1980) calls an *action projection;* the caller projects an action that will follow if the nurse gives an affirmative answer to the question (But although the caller's utterance, on the surface level, may be formed as a question, callers do not necessarily await nurses' responses before moving on to their next utterances. Therefore the callers' preliminary questions need not necessarily be viewed as "questions to ask questions", but as *announcements* of questions that will follow); (b) Not only do the callers ask if they can ask questions, they also portray the general topics of the questions – "about ticks", "an eye" and "medical character". That is, they present their problems. After the callers' question projections, the nurses utter affirmative responses/continuers (arrows 2). (In a number of cases, the nurses interrupted the callers at this point to inform them about the recordings and to ask for permission to continue recording.) In no instance did the nurses hinder callers from beginning to provide background information.

What then follows is not the question that the callers projected in their first utterances. Instead they present their problems just as when they request to see doctors (arrows 3). The callers give at least three pieces of information: (a) Who has the problem – "*my daughter*" in extract (10) and "*I*" in extract (11); (b) Descriptions of their problems. In extract (10), the problem is described as a "*tick*" problem and the caller gives a specification of which part of the body the tick is attached to – "*groin*". In extract (11), the problem is depicted as "*bloodshot eye*" followed by a description of what it looks like: "*looks like some kind of rupture*"; (c) How long the problem has lasted – "had a tick in her groin yesterday" in extract (10).

Finally, the callers ask their questions (arrows 4). In extract (10), the caller asks "*I'll just ask how=huhh I got it away but hh (1.6) so that it isn't anything dangerous(hh)?*". Observe that the caller's question is rather open: The caller does not set up alternatives (for instance yes/no), but merely asks "how to get it away". That is, the caller asks for general help with the problem. Similarly, in extract (11), the caller says "*what measures should you take*(*hh*)?", which is an open question requesting general help. Thus, it is not the case that callers' question projections necessarily project *specific* questions. The question format is not necessarily used by callers who want to ask specific *questions.* It is just one of three formats used to present problems.

Which callers use this format? Let us first note that the problems presented in the above extracts were *explicitly non-acute.* In both cases, the problems were presented as minor. In extract (10), the caller began her pre-question by saying,

"*I would just like to*" (etc.) and later "*I'll just ask*". In extract (11), the caller began by saying, "*I have a simple question*". That is, in both cases, the callers marked their upcoming questions as being about minor problems. Note also that the caller in extract (10) portrays the overall problem as *under control*. The caller says that she wants advice about how she, herself, can remove the tick. She conveys that she has the problem under control, but that she merely needs instructions on one aspect of it – how to remove the tick.

The non-acuteness of the problems presented with the question format becomes even more evident when comparing them with how problems are presented with the narrative format. When I show instances of callers' narratives below, we will see that they regularly design their narratives with "complications": callers not only describe a problem but they also say how this problem has become worse. (For instance, a caller that used a narrative format said, "I have a son he has been coughing now for a month". This description is, per se, a problem worthy of medical attention. But this caller continues to describe the problem by saying that "I was down at the general practitioner's surgery about ten days ago but it hasn't got better", and then moves on to construct a complication, "and he coughs so he almost vomits".) The complication is a means to show that the caller not only has a problem, but has a problem that is becoming worse (which implies need for medical attention). Complications were not built into the format under examination here, that is, callers' questions. In sum, the question format allows patients to give longer presentations of their problems than does the request format. The question format also allows patients to present their problems as non-acute and as only requiring minor information.

2.3 Narratives

In 123 of the 209 telephone calls the callers used the narrative format. We see a short but clear example of this in the following extract:

(12) 1 VCTEL 3

1 C: ˙hhh du ja har en kille här han har hållt på
˙hhh you see I have a boy here he has been
2 hh=hostat nu i en månad(hhh).
hh=coughing for a month now(hhh).
3 N: jA::?
yES::?
4 C: tch! ja va nere på läkarstationen de:
tch! I was down at the primary care centre it:
5 va:: de e en tie dar sen ∫ungefär
was:: it was about ten days ago ∫approximately

```
6        men dä har inte blitt bättre ˙hhhh
         but it hasn't got better ˙hhhh
7        o han hostar så han spyr iblanhh,
         and he coughs so that he vomits sometimes(hh),
8   N:   ha? (pt)=˙hh när är han född?
         uhu? (pt)=˙hh when was he born?
```

Like the other two formats, these narratives have a distinct sequential shape: (i) The caller begins her narrative by giving an *initial presentation* of the problem – "˙hhh you see I have a son here he has been hh=coughing now for a month" (1–2); (ii) The nurse utters a *continuer* indicating that she functions as a listener to the telling in progress (3); (iii) The caller *details* the problem (4–7); (iv) The nurse begins to ask questions about the problem. In this particular instance, the telling is rather short, but the extract illustrates the overall sequential pattern that narratives usually take. In what follows, I will analyse different parts of it in more detail.

Initial presentation of the problem

Just as with requests and questions, callers who use the narrative format regularly begin by giving an initial presentation of the problem. They provide information about who has the problem, how long the problem has lasted, and a short description of it. This package is rather similar to the *story preface-package* that precedes many conversational stories, for instance:

(13) [Sacks 1974]

```
1   Ken:    You wanna hear muh-eh my sister told
2           me a story last night.
```

A story preface usually consists of the following parts (Sacks 1974): (a) an *offer* to tell something, which creates an opportunity in the following slot for the listener to accept or reject the telling (depending on the response a story may or may not occur); (b) an initial *characterization* of the story, which, among other things, conveys to the listener how the story should be listened and responded to (a "sad story", "a happy story", etc.); (c) *when the told about event happened*; and (d) a mentioning of the *source of the story*, which, among other things, gives the listener the means to decide if he has heard it before. When a story preface is followed by an accepting response from the listener, the turn-taking rules governing everyday conversation (Sacks, Schegloff, & Jefferson 1974) are lifted and a situation is created where one person, the actor who has proposed to tell the story, can tell the story until it has reached its recognizable end. Subsequently, the turn-taking rules for ordinary conversation are made relevant again and listeners can laugh, comment on the story, ask questions, tell their own stories, etc.

The callers' initial presentations of their problems may be viewed as story prefaces of a kind, since they characterize their problems, describe when their problems started and who has them, the caller or someone else. But there is an important interactional difference between story prefaces in conversational story-telling and problem presentations in telephone calls to Swedish primary care: The story preface in ordinary conversation – if accepted by the listener – provides the teller with space to tell a story. The story preface is the first step in negotiating for the right to tell the story. This is not the case in telephone calls to primary care, since the slot in which these initial characterizations are made is pre-defined as the space in which a caller is expected to present a problem. The interactional work that initial characterizations of problems perform is not to negotiate for the *right* to present problems.

But initial characterizations do other important interactional work: (a) They convey that *a longer* telling is in progress (the caller only gives a gloss description of a problem and tells how long it has lasted, not what the *present* problem consists of); (b) The initial description also briefly characterizes the problem (who has it, for how long, and a descriptive term), giving the nurse a means for deciding *how to listen* to the story. That is, the initial descriptive term ("coughing") allows the nurse to hear the following information "I was down at the primary care centre it was about ten days ago approximately but it hasn't got better" as information about "coughing not getting better". The term "coughing" enables the nurse to decode what is of relevance in the caller's subsequent problem presentation. In sum, the interactional work that initial presentations achieve has not so much to do with negotiating the right to tell as with preparing the listener, the nurse, for listening properly to the telling.

Constructing a complication

As mentioned above, nurses regularly utter continuers after callers' initial problem presentations. These continuers convey that they are prepared and willing to listen.

Then callers start to *detail* their problems. One important feature of these detailings is that, using them, callers *construct complications.* One first observation about the caller's presentation of her problem in extract (12) is that it is *chronological.* In her initial presentation she begins by saying that her son "*has been coughing for a month*". She then continues by telling that she was at the primary care centre "*about ten days ago*" and how the problem has developed since then; "*it hasn't got better*". Finally she describes the present state of the problem;

"*he coughs so that he vomits sometimes*". Thus, we see that the mother's narrative moves from the past, when the problem began, to the present. A second observation about this problem presentation (extract (12)) is that *the problem is formulated several times.* The narrative consists of a series of formulations of the problem: "*he has been coughing for a month now*", "*it hasn't got better*" and "*he coughs so that he vomits sometimes*". We see a similar pattern in the following extract:

(14) 8 VCTEL 31

```
1   C:  ˙hh e: jo ja ringer för att (.) ja har
        ˙hh e: yes I'm calling because (.) I have
2       problem med ont i magen.
        stomach-ache problems.
3   N:  mm?
4   C:  och e:: de e ju lite så galls-
        and e:: it is you know a little so galls-
5       ʃja e opererad för gallste::n
        ʃI've been operated for gallsto:ne
6       [ oh nu sitter de liksom =
        [ and now it is located just about =
7   N:  [mm¿
8   C:  = under e: ˙h höger h revbensbåge,
        = under e: ˙h right h rib,
9   N:  mm. mm.
10  C:  oh de=e ja har haft ont se::ne
        and it=is I have had pain sin::ce
11      >˙hh< de började väl lite smått i lördags.
        >˙hh< it started I guess a little last saturday.
12      =men ʃja ha ja kan va lite öm så där
        =but I ʃhav- I can be a little tender like that
13      ibland.
        sometimes.
14  N:  mm[.
15  C:    [lite molande så.
          [aching a little bit like that.
16  N:  mm¿
17  C:  ʃså ja inte haft nåt (.)
        ʃso I haven't had any (.)
18      liksom riktit anfall nu(hh).
        like real attack now(hh).
19  N:  n(tch!)äe?
        n(tch!)o?
20  C:  men däremot så: (.) e de rejält
        but however so: (.) it is really
21      molande värk(h) ˙hh oh de liksom
        aching pain(h) ˙hh and(h) it like
```

```
22       försvinner inte nån gånghh,
         doesn't disappear ever,
23       (.)
24  N:   näe. näe. (.) ˙hh E- när dom opererade dej
         no. no. (.) ˙hh E- when they operated on you
```

We can see that the caller, just as in extract (12), formulates the problem several times, as "*stomach-ache problems*" (2), "*it is you know a little so galls-*" (4), "*it is located just about under e: ˙h right h rib*" (6, 8), "*I have had pain sin:ce*" (10), "*I can be a little tender like that sometimes*" (12–13) and "*it is really aching pain and it like doesn't disappear ever*" (20–22). A third observation about these narratives is that the first descriptions of the problems usually portray an *imprecise minor problem* and that the following descriptions move towards increasing severity. In extract (14), the caller began by describing her problem with the imprecise "*stomach-ache problems*" (this description is very general and in need of specification. It does not exclude that the problem may be minor). When the caller utters her next formulation of the problem, "*it is you know a little so galls-*" (gallstone) and now "*it is located just about under e right rib*", this is a more precise description of her problem. Then the caller goes on to formulate the problem as "*I can be a little tender*" and "*aching a little bit*" and finally "*it is really aching pain and it like doesn't disappear*". But the later formulations do not only present the problem as more severe than the earlier ones. The last formulation of the problem, since it describes the caller's *present* health state, is also describing a *tendency in the present* – a deterioration of the caller's health. That is, not only does the caller have a health problem that has deteriorated, but it is deteriorating at this moment. This is the complication.

When the caller reaches the part of the narrative where the complication is revealed, she has made two interactional achievements. The first is locally interactional and has to do with the issue of who should speak next: When callers have uttered a complication nurses regularly view the narratives as having reached an end point. Now nurses start to question the callers. In this respect, these complications are similar to the "point" or the "punch line" of a story told in ordinary conversation. (Of course there are instances in which the nurses do not take over and where callers do reparative work to give the turn to the nurse, but this issue is too complex to be dealt with here.) (Also see Ruusuvuori 2000 for an analysis of similar processes in doctor-patient interactions in Finnish primary care.) The second interactional job that the complication does is to appeal to a maxim used in medicine: "one should always stabilise health conditions". There seems to be a rule in medicine stating, roughly, "A stabile pathological condition is not as acute as a less pathological condition that is deteriorating." By presenting his state of health as deteriorating, the caller increases his chances of receiving quick medical attention.

While most narratives result in complications, not all do:

(15) 9 VCTEL 13

```
1   C:  e jo de e så att e- efter nä ja fått
        e yes it's like this that e- after when I had
2       barn då: ˙hh o sen säj mån efteråt¿
        my baby then: ˙hh and then say a month after¿
3       (.) e: så ˙hh nä ja gå på bäcken nä
        (.) e: so ˙hh when I go to the toilet when
4       ja gå o bajsar så ˙h kommer de blo:d(h)?
        I go to shit then ˙h there is blo:d(h)?
5       (1.4)
6   N:                har du blod i avför[ingen,
        do you have blood in your excre[ment,
7   C:                                   [a :
                                         [ yes:
8       (.)
9   N:  ja hur länge har du haft de,
        yes how long have you had that,
```

In this extract we see that the caller gives an initial presentation of the problem (1–4). She tells who has the problem, when it started, and gives a short description of it; "*there is blo:d*". Then follows a 1.4-second pause. The nurse does not utter any continuer, nor does the caller continue her presentation of the problem. After the pause, the nurse utters a question: "*do you have blood in your excrement,*". Observe that this is not primarily a diagnostic question (which comes on line 9), but a news-marked receipt (Heritage 1984b). The nurse repeats the central part of the caller's previous utterance ("*blood in your excrement*"). She displays surprise and, perhaps, that she is a bit shocked. After the caller's affirmative response (7), the nurse begins to ask diagnostic questions (9). Thus we see that the nurse in this instance treats the caller's initial problem presentation as sufficient for her purposes; the problem presented is serious enough. No complication needs to be constructed by the caller.

It seems that complications need not be used when problems, in themselves, are very serious. Callers and nurses treat them as legitimate medical problems even without complications. Complications tend to be used when the seriousness of callers' problems is *negotiable* – when the need to see a doctor can be discussed.

Resisting the problem

When callers construct complications, they describe their problems as deteriorating. But, at the same time, callers often do something that seems to be quite the opposite: they *resist their problems.* They do so in various ways:

(i) Problem-resistance can be achieved by saying that the problem had *first been incorporated into a frame of thought that was not medically problematic* (before it was viewed as a medical problem):

(16) 2 VCTEL 6

```
1   C:  hh NEJ JA HAR- ja ringde hela
        hh NI I HAV- I called the whole
2       förmiddan men de har vart upptaget
        morning but it has been engaged
3       så mycket då va,
        so much then you see,
4   N:  ja:. h de ringer mycke, hh
        yes:. h it rings a lot, hh
5   C:  ha, ˙hh e: ja ha då en längre tid nu
        yes, ˙hh e: I have then for a longer time now
6       haft ont(hh) här under hAKAN(h) (.) ner
        had pains(hh) here under the chIN(h) (.) down
7       mot e >adamsäpplet om man säjer så vA?
        towards e >the adam's apple if you say so you sEE?
8   N:   ja[?
        yes[?
9   C:      [de e ömt oh SÅ ((harkling))=
            [it is tender and(h) SO ON ((clears the throat))=
10      =o känns varje gång ja sväljer.
        =and I feel it every time I swallow.
11      (1.0)
12      (pt) o ja har har tänkt de e väl nåt
        (pt) and I thought that it is I guess something
13      me halsen men- ˙hh de går ju inte över
        about the throat but- ˙it doesn't you know disappear
14      så att- (.) ((harkling)) nästan blivit
        so that- (.) ((clears the throat)) it has almost got
15      värre tycker ja de sista: veckorna då va?
        worse I think for the last: weeks then you see?
16      (1.1)
17  N:  hh e- syns de något utanpå så           hä[r eller,
        ˙hh e- is it visible on the outside like th[is or,
```

In this extract, the caller describes how he has had pains under his chin (5–7) and how it hurts every time he swallows (9–10). Then he adds "*I thought that it is I guess something about the throat*" (12–13). That is, the caller conveys to the nurse he has tried to understand his observation of the pain under the chin as a pain of the kind you get now and then; a "normal" pain not requiring medical attention. Thereby the caller manages to present himself as a person who seeks medical attention only when problems are serious enough.

A similar pattern was found by Jefferson (1984b) in her study of how people describe situations such as airplane hijackings, car accidents, murders, etc. She found that a common pattern in these tellings was the one she called "at first I thought X but then realized". Jefferson exemplifies this with the following extract, where the late US President Kennedy's driver describes what happened when Kennedy was murdered: "Well, we were going down Elm Street, I heard a noise that I thought was a backfire of one of the motorcycle policemen.... And then I heard it again. And I glanced over my shoulder. And I saw Governer Connally like he was starting to fall. Then I realized there was something wrong." In other words, the driver says that he first made a "normal" interpretation of the sound (as emanating from a motorcycle) before realising that it was gunfire. By displaying this interpretation, the driver succeeds in presenting himself as a person who makes normal interpretations of what happens around him. He is not the kind of person who interprets any loud noise as a gunshot. In a similar manner, callers to primary care manage to describe themselves as trustworthy witnesses to their own health problems. (Observe that they are not resisting their problems in the present, *now.* They are merely telling the nurses that they have tried to manage their problems in the *past.*)

(ii) Another way in which callers resisted problems was by saying that they have observed their problems and have *tried to manage and do something about the problems themselves:*

(17) 4 VCTEL 6

```
1   C:  ja skulle vilja samtala me (nån) om mitt ben
        I would like to talk to (someone) about my leg
2       för ja har ˙hhh gjort nånting konstit med de
        because I have ˙hhh done something strange with it
3       oh sen har ja- ˙hhh vatt hemma fjorton dar
        and(h) then I have- ˙hhh been home a fortnight
4       sen har ja provat åh jobba med de o ja
        then I have tried to work with it and I
5       får inte rätt på de hänger inte me mej riktit,
        can't get it right it doesn't follow me properly,
6   N:  nehe?
        noho?
7       ˙hh[h-
8   C:     [oh ja bär möbler förstår du så att
           [and(h) I carry furniture you see so that
```

```
9        ja frestar de rätt så kraftit,
         I am stressing it pretty hard,
```

The caller says that he took sick leave for two weeks and then tried to go back to work. He has tried to live a normal life as long as possible despite his problem. But there were also instances in which callers had tried more actively to manage their problems. In one instance a caller said that "*I discovered a tick and my husband helped me to take it away BUT the head is left, and he tried and tried and tried to take it out and we like there came blood and so on but he couldn't get it out*". This caller had tried her best to take care of the problem herself before calling the nurse.

(iii) Another way to do problem resistance is by *describing non-pathological states:*

(18) 12 VCTEL 7

```
1   C:  hh e jo e: de ä så att han har
        hh e yes e: it's like this he has
2       e: fått ont i sitt ö:ga. ˙hh[e-
        e: got pain in his e:ye. ˙hh[e-
3   N:                             [°han har ont
                                   [°he has pain
4         i ö:ga ja[?
        in e:ye yes[?
5   C:             [ja, han har fått de- haft
                   [yes, he has got it- had
6       de sen i natt.
        it since last night.
7       (0.8)
8   N:  [ja:?
        [yes:?
9   C:  [och e: ja de kliar o(h) de e:: hh rättså
        [and e: yes it itches and(h) it is:: hh pretty
10      rött o som man säjer irriterat runtom eller så,
        red and as you say irritated around or so,
11  N:   j[a?
        ye[s?
12  C:    [men de e inte jätteilsket i ögonvitan.
          [but it isn't really bad in the white of the eye.
13  N:  nej, ne[j,
          no, n[o,
```

The caller begins by giving rather detailed descriptions of what the eye looks like: he has had it since last night, it is itching, is pretty red and irritated. Then she describes a feature of the eye that is non-problematic; "*it isn't really bad in the white of the eye*". Thus the caller mitigates the problem she described previously by adding a detail that does not seem to be a part of the problem.

(iv) One way to do problem resistance is by saying that the problem has *disappeared for a while but has returned.* In the following extract the caller says that her son has a high fever. Then she adds that it goes down in the mornings:

(19) 1 VCTEL 2

```
1    C:  e: ja ringer för ja har min lille
         e: I'm calling because I have my little
2        kille här som heter jakoʃb som är sex år
         boy here whose name is jakoʃb who is six years old
3        <han har haft hög feber> sen i onsdas?
         <he has had high fever> since last wednesday?
4    N:  mm?=
5    C:  =bara feber och ont i halsʃen o: väldigt
         =just fever and a sore throʃat and: very
6        röd o svullen o vita prickar i halsen
         red and swollen and white dots in the throat
7        ·hhhh och e:m han har legat mellan trettinie
         ·hhhh and e:m he has had between thirty-nine
8        och förti och sen det har vart nere lite
         and forty and then it has been down a little
9        på morronen och sen så springer det upp då¿
         in the morning and then it runs up then¿
10   N:  mm?
11   C:  så ja v- li:te orolig för den här halsen
         so I wa- li:ttle worried about that throat
12       då för han har flera gånger åkt på halsfluss
         then because he has had tonsillitis several times
13       då,
         then,
14   N:  ja:? (.) feber ida?
         yes:? (.) fever today?
```

In sum, we see that callers often embed problem resistance into their narratives. Similar patterns of problem resistance were found by Wooffitt (1992) when he analysed interviews with people who claim to have had various paranormal experiences, for instance telepathy, clairvoyance, out of body experiences or direct contact with spirits or extraterrestrial beings. Wooffitt focussed his analysis on the strategies these people use to convince conversational partners about their actually having experienced these things. Their problem is "how to convince others that they are trustworthy witnesses" or "believable". This is a central interactional problem for them and a problem that permeates every single aspect of conversation when they tell others about their paranormal experiences.

In one of Wooffitt's interviews, a professional medium, when describing her first contact with a deceased person, says that "every time I walked into the sitting room, (0.3) er:m. (0.7) right by the window (0.3) and the same place always I heard a lovely s:ound like de de dede dedede dededah just a happy (.) little tune (0.5) a:nd >of course< I tore apart ma window I tore apart the window frame I >did Everything< to find out what the hell's causing that cos nobody else ever heard it ˙hhh (0.2) >y'know< (0.) there could be ten people in the room nobody'd hear it but me< (0.7) er:m and I wanted to know what was the: (.) material cause of this" (Wooffitt 1992: 74). We see that the medium in various ways tries to convey how she tried to resist understanding her experience as a paranormal event: (i) She describes the sound that she heard as a "lovely sound", suggesting that her first interpretation of it was far from something frightening, mysterious, or unnatural, which it probably would have been had she first interpreted it as paranormal. Thus she uses the "first I thought" format. Above we saw that the Swedish callers also used this format; (ii) She implies, by saying that "and the same place always I heard", that she did nothing about the problem for a while. She lived with the sound before she took measures; (iii) The medium says that once she had started to take measures, she tried to do something about it herself. She says that "I tore apart ma window I tore apart the window frame I >did Everything< to find out what the hell's causing that". She tried to find a "natural" rather than "supernatural" explanation for the sound. But she didn't search for help from others. She tried to do something about the problem herself, which is similar to what the Swedish callers sometimes said they did.

Let me summarise: By constructing complications, callers emphasise how their problem has deteriorated. Thereby they manage to design their problem to fit the logic of medicine, i.e., that a deteriorating health state is more worthy of medical attention than is a health state under control. But designing a problem in this way is risky: The nurse may view the caller's description of the problem as exaggerated and the caller's ability to assess medical conditions as poor (or even view the caller as a hypochondriac). Problem resistance counteracts this: callers who resist their problems present themselves as persons who do not seek medical care for minor problems. They present themselves as persons who have a moderate view of their own problems and seek medical care only when their problems have become more severe. They can present themselves as trustworthy witnesses to their own health states.

But problem resistance has another interactional function in terms of its position in the narratives: Callers' problem resistances usually occur before the complications are expressed. Since problem resistance reflects callers' attempts to non-medicalise their problems, complications are in sharp contrast to them. The utterances in which problems are resisted create the background against which complications can be contrasted.

Letting go

As we have seen, by using narratives callers can tell about their problems in a chronological order, allowing them to set up complications and do problem resistance. But not all callers resisted their problems. Instead, in some instances, callers let their emotions show and explicitly displayed worries about health problems, as in the following extract:

(20) 5 VCTEL 1

```
1   C:  ja har en dotter som ä::.
        I have a daughter who is::.
2       va ska ja säj? ˙hhh hon har en
        what should I say? ˙hhh she has a
3       stor=stor knuta bakom örat (.) som
        large=large bump behind her ear (.) that
4       som gör väldigt ont=man kan inte röra vid den
        that hurts a lot=you can't touch it
5       o(h) hon kan inte vrida huvudet? ((sjungande röst))
        and(h) she can't turn her head? ((singing voice))
6   N:  °ne::[j,
        °no: [:,
7   C:       [och hon har hatt e:: feber kallsvettig
             [and she has hąd e:: fever cold sweats
8       och så går de över och så kommer de igen¿
        and then it gets better and then it's back again¿
9   N:  mja::, e- e- e- de här själva öronmusslan,
        nyes::, is- is- is- the ear-conch itself
10      =e den osså engacherad?
        =is that also involved?
```

The caller expresses her worries in a number of ways: (i) She grades the seriousness of the problem. She uses the adjective "*large*" before "*bump*" (3) and "*a lot*" before "hurts" (switched word order in English); (ii) "Lot" is emphasised; (iii) She "sings" out "*and she can't turn her head?*", as if she is about to loose control of herself and start to cry; (iv) The caller does not await any continuer from the nurse after her initial problem presentation. Instead she moves directly into exposing the problem. Observe also that there is no problem resistance at all in this extract. These displays of worries, like problem resistance, are ways to display the callers' *morale*. Callers who resist their problems present themselves as "rational" and as "having things under control", whereas callers who display worries and do no problem resistance present themselves as "emotional". It seems that these are the two primary ways of presenting morale.

Invoking patient status

Another technique that callers used to increase their chances of receiving medical aid/seeing a doctor was to, implicitly or explicitly, present themselves as *patients:*

(i) Callers could present themselves as patients by *referring to a doctor they had met who had said that if X happens then do Y.* In one instance, a caller said "*then I had some infection in the muscle* (.) *in the muscles in the eye.* (N: *yes*?) *so I got some cream from him and if it hadn't stopped after a week then I should contact you*" (8 VCTEL 1). (This caller had not mentioned the doctor previously in the narrative, which makes her use of "him" rather odd. But "him" is understandable as "the doctor" when the caller says that she got some cream from him, as the doctor can prescribe ointments.) By referring to the doctor, this caller conveys to the nurse that she is a patient and that the responsibility for her call is the doctor's. She is just being a compliant patient who follows the doctor's advice.

In some cases it was unclear whether the doctor had actually advised the caller to contact the centre: "*I e: spoke to a orthopaedist the day before yesterday* (N: *yes*?) *I work at the ambulance in the city and I spoke to her there because I have s such a pain really in ˙hhh the right toe on the big toe's joint*?" (etc.) (12 VCTEL 36). In this instance, it is unclear whether the doctor at the caller's workplace has actually examined the toe and given any advice.

(ii) Callers might invoke their status as patients by saying that they have visited a caregiver in the past for *a medical problem that has now returned.* In one instance, the mother of a teenage daughter said that "*she has warts on her index finger that don't want to go away,=we have been down there and frozen them e:* (.) *two or three times we were down there,*" (12 VCTEL 28). (The mother refers to a procedure using carbon dioxide to freeze warts.)

(iii) Callers could invoke patient status by *complaining* about something that has been promised or done to them in the past. Most complaints concerned the dates patients were promised that they would receive test results. In one instance, a caller said that "*yes it's like this I visited ˙hhh you hh yes: in the beginning of march ˙hh* (N: *mm*?) *and then a referral was sent to the x-ray to x-ray the nasal sinuses?* (N: *mm*?) *and I was there* (.) *it is a fortnight toʃday:?* (N: *mm?*) *and, I am waiting for e the answer e the result would go to you.* (N: *m*,) *and then I would get a messAGE?* (.) *but I'm waiting- should it* (.) *take s*(*h*)*o long time* (*he he*) *I was about to say,*" (2 VCTEL 20).

(iv) Finally, callers invoked patient status by using "I should" constructions, as in the following instance: "*hello ˙hh you know I should come and(h) take a new e: blood sample*" (followed by the narrative). By using "I should", the callers rather efficiently referred to obligations made by some other authority. "I should" conveys to the nurse that the patient is merely following someone else's instructions. Observe that these "I should" constructions do not contain the other narrative features discussed above: These patients do not provide the 3 pieces of information routinely given in initial problem presentations (who has the problem, its nature and duration). Neither do they end with complications. Nor do we find that patients reveal their emotions when using this format. The reason for these differences is, of course, that the patients who use this format are not new patients. They are merely calling back to the centre in order to fulfil one or another obligation connected to their status as established *patients*.

3. Discussion

We have seen that callers present their problems using one of three formats: requests, questions, or narratives. The three formats have different interactional consequences for the talk that follows:

i. They provide different amounts of space for callers to present problems using their *own* words: When requesting, callers' own presentations of their problems are very short. Of course all callers continue detailing their problems later in the calls, but these detailings are steered by nurses' questions. After a caller has requested to see a doctor, and a nurse has begun diagnosing, the caller's following problem presentation must be adjusted to the frame provided by the nurse's questions. When callers choose to use the narrative format they create interactional space sufficient for lengthy detailings of their problems. As soon as a caller has begun a narrative (as a telling of health state in the distant past), the nurse must wait until the narrative describes something that happens in the present. Narratives provide callers more space in which to present problems in their own words than do requests. The question format also provides space for callers to detail their problems, since the pre-questions (which are regularly followed by an acknowledgement) create a situation in which background information can be given until the questions are stated. But callers that have a great deal to tell tend not to use this format, but choose the narrative format instead. This may have to do with the seriousness of the problem that the caller wishes to present: Problems presented in the question format tend to be explicitly minor, whereas problems presented in the narrative format tend to be more severe. Callers who wish to present more severe problems may choose the narrative format because it provides other possibilities to construct complications, display worries, etc.

ii. A large number of callers are concerned with the following interactional issue: How do I present my problem as serious enough to be given attention from the medical institution (for instance, a time boooked with the doctor)? This concern seems to drive many callers to present the problems as deteriorating. But there is a possible risk in being too focussed on the problem, too concerned about it. The nurse may start to question or simply not believe what is told by the caller. She may even view the caller as a hypochondriac. It may also be the case that this issue is especially relevant when the speakers are two persons who do not know each other. Therefore callers are faced with a second issue: How do I present myself as a trustworthy witness of the problem I describe? How do I describe my problem in a believable way? As a result of this, we see how callers do problem-resistance. The narrative format seems especially suitable for giving these "balanced" or "matter of fact" presentations of problems. They provide callers with space enough to give full-blown matter-of-fact descriptions of problems. As we have seen, requests to see doctors do not provide callers the space to do this, and questions tend to be used to describe problems as not necessarily in need of a a doctor's attention, as minor.

iii. The use of the three formats have consequences for the identities ascribed to the nurses: When callers request to see a doctor, nurses are viewed as "secretaries", as bureaucratic steps leading to an appointment with a doctor. This view of nurses becomes especially visible when callers only utter diagnoses (without requests). In these cases, callers sometimes present themselves as having interactional rights to diagnostic knowledge equal to that of "the secretaries". We saw above that nurses opposed callers' rights to use diagnoses. At the same time, thus, they opposed the ascribed identity. The question format ascribes another identity to the nurses; when callers use it, they approach the nurses as "medical experts" (observe that these 20 cases are the only ones in which the callers *explicitly* assigned medical expertise to the nurses. Observe also that in all these cases the problems were *minor*. This reflects a more fundamental view of nurses as persons who are competent only to assist in treatment of minor problems). The identity of the nurse is left more open in all those cases in which callers used narratives. The caller's problem is merely presented to the nurse, who is expected to decide what should be done next. When narratives are used, it seems to be an open issue as to whether the nurse is an "expert" or a "secretary". It is up to herself to decide who she is.

Note

1. Observe that not all 66 request sequences look *exactly* like the examples presented below. There are a few instances that run differently, but for other interactional reasons. An overwhelming majority of all cases are *very* similar to the ones presented below. It wouldn't be possible to present all varieties in this paper.

References

Coupland, J., Coupland, N., & Robinson, J. (1992). "'How are you": Negotiating phatic communion'. *Language in Society, 21*, 207–230.

Coupland, J., Robinson, J., & Coupland, N. (1994). 'Frame negotiation in doctor-elderly patient consultations.' *Discourse & Society, 5*(1), 89–124.

Drew, P. (1991). 'Asymmetries of knowledge in conversational interactions'. In I. Markova & K. Foppa (Eds.), *Asymmetries in Dialogue* (pp. 21–48). Hemel Hempstead: Harvester Wheatsheaf.

Spri (1987). *Distriktssköterskan.* Spri-rapport 214.

Heritage, J. (1984a). *Garfinkel and Ethnomethodology*. Cambridge: Polity Press.

Heritage, J. (1984b). 'A change-of-state token and aspects of its sequential placement'. In J. M. Atkinson & J. Heritage (Eds.), *Structures of Social Action: Studies in Conversation analysis* (pp. 299–345). Cambridge: Cambridge University Press.

Jefferson, G. (1980). 'On "trouble-premonitory" response to inquiry'. *Sociological Inquiry, 50*(3/4), 153–185.

Jefferson, G. (1984a). 'Stepwise transition from talk about a trouble to inappropriately next-positioned matters'. In J. M. Atkinson & J. Heritage (Eds.), *Structures of Social Action: Studies in Conversation analysis* (pp. 194–222). Cambridge: Cambridge University Press.

Jefferson, G. (1984b). 'At first I thought: A normalizing device for extraordinary events'. Tilburg: Katholieke Hogeschool (unpublished manuscript).

Jefferson, G. (1985). 'On the interactional unpackaging of a gloss'. *Language in Society, 14*, 435–466.

Jefferson, G. (1988). 'On the sequential organisation of troubles-talk in ordinary conversation'. *Social Problems, 35*(4), 418–441.

Jefferson, G. & Lee, J. R. E. (1980). *The analysis of conversations in which "troubles" and "anxieties" are expressed.* Final Report to the (British) SSRC, Report Nos HR 4805/1–2.

Jefferson, G. & Lee, J. R. E. (1992). 'The rejection of advice: managing the problematic convergence of a "troubles-telling" and a "service encounter"'. In P. Drew & J. Heritage (Eds.), *Talk at Work* (pp. 521–548). Cambridge: Cambridge University Press.

Landstingsförbundet (2001). *Basårsstatistik 2000, Verksamhet, del 1.* Samhällsekonomiska sekretariatet at Landstingsförbundet.

Leppänen, V. (1998). *Structures of District Nurse – Patient Interaction.* Department of Sociology at Lund University, Lund, Sweden (Doctoral dissertation).

Marklund, B. (1990). *Medicinsk rådgivning på vårdcentraler och apotek. Omfattning, innehåll och utvärdering med särskild tonvikt på vårdcentralernas telefonrådgivning.* Department of General Practice at the University of Gothenburg, Gothenburg (Doctoral dissertation).

Ruusuvuori, J. (2000). *Control in Medical Interaction: Practices of Giving and Receiving the Reason for the Visit in Primary Health Care.* Acta Electronica Tamperensis 16, Tampere (Doctoral dissertation).

Sacks, H. (1974). 'An analysis of the course of a joke's telling in conversation'. In R. Bauman & J. Scherzer (Eds.), *Explorations in the Ethnography of Speaking* (pp. 337–353). Cambridge: Cambridge University Press.

Sacks, H., Schegloff, E., & Jefferson, G. (1974). 'A simplest systematics for the organisation of turn-taking for conversation'. *Language, 50*, 696–735.

Schegloff, E. A. (1980). 'Preliminaries to preliminaries: "Can I ask you a question"'. *Sociological Inquiry, 50*, 104–152.

Schwartz, H. (1976). 'On recognizing mistakes: A case of practical reasoning in psychotherapy'. *Philosophy of the Social Sciences, 6*, 55–73.

Smith, D. E. (1978). '"K is mentally ill": The anatomy of a factual account'. *Sociology, 12*, 23–53.

Whalen, M. & Zimmerman, D. (1987). 'Sequential and institutional contexts in calls for help'. *Social Psychology Quarterly, 50*, 172–185.

Whalen, M. & Zimmerman, D. (1990). 'Describing trouble: Practical epistemology in citizen calls to the police'. *Language in Society, 19*, 465–492.

Wooffitt, R. (1992). *Telling Tales of the Unexpected.* Hemel Hempstead: Harvester Wheatsheaf.

Zimmerman, D. (1992). 'The interactional organization of calls for emergency assistance'. In P. Drew & J. Heritage (Eds.), *Talk at Work*. Cambridge: Cambridge University Press.

Chapter 9

Constructing and negotiating advice in calls to a poison information center

Håkan Landqvist

1. Introduction

In his article 'Institutional discourse', Agar (1985) claims that the giving of directives is one of three things that institutional discourse must accomplish (the other two being diagnoses and reports). This is not meant to imply that what Agar (1985: 155) refers to as directives are a necessary constituent in *all* forms of institutional discourse. In calls from the public to a Poison Information Center (PIC), however, the advice regarding treatment (which in these calls correspond to Agar's term 'directive') is of fundamental importance. The advice can be seen as the second pair part to the initial, explicit or implicit question that embodies the reason for calling (cf. Zimmerman 1984: 219f. on complaint/request and remedy). However, the giving of advice is not an easy enterprise. Given that a paramount concern for the advice-giving party (henceforth referred to as the pharmacist) is to make the other party, the caller, comply with the advice, certain specific circumstances are likely to have an influence on advice-giving. First, the advice is a deferred action request (Lindström 1999), i.e. the action that the advice giver wants the advice recipient to take is meant to be carried out after the call has ended. That means that the pharmacist has no way of monitoring that the caller will actually heed the advice. Secondly, the advice giver has no means of actually forcing an unwilling advice recipient. Thirdly, even though the pharmacist cannot use force, this does not mean that the calls are symmetrical; in reality the situation is asymmetrical regarding both knowledge and authority. The pharmacist can therefore try to assure compliance by walking a tightrope between, on the one hand, displaying knowledge and exerting authority and, on the other, maintaining friendly (or at least civil) relations with the caller.

Opening/Identification
Reason for call/Request for information
Interrogative series
Advice
Closing

Figure 1. Organization of a PIC call[1]

This chapter will present some ways in which advice-giving is achieved in calls to the Swedish Poison Information Center. In trying to explain the detailed construction, I will concentrate on the seriousness of the incident, and the caller's demonstrated willingness (or lack thereof) to co-operate. First, however, some brief information about the institution and the calls is warranted.

In Sweden there is only one poison information center. The main purpose of the institution is to give accurate advice to citizen callers and to medical institutions concerning how to deal with what might be a poisoning (Persson 1990). This advice is given by pharmacists who are experts in their field.

The PIC corpus, which was gathered in 1995, consists of 377 calls from laypersons to the PIC.[2] All calls are authentic and would have taken place (and would even have been recorded) regardless of whether they were to be part of the corpus or not. The calls last on average 2 min 44 sec. During the recording of all daytime calls, I was present, sitting in the background and taking field notes. All calls are routinely categorized by the pharmacist according to – among other things – the estimated risk of the incident. The risk categories are *clear risk, some risk* and *no or small risk*. A subset, consisting of 69 calls, from that corpus constitutes the material for the present study. This, in turn, comprises three subcorpora: the *clear risk* corpus (C-corpus) with 20 calls, the *some risk* corpus (S-corpus) with 26 calls and the *no or small risk* corpus (N-corpus) with 23 calls.

A call to the Swedish PIC is typically organized as outlined in Figure 1.

As an example of the sequential context in which the advice sequences are situated, I will offer the following call (example (1)). Though unusually short, the call is nevertheless complete with all phases, from the opening phase and the following reason for the call, through an interrogative series to an advice phase and finally a closing phase.

(1) #19569[3]

Date: 1995-09-28 Time: 6.29 PM
Length of call: 01.20 min
Estimated risk: none

Ph = Pharmacist
C = Caller (woman, possibly in her fifties)

1 Ph: *Giftinformation, jourhavande apotekare¿*
Poison information pharmacist on-duty
2 C: *.hh Ja hejsan, ja skulle höra (0.4)*
yes hello I should hear (0.4)
.hh Yes hello I'd like to hear (0.4)
3 *hur man gör om man har fått bensin i sej.*
what one does if one has got gasoline in oneself
what to do if one accidentally has consumed gasoline
4 *(1.1)*
5 Ph: *Ja:, e de du som ha fått de, eller?=*
Yes is it you who has got it or
We:ll is it you who's done that or=
6 C: *=Näej, de e min man.*
=No it is my husband
7 Ph: *Ja::. (0.6) hur mycke kan de röra sej om¿*
Yes how much can it be about
We:ll: (0.6) how much can it be
8 C: *Ja:, de e väl inte nå stora mängder,*
Yes it is *väl* not any big quantities
We:ll I'd say there are no big quantities
9 *men har ju fått kanske nå munnar i sej så där.*
but has *ju* got maybe some mouthfuls in himself so there
but he has of course got maybe like some mouthfuls in him
10 ***(0.6)***
11 Ph: *Men hur, (.) har han slangat då, eller¿=*
But how (.) has he been siphoning off gas then or
12 C: *=Ja::, å fått de: (0.2) fel håll då.*
Yes:: and got it (0.2) the wrong way then
13 Ph: *Mm:, (0.3) har han: hostat eller kräkts, eller nåt sånt.*
***Mm:* (0.3) has he been coughing or vomiting or anything like that**
14 C: *Näej. de har (han) j- han rapar väl upp de, så här.*
No that has (he) *j-* he belches *väl* up it like this
No (he) has *j-* I suppose he belches it up like this
15 *(0.2) har väl hänt för nån halvtimme se:n,*
(0.2) has *väl* happened for some half hour ago
(0.2) has I suppose happened about a half hour ago
16 Ph: *Oke:j. nä för enstaka klu:nk eller så, (0.5)*
Okay no because a single gulp or so (0.5)

17 *då e de ju bara å avvakta,*
then it's really just to wait and see
18 *å då ska han ju ta nåt fett,*
and then shall he *ju* take something fat
and then he should really eat something with a lot of fat
19 *(0.7) eh grädde eller gräddglass*
(0.7) uh cream or ice cream
20 *eller man kan ta en massa smö:r på en smörgås också¿*
or one can take a lot of butter on a sandwich also
or one can put a lot of butter on a sandwich also
21 C: *E de bra:.*
Is that good
22 Ph: *Jaa.*
Yes
23 C: *Mm:¿*
24 Ph: *De e inte särskilt giftit å få i: sej*
It is not especially poisonous to get into oneself
It isn't especially poisonous to swallow
25 *i [mindre mäng]der,==*
in [lesser quanti]ties
26 C: *[De e inte de¿]*
[It is not that]
[It isn't]
27 Ph: *= utan den stora risken e (ju)*
but instead the big risk is (really)
28 *att de kommer ner i luftvägarna,*
that it comes down into the respiratory tract
29 C: *Jaha:,*
I see
30 Ph: *Å ger en lunginflammatio:n.*
And causes pneumonia
31 C: *Jaha du:. (0.2) mm:.*
jaha you (0.2) mm
Oh I see (0.2) mm:
32 Ph: *Sen om man får i sej mycke*
then if one gets in oneself a lot
Then if one drinks a lot of it
33 *så kan man ju bli förgiftad*
so can one *ju* become poisoned
one can really get poisoned

34 *av de här också då,*
by this also then
35 *till skillnad från .hh lacknafta å sånt,*
to difference from ligroin and such
as opposed to ligroin and such
36 C: *Ja:. .hh men man kan avvakta då tycker du å:,=*
yes but one can wait and see then think you and
Yes .hh but you think one can wait and see then and=
37 Ph: *=Ja just de,*
yes exactly that
=Yes exactly
38 *å å men e de så att han skulle kräkas*
and and but is it so that he would be vomiting
and and but if he would be vomiting
39 *eller ()- få en sån där krafti hackhosta då¿*
or get one like that strong hacking cough then
or have one of those strong hacking coughs then
40 *.hh eller om han känner sej rejält påverkad,*
or if he feels himself much affected
.hh or if he felt much affected
41 *yr eller slö eller [nånting sånt där],*
dizzy or listless or [anything like that]
42 C: *[Ja:, han känner väl] sej*
[yes he feels *väl*] himself
[Well: I suppose he's feeling
43 *lite yr så där tycker han.*
a little dizzy like that he thinks
44 *(1.8)*
45 Ph: *Ja::.=*
yes
We::ll=
46 C: *=Mm:¿*
47 Ph: *Nä asså e de eh påtaglit,*
no so then is it obvious
no then if it's obvious
48 *då ska man (ju) in ti sjukhus,*
then shall one (*ju*) in to hospital
then really one should go to a hospital
49 *[(e de)] e de en aning y:r så kan man väl avvakta.*
[(is it)] is it slightly dizzy so can one *väl* wait and see

[(is it)] **is it slightly dizzy I suppose one can wait and see**

50 C: *[Ja:.]*
[Ye:s]

51 C: *Mm:. (0.6) pt .h ja men de e bra, då vet vi.*
mm (0.6) yes but it is good then know we
Mm (0.6) pt .h well that's good then we know

52 Ph: ***Mm:?***

53 C: *Tack ska du ha.*
Thanks shall you have
Thank you

54 Ph: *Tack he [j.]*
Thanks by[e]

55 C: *[Ta]ck hej.*
[Tha]nks bye

First, I will give an outline of the entire call in order to show how the advice is sequentially fitted into the call. I will expand on the analysis of the advice-giving sequences.

The pharmacist produces a standard opening turn, which gives the call a formal footing, to use a term from Goffman (1974), by first identifying the institution and then referring to his own professional role as pharmacist on-duty (l. 1). The caller responds with a greeting that is somewhat informal and proceeds immediately to the question, which is the reason for the call. The question is formulated as a hypothetical one (see Sjöberg 1999) (ll. 2–3). Instead of answering the question, the pharmacist produces a question of his own, which is about whether the caller is referring to something that has happened to *herself* or not. And the caller replies that the person who has ingested gasoline is her husband. When the parties have established who is concerned, the pharmacist asks about the quantity of the intake (l. 7). In response, the caller produces a rather vague description of the quantity, characterizing it as "*no big quantities*" (l. 8), and then making it somewhat more precise in saying "*maybe like some mouthfuls*" (l. 9). Since the caller has not so far, on her own initiative, produced any explanation regarding why the husband has consumed gas, the pharmacist now addresses the relevance of that information in ll. 10–11. He does so by waiting for an explanation and then, when none is offered, he starts a question with *but how*. The open question is however changed into an offered hypothesis as to whether the husband has been siphoning off gas. The caller confirms this hypothesis (l. 12). The pharmacist then continues by asking whether the husband has been coughing or vomiting or the like.

The reason for the question – which in this call is never given explicitly – is that petroleum products can get into the respiratory tract during coughing or vomiting, and this can cause pneumonia. The caller gives a negative answer (l. 14) and then she volunteers information that the incident took place about a half-hour earlier.

Lines 5–15 comprise the interrogative series, which in this call is unusually short. During this phase the pharmacist asks for the information that is necessary in order to judge the seriousness of the possible poisoning. Very often, this phase is followed by a temporary break in the conversation, when the pharmacist puts the caller on hold while consulting a database or archive for information about the toxicity of the relevant product or substance. The pharmacist in this particular call does not need to do this, because the consumption of a petroleum product is a relatively common subject in calls to the PIC, and he already knows about possible risks and about what to do to minimize such risks. Therefore, he initiates the advice sequence right away (l. 16–).

He introduces the advice by referring to the consumption of a single gulp (l. 16), and continues the utterance by telling what to do in that case ("*then it's really just to wait and see*," l. 17). He then further continues the advice turn by giving additional advice, saying "*and then he should really eat something with a lot of fat*," l. 18. After a short pause he adds some examples of which things to eat, namely cream or ice cream or a lot of butter on a sandwich. At the earlier transition relevant points (TRPs) in this turn, he has shown with his intonation that he was not yet finished. At the end of the turn, however, his intonation is rising, which seems to indicate that the turn is complete.

The caller takes the turn and poses the question "*Is that good*" (l. 21). It does seem rather redundant to ask this if the function of the question is purely to get information whether the proposed measures are good or not. After all, the pharmacist would not give such advice if he didn't think they were useful. So the question probably has more of an affiliating function, making the caller more active in the decision process. The pharmacist naturally confirms the question (l. 22). The caller responds with a rising *Mm:¿*. Such a response is very common in Swedish conversation and its meaning can vary. Normally it's equivalent to *yes*. And that is also the meaning I suggest that it has on this occasion. It confirms the reception of information and indicates that the speaker at the moment has nothing more to add.

To summarize what happens in this advice sequence: The pharmacist outlines that the immediately-following advice pertains to measures to be taken when somebody has consumed a small amount of gasoline. In the same turn, he advises that the husband can just wait and see, and that he should eat something with a lot of fat, and gives three examples of possible things to eat. The caller asks if that would be good things to eat, a question that I would argue primarily has affiliative functions. The pharmacist gives a confirmation, and the caller gives a minimal response that acknowledges the information.

Now, the pharmacist explains the risks that are associated with the consumption of gasoline. The explanation sequence starts at line 24 and ends at line 36. My purpose here is not to go into the full, but just to summarize the sequence. The pharmacist tells the caller that it isn't especially poisonous to consume smaller quantities of gas, but that the main risk would be if it were to get into the respiratory tract and then cause pneumonia. He goes on to say that if one consumes bigger quantities, one can get poisoned, and he contrasts this with the effect of another chemical, ligroin, which is not poisonous at all.

At this point (l. 36), the caller acknowledges the information and asks a question in which she summarizes the received advice. The question is in the form of a formulation of the gist of the advice, and it is subjected to the pharmacist to either confirm or reject.

The pharmacist gives a latching confirmation, "=*Yes exactly*" (l. 37), and then goes on to name circumstances that would demand another type of advice (ll. 38–41). These circumstances involve more serious symptoms like vomiting, having a strong hacking cough, or feeling much affected, dizzy or listless. This is the beginning of what I call the "just-in-case advice." Sequences with just-in-case advice are common in calls where the risk does not seem to be too serious and where the proposed actions therefore are not very far-reaching. (In the S-corpus, 9 of 26 calls contain a just-in-case advice sequence, and of the 23 calls in the N-corpus, 5 contain one such sequence, and another 2 calls contain 2 and 3 just-in-case sequences. There are no instances of this kind of advice sequence in any of the 20 calls in the C-corpus.) Before I return to this particular call I will show a typical progression of such a sequence:

(2) #17357

Date: 1995-09-01 Time: 7–8 PM
Length of call: 06.53 min
Estimated risk: some
Ph = Pharmacist
C = Caller (woman, possibly in her thirties)

170 Ph: *å skulle de va så att han vaknar å kräks ytterliare*
and should it be so that he wakes up and vomits more
and if he should wake up and vomit more
171 *>eller börja< hosta under natten?*

or start coughing during the night
172 C: *M[m.*
173 Ph: *=[Eller att han skulle vakna me fe:ber imorron bitti,*
[or that he should wake up with fever tomorrow morning
174 *å verka påverkad på andningen på något vi:s,*
and seem to be affected on the breathing in any way
and seem to have his breathing affected in any way
175 *då ska du åka in ti sjukhus me honom¿*
then shall you go in to hospital with him
then you need to take him to a hospital
176 C: *Ja just de.=*
Yes exactly that
Yes precisely
177 Ph: *=.Ja:.*
.Yes
178 (*0.9*)
179 C: *Jodå.*
Right
180 Ph: *Mm:, men då vet du¿*
Mm but then you know

Before this sequence starts, the pharmacist is telling the caller to keep the child under close watch. Such an utterance normally precedes the just-in-case advice. The sequence begins with the just-in-case part, here represented by the expression "*and if he should*" (l. 170). Then comes a list with more serious symptoms (ll. 170–171, 173–174). The list of symptoms is followed by the advice regarding what measures to take when faced with one or more of the symptoms (l. 175). The caller produces a response that indicates that she is in agreement with the advice ("*Yes precisely,*" l. 176). The last utterance is immediately followed by a latching *.yes* from the pharmacist (l. 177). The *.yes* indicates that there is nothing more to be said, and thus seems to be an attempt to move into a closing sequence (which is one function a *yes* uttered while inhaling can have in Swedish conversation; see Landqvist 2001; cf. Hakulinen 1993:61 on the function of the Finnish equivalent *.joo*).

In example (2) the caller does not take any initiatives of her own until she agrees with the advice that the pharmacist proposes. Her only other response in example (2) is a continuer at l. 172. These are the typical reactions of the caller. Now, let's see what happens in the corresponding just-in-case advice sequence in example (1). Before the pharmacist has come to the advice that concerns more serious circumstances, the caller confirms in overlap one of the mentioned symptoms, namely that her husband is feeling a little dizzy (ll. 42–43). That piece of unexpected information causes a stop in the pharmacist's ongoing advice-giving and causes him to pause for as long as 1.8 seconds. Thereafter, he produces a very indecisive *We::ll.*, which immediately is followed by a confirming =*Mm:¿* from the caller.

This is where the next advice sequence starts, which is initiated by the new information regarding the husband's symptoms. The pharmacist now splits the advice into two different prescribed lines of action, depending on how strong the symptoms are. First, he mentions what to do if there are strong symptoms and says that "*then really one should go to a hospital*" (l. 48). Simultaneously with an acknowledging feedback from the caller, he goes on to say what to do if the dizziness is only slight. Then one can wait and see (l. 49). The caller responds with a *Mm:.*, which constitutes the end of this advice-giving sequence and therefore also makes the closing of the call relevant, since the goal of the call now is accomplished (cf. Schegloff & Sacks' *Opening up closings*, 1973: 306, Note 10, 308 and Nordberg's *Closings in alarm calls*, 1999: 67f.). The caller initiates the closing sequence by giving a positive evaluation (l. 51). (Whether it is the advice or the call as a whole that is evaluated here is unclear.) The pharmacist responds to the evaluation with a minimal *Mm:?*, and the caller thanks the pharmacist. After that the parties give their closing greetings and the call is closed.

We have seen that in the advice-giving sequences the caller is active in the reception or formulation of advice. In the first sequence, she not only accepts the advice with an acknowledgement such as *okay*, but she asks whether the proposed action alternatives are good. This seems to be a way of affiliating with the advice giver.

In the second sequence, she squeezes in a piece of information about her husband's symptoms that proves to be of such importance to the advice-giving that the pharmacist changes the course of his recommendations. However, it should be noted that worst case scenarios are regularly addressed just in case the situation should be worse than expected, even in calls where the parties seem to agree that the case probably is not very serious.

This example also shows that advice-giving can be a rather complex activity, where new information of great importance to the advice may surface even during the advice-giving activity.

2. The first advice sequence

In example (1) it was shown that a call can – and normally does – contain more than one advice sequence. The first sequence is almost always positioned within or after the interrogative series. In the S-corpus, the most-often encountered position for the first advice sequence is right after the parties have resumed the call after a break, during which the pharmacist has consulted a database or archive for information on the medical effects of the substance. This seems to be the natural position for the main advice sequence. In 9 of the 26 calls in the S-corpus, the first advice is given in this position. Before the break, the pharmacist announces that she is about to put the caller on hold in order to do a check-up on possible effects of the substance. At this point, s/he usually also asks for the caller's phone number "just in case the call is cut-off." The pharmacist's way of introducing the temporary break (ll. 26–27) makes advice relevant right after the break. See example (3).

(3) #16629

Date: 1995-08-25 Time: 11.00 AM
Length of call: 05.18 min
Estimated risk: some
Ph = Pharmacist
C = Caller (woman, possibly in her thirties)

25 C: *Va kan liksom,=*
What can sort of
26 Ph: → *=Mm:, vi ska se¿ ja ska ta fram, (1.0)*
Mm we'll see I'll take forth (1.0)
Mm we'll see I'll get ahold of (1.0)
27 → *↓information om du dröjer ↑kvar i luren?*
information if you stay in the receiver
the information if you hang on
28 C: *Ja?*
Yes
29 Ph: *Kan ja- (0.3) ta ditt telefonnummer varifrån du ringer¿*
Can I- (0.3) take your phone number wherefrom you're calling
Can I- (0.3) write down your phone number where you're calling from
30 C: *Ja %de e: sju sex fem*
Yes it is: seven sex five
31 Ph: *Mm*
32 C: *Förtitre trettitvå%*

Fortythree thirtytwo

33 Ph: *Här i Stockholm.*

Here in Stockholm

34 C: *Ja¿*

Yes

35 Ph: *Mm: vänta kva:r.*

Mm: hang o:n

36 C: *Ja¿*

Okay

37 → (*9.2*) ((break during which the pharmacist asks a doctor at the PIC about some written information about wasp stings))

38 Ph: *(Hall)å ja¿=*

(Hell)o yes

Yes hello

39 C: *M:m?=*

40 Ph: → *Mm? .h de (.) du kan prova att göra eh de e på själva då*

Mm .h it (.) you can try to do uh it is on the very then

Mm .h what (.) you can try to do uh that's exactly on the

41 → *stickstället, den: lokala reaktionen kan man lindra, (0.4)*

place of the sting the local reaction can one soothe (0.4)

42 → *genom att man (0.2) tar (.) en (1.0) eh tablett (0.6)*

through that one (0.2) takes (.) a (1.0) uh pill (0.6)

by (0.2) taking (.) a (1.0) uh pill (0.6)

43 → *Magnecy^l^, nån tablett som innehåller acetylsalicylsyra å de*

of aspirin some pill that contains acetyl salicyl acid and the

44 *vanliaste man har (-) brukar ha hemma e Magnecyl.*

most common one has (-) usually has at home is aspirin

45 C: *Jaa¿=*

Yes

46 Ph: *=Tror du att du har de¿*

Do you think that you have it

47 C: *Nää ja har nog inte de. =ja har nog bara Alvedon när de*

No I have possibly not it I have possibly only Alvedon when it

No I don't think so =I've probably only got Alvedon when it

48 *gäller såna där.*

comes to those

In the calls where the effects might be more severe, i.e. in the C-corpus, the most common position for the delivery of the first advice is another, namely immediately after the caller has described the incident or the symptoms. In this corpus, there are 23 instances of advice that constitute the first advice in their respective calls.[4] Of these 23, 11 first instances of advice occur in the currently mentioned position. Example (4) shows one of them (l. 6).

(4) #16785

Date: 1995-08-27 Time: 3.30 PM
Length of call: 00.49 min
Estimated risk: clear
Ph = Pharmacist
C = Caller (woman, possibly in her thirties)

4 C: *Ja hej min son har blivi fått eh ett getingstick*
Yes hello my son has become got uh a wasp sting
5 *på tungan.*
on the tongue
on his tongue
6 Ph: → *Jaha, då ska du åka in me honom ti sjukhus på en gång.*
I see then shall you go in with him to hospital at once
I see then you need to take him to a hospital at once
7 C: *Jaha:?*
Yeah?
8 Ph: *Hur gammal e han¿*
How old is he
9 C: *Han e tre:.*
He is three
10 Ph: *Ha:. (.) och de hände alles nyss=*
Ha (.) and it happened just now
11 C: *=Jaa.*
Yes
12 Ph: *Jaa. var ringer du ifrån?*
Yes where are calling you from
Yes where are you calling from
13 C: *Från Förort*
From Suburb
14 (***1.0***)
15 Ph: *Eh::m (0.4) ja du v- de e: asså (.)*
Uhm (0.4) yes you kn- it is therefore
Uhm (0.4) yes you kn- that's
16 *faran me e att man kan (0.4) om man får re i munnen,*
the danger with is that one can (0.4) if one gets it in the mouth
17 *de er att de: de kan bli: >sån krafti svullnad,*

it is that it it can become such strong swollenness
that is that there can be such a strong swollenness
18 *så att de kan bli svåriheter att få luft<.*
so that it can become difficulties to get air
so that it can be difficult to breathe
19 C: *Jaa, [jo].*
Yes [yes]
20 Ph: *[Så] de e (därför) därför måste man åka in me honom.*
[So] it is therefore therefore must one go in with him
[So] that's why that's why one must take him to a hospital

Early advice as the one in example (4) seems to be placed immediately after the caller in the previous turn has supplied enough information for the pharmacist to realize the possible consequences of the incident. In the current example, the caller's information about the place of the sting is sufficient for the pharmacist to immediately give definite advice. The caller responds with a *yeah* with rising intonation. This seems to be a news receipt (Heritage 1984). The intonation might convey a slight questioning of the advice. Such a questioning is, however, not reflected in the pharmacist's next turn. Instead of explaining the reasons for the advice at this stage, she poses a couple of questions (two diagnostic questions, ll. 8, 10, and one regarding the location of the caller, l. 12). Only after the question pairs and after a rather long pause (2.0 seconds) the pharmacist goes on to explain why she has given the troublesome advice (ll. 15–20). It is more common that advice, which proposes that the caller should take such difficult and time-consuming measures, is immediately supported by an account as in example (5).

(5) #18011

Date: 1995-09-09 Time: 11.05 AM
Length of call: 01.29 min
Estimated risk: some
Ph = Pharmacist
C = Caller (man, possibly in his fifties)

27 Ph: *Ja hallå?*
Yes hello
28 C: *Ja?*
Yes
29 Ph: *J- hallå?*
Y- hello
30 C: *Ja¿*

Yes
31 Ph: → *Ja även fast att den e spädd såhär*
Yes even though it is diluted like this
Yes even though it is diluted like this
32 *så innebär de en risk för ögonen. (0.6)*
so means it a risk for the eyes (0.6)
it would put the eyes at risk (0.6)
33 *så att hon måste ju ändå kontakta en ögonläkare,*
so that she must *ju* all the same contact an eye doctor
so that she really has to get in touch with an eye doctor
34 *även om hon har spolat rejält.*
even if she has rinsed thoroughly
35 *(0.6) nu vet ja inte varifrån i landet du ringer.=*
(0.6) now know I not wherefrom in the country you are calling
(0.6) now I don't know where in the country you're calling from
36 C: *=Ja ringer från eh Ly- Lysekil.*
I'm calling from uh Ly- Lysekil

Calls with a higher as opposed to a lower risk tend to differ not only regarding the sequential placement of the first advice. There are also differences in the way advice is presented. Before I elaborate on the differences, however, it should be noted that the most common shape of an advice is with the use of the modal verb *ska* ('shall') and that this advice shape is more or less equally common in both the C- and the S-corpus (32.4% and 28.3% of the advice instances respectively). As for other ways of presenting the advice, it is notable that the C-corpus has a much higher percentage of advice in the form of imperatives (14.8% compared to 2.0% in the S-corpus) and of those with the verb *måste* ('must,' 'have to') (12.0% compared to 2.6% in the S-corpus). Furthermore, advice in the form of questions (*Can you take him to a hospital?*) appear only in the C–corpus. On the other hand, the S-corpus has a higher percentage of advice with the verbs *får / får se till att* (roughly meaning 'have to') (10.5% compared to 0.9% in the C-corpus), *kan* ('can') (15.1% compared to 2.8% in the C-corpus) and *bör* ('should') (5.9% compared to 0.9% in the C-corpus).

There is a difference in strength between advice constructed with the verbs *can* and *should* as opposed to those constructed with imperatives or the verb *must*. The former types imply that the caller would be freer to ignore the advice, while the latter types imply a clear necessity. This indicates that the pharmacist uses stronger formulations in calls where the risk of poisoning is greater. One reason for this might be that in these calls the caller is told to take measures that may involve some inconvenience on her/his part (e.g. going to the hospital), and this advice may result in more resistance from the caller.

3. Caller's response and its consequences

The caller's responses to the advice vary. When the caller gives a response that indicates compliance, e.g. an explicit acceptance or a question that is in line with the advice, the parties can move on to further details of the advice or to the closing of the call. This is the case in example (6).

(6) #17405

Date: 1995-09-02 Time: 10.30 AM
Length of call: 03.36 min
Estimated risk: some
Ph = Pharmacist
C = Caller (woman, possibly in her thirties)

126	Ph:		*Så att d- då tycker ja ändå att du ska åka in*
			So that th- then think I anyway that you shall go in
			So that th- then I think anyway that you should go to the hospital
127			*för [säkerhets skull] för en kontroll.*
			for [security's sake] for an examination
			to [be on the safe side] for an examination
128	C:		*[För säk- ja:]*
			[For sec- yes]
			[To be on the s- yes]
129	C:		*.Ja:,*
			.Yes
130	Ph:		*.Ja,=*
			.Yes=
131	C:	→	*=Ja men då gör [ja de].*
			Yes but then do [I that]
			=Yes but then I'll do [that]
132	Ph:		*[Mm:?]*
133	C:		*Va bra:,=*
			What good
			Good
134	Ph:		*=(J [a)]*
			=(Y[es)]

135 C: [+T]ack+ *så jättemycke:=*
[Th]ank you so very much=
[Th]ank you very much=
136 Ph: *=He:j?*
=Bye
137 C: *He:j¿*
Bye

Here the caller gives an explicit commitment to following the advice (l. 131). That is followed by an acknowledgment token (*Mm:?*) by the pharmacist, who in that way indicates that she has nothing more to add. The caller responds by giving a general evaluation (l. 133), which is the beginning of the closing sequence.

Another way of accepting advice is to produce an *okay* in the turn following the advice; see example (7). This is a somewhat less clear way of showing acceptance of the advice than by producing a marked acknowledgment. The reason why an *okay* is less clear is because of the other functions that such an utterance can have, e.g. initiating the closing of the call or showing that one has nothing more to add (see Schegloff & Sacks 1973; Levinson 1983: 316–386 and Beach 1993 with references).

(7) #16603

Date: 1995-08-25 Time: 01–02 AM
Length of call: 06.01 min
Estimated risk: clear
Ph = Pharmacist
C = Caller (woman, possibly in her thirties)

205 Ph: *Så så prata me henne, å prata me dom på nittitusen.*
So so talk with her and talk with them at ninety thousand[5]
So so talk to her and talk to them at 9-1-1
206 C: *Mm.*
207 Ph: *Gör de me en gång.*
Do it at once
208 C: → *Okej.*
Okay
209 Ph: *De e bra.*
It is good
That's good

The *okay* (l. 208) is followed by an evaluation from the pharmacist that treats the previous turn as an adequate advice response.

The above excerpt from a call also illustrates another response to advice. The fragment begins with an advice token (l. 205) to which the caller responds with a very weak response, a *mm* (l. 206). However, that is not treated by the pharmacist as an adequate response to advice, as can be seen in the next turn where the pharmacist reinforces the advice (l. 207). By the term 'weak response' I mean different kinds of responses, from minimal responses like *mm* or *yes*, to next turn repair initiators (Levinson 1983: 339) and, finally, to outright rejections of the advice. If the caller only gives weak responses (or outright rejects the advice), the advice-giving is continued. This is effectively illustrated in an earlier fragment of the above call. But first, some background information is due. This caller has phoned the PIC because a friend of hers has taken an overdose of what is referred to as "Helsingör pills," i.e. a type of weight reduction pill. The problem is that the friend has locked herself in a room and won't let the caller in, so the caller thinks that it will be impossible to follow the pharmacist's advice to take her friend to a hospital.

(8) #16603

Date: 1995-08-25 Time: 01–02 AM
Length of call: 06.01 min
Estimated risk: clear
Ph = Pharmacist
C = Caller (woman, possibly in her thirties)

67	C:		*E de nånting ja kan göra här själv (hh)/(eh),*
			Is it anything I can do here myself (hh)/(eh)
			Is there anything I can do here myself (hh)/(eh)
68	Ph:		*Nä de e inte så mycke som du kan göra <själv hemma>,*
			No it is not so much that you can do by yourself at home
			No there isn't very much that you can do by yourself at home
69			*utan de e ju farlit bland annat för hjärtat då då:¿*
			instead it is *ju* dangerous among other things for the heart then then
			instead really it's dangerous among other things for the heart
70	C:	→	***Mm:.***
71			(*1.3*)
72	Ph:		*Så att- så att hon sku-*
			So that so that she shou-
73			*ja menar de här e ju en stor dos, =å hon e:h*
			I mean this is really a big dose, =and she u:h

74 *ja tycker absolut att hon måste in ti sjukhus.*
I think absolutely that she must in to hospital
I definitely think she's got to go to the hospital
75 C: → *Mm:.*
76 Ph: → *Så att hon kan få behandling där.*
So that she can get treatment there
77 C: → *Mm:.*
78 (*2.9*)
79 C: → *Ja.*[6] ((*fronted a*))
'Kay
80 Ph: *Å de de e liksom inte sådär att man man man kan göra*
And it it is like not that way that one one one can do
And it's kinda not that way that you can do
81 *nånting specifikt hemma, utan de behövs den här, (0.5)*
anything specific at home but instead it is needed this one (0.5)
anything specific at home instead it's necessary (0.5)
82 *de man kan göra på sjukhus så att säja.*
what one can do at a hospital so to speak
83 C: → *Mm:.*
84 Ph: *Så att de, (1.1) kan du inte ringa till dom igen då på*
So that it (1.1) can you not call to them again then on
So that it (1.1) can't you call them again then on
85 *nittitusen å diskutera me dom hur man (.) ka- ska göra¿*
ninety thousand and discuss with them how one (.) ca- shall do
9-1-1 and talk to them about what to do

The pharmacist gives a negative answer to the caller's question whether there is something the caller can do without taking her friend to a hospital, and reinforces the need to go to hospital (ll. 68–74). The advice "*I definitely think she's got to go to the hospital*" (l. 74) is very strong. The caller's response is just an acknowledgment token (*Mm:.*, l. 75), that says nothing about the caller's willingness to comply with the advice.[7] As Lindström (1999) has shown in her dissertation 'Language as social action. Grammar, prosody, and interaction in Swedish conversation,' minimal responses like *yes* or *mm* are not accepted as a preferred response to a deferred action request. Such behavior can even be understood as rejection-implicative (cf. Davidson 1984). Here, the pharmacist pursues a more elaborate sign of compliance by adding a turn fragment ("*So that she can get treatment there*," l. 76) which gives the caller a new chance to give a preferred answer. Instead, the caller repeats her former acknowledgment token (*Mm:.*, l. 77). The long silence afterwards indicates that the caller treats this answer as adequate and is waiting for the pharmacist to take the turn again. The pharmacist, however, refuses to do this and indicates in that way that she does not accept the caller's response as an adequate answer to her advice. Finally, after 2.9 seconds, the caller tries another response, this time a "'*Kay*"[8] (l. 79). But this response is not treated as enough, and the pharmacist repeats the already given information that there is nothing the caller can do at home and that her friend needs to go to a hospital (ll. 80–82). The caller responds once more with a *Mm:.* (l. 83) and the pharmacist goes on to make a more concrete and detailed action-forwarding utterance, this time in the form of an emphatic question.

The inadequacy of a minimal response such as *yes* to advice is clearly illustrated also in another call, example (9), which incidentally had immediately preceded the above call. Throughout the call, the caller has given only minimal responses such as *yes* or *mm* to advice.

(9) #16602

Date: 1995-08-25 Time: 00–01 AM
Length of call: 03.53 min
Estimated risk: clear
Ph = Pharmacist
C = Caller (woman, 23 years old)

120 Ph: *[A-] att ja vill faktist att du ska*
[Th-] that I want actually that you shall
[Th-] that I actually want you to
121 *ringa tibaka ti nittitusen me en gång.*
call back to ninety thousand at once
call 911/112 again at once
122 C: *Jaa. (1.0) mm.*
Yes (1.0) mm
123 Ph: → *Nej, säj inte ba^ra^ ja,*
No say not ju^st^ yes
No don't just say yes
124 → *utan [du m]åste göra de ock [så].*
instead [you h]ave to do it al [so]
instead [you re]ally have to do [it]
125 C: *[Ja.]* *[ja] ska göra de.*
[Yes] **[I] will do it**
126 Ph: *Lova mej att du gör de.*
Promise me that you do it
Promise me that you will do it

The caller (l. 122) once again responds to an (upgraded) advice with a simple *yes*, and when that does not result in the pharmacist taking the turn, produces another minimal response token, *mm*. The pharmacist's following metacomment (ll. 123–124) clearly shows her view regarding the inadequacy of a *yes* or *mm* as a response to an advice.[9]

Minimal responses to advice can thus indicate that the caller is not interested in following the advice. Another type of response that also seems to have this quality is a request for confirmation (the term is borrowed from West 1984: 109), as in example (10).

(10) #17393

Date: 1995-09-02 Time: 06–07 AM
Length of call: 02.06 min
Estimated risk: clear
Ph = Pharmacist
C = Caller (man, possibly in his thirties)

43	C:		*Va gör man då¿*
			What does one then
			What would one do then
44	Ph:		*Ja:, de man kan de e inte så mycke man kan göra själv¿*
			Yes what one can it is not so much one can do oneself
			Well what one can there isn't much one can do by oneself
45			*utan man ska åka in ti sjukhu:s¿*
			instead one shall go in to hospital
			instead one should go to the hospital
46	C:	→	*Man får göra de?*
			One has to do that
47	Ph:		*Ja: de ska man, för att (dom vill också) lyssna på dina*
			Yes that shall one 'cause (they want also) to listen to your
			Yes one should 'cause (they also want to) listen to your
48			*lunger och att, .hh man behandlar de som en*
			lungs and that .hh one treats it like a
			lungs and .hh one treats it like
49			*lunginflammation ungefär kanske inte att man får*
			pneumonia more or less maybe not that one gets
50			*antibiotika men man .hh behandlar dom symtomen som e:? (0.3)*
			antibiotics but one .hh treats those symptoms which are (0.3)
			antibiotics but they .hh treat those symptoms that you have (0.3)
51			*.hh så att du ska åka in ti sjukhus de måste du göra:¿ (0.4)*
			.hh so that you shall go in to hospital that must you do (0.4)

.hh so you should go to the hospital that you got to do (0.4)
52 *.ja.*
.yes
53 C: → *>Men va heter< de:, öh nu e de så (>att ja har ett<) jobb*
But what is called it uh now is it so (that I have a) job
But what's it called uh now the thing is (that I got a) job
54 → *från klockan två.*
from two o'clock
55 Ph: ***.Mm,***
56 C: → *Som ba- ((lätt hostning)) de e bara ja som kan sköta de här.*
that on- ((coughs)) it is only I who can do this

The caller's request for confirmation (l. 46) following advice from the pharmacist is an ambiguous response to the advice. It can be a neutral way of marking the reception of information (cf. Börestam Uhlmann 1994), but it can also be a way of probing the advice, by asking whether it really is necessary to follow the forwarded action. When interpreted in the latter sense, it can easily be seen as a pre-disagreement token. Compare the way in which a questioning repeat can function as a delay device when there is disagreement with an assessment and agreement is the preferred response (Pomerantz 1984:71, example (39); the transcription conventions have been adopted to the conventions I use elsewhere in the present article):

(11) (TG:1)

1 B: Why whhat'sa mattuh with y-Yih sou[nd=
2 A: [Nothing.
3 B: = HA:PPY, hh
4 A: → I sound ha:p[py?
5 B: [Ye:uh.
6 (0.3)
7 A: No:,

Pomerantz's article is about second assessments to a previous assessment, instead of responses to advice. Nevertheless, I think there is striking similarity between the way in which the relevant turns are constructed. According to Pomerantz the function of the questioning repeat in this case is to delay the disagreement. The request for confirmation in my present example has a similar function. The pharmacist's response (ll. 47–52), where she confirms the advice and justifies it, supports this analysis. Additional support for the analysis is provided by the caller's following argument against the advice (ll. 53–54, 56).

The argument against accepting the advice (ll. 53–54, 56) is an attempt to reject the advice in a somewhat indirect way. Such a rejection is of course the most explicit sign of not committing to the advice. Outright rejections are seldom seen in these calls, a fact that probably has to do with the asymmetric relation between the two parties. Where the rejection to the advice is given it is typically presented in a dispreferred format (the counter-argument in ex. (10) is initiated by delay devices such as the expression "*but what's it called*" and the hesitation mark *öh*) and formed in a way that expresses a "no fault quality" of the rejection (in the current example the reference to a job that only the caller is capable of doing, and which he therefore has to do).

A caller who finds the given advice too troublesome can present a lower bid, and in that way try to avoid the proposed, more inconvenient course of action. This is the case in example (12), which is from a call by a woman who has accidentally eaten too many bitter almonds. The pharmacist has already given the advice to go to a hospital three times earlier in the call, but it has been met with different kinds of more indirect resistance, e.g. a request for repetition and a question that functions as an argument against the proposed course of action ("yes but don't you think...").

(12) #16581

Date: 1995-08-24 Time: 08–09 PM
Length of call: 06.08 min
Estimated risk: some
Ph = Pharmacist
C = Caller (woman, possibly in her thirties)

172	Ph:	*S[å ja tyck-]*
		S[o I thi-]
173	C:	*[Ja men va] va kan dom göra på sjukhuset då¿*
		[Yes but what] what can they do at the hospital then
174	Ph:	*Ja dom dom kan ju behandla dej, dom symtom du har.*
		Yes they they can *ju* treat you the symptoms you have
		Well they they can of course treat you the symptoms that you have
175	C:	*Mm.*
176	Ph:	*Och man man kan behöva ta en del prover,*
		And one one can have to take some tests
		And one one might have to take some tests
177		*för å se om man behöver sätta in behandling.*
		for to see if one has to put in treatment
		in order to see if it's necessary to start a treatment

178 C: °Okej.°
Okay
179 Ph: *Så att ja tycker att du ska åka in ti sjukhus.*
So that I think that you shall go in to hospital
So I think you should go to the hospital
180 C: → *.hh hm: a ja ja väntar lite, å ser om ja känner mej sämre,*
.hh hm: yes I I wait a little and see if I feel myself worse
.hh hm: yes I I will wait a little and see if I feel worse
181 *för att nu: (0.3) de e en trekvart sen ja åt.*
'cause that now (0.3) it is one threequarter of an hour since I ate
cause now (0.3) it's been forty-five minutes since I ate
182 Ph: *Ja:.=*
Yes
183 C: *=>Så de borde väl<,*
So it should *väl*
So I guess it should
184 (*1.0*)
185 Ph: *Nej ja tycker inte man ska vänta ytter[liare*
No I think not one shall wait fur[ther
No I don't think you should wait any lo[nger
186 C: *[Tycker] du inte de¿=*
[Think] you not that
[Don't] you think so

In response to the once more repeated advice (l. 179) the caller states that she will wait a little and see if she starts to feel worse (l. 180), i.e. instead of agreeing to go to the hospital right away, she proposes to do it only if she feels worse. This is what I call "giving a lower bid," to use card-playing terminology. The lower bid is backed up by an account concerning the small amount of time that has passed since the incident occurred (ll. 181, 183). The mitigation *a little* and the account give the turn a dispreferred format (Pomerantz 1984). The pharmacist responds by maintaining her advice without any mitigating devices,[10] i.e. a dispreferred action presented in a preferred format.

4. Conclusion

It is evident that the caller in the calls to the PIC is not merely a passive recipient of advice. Rather, s/he contributes actively in the construction of advice. This has been shown in example (1), where the caller in the middle of the advice-giving contributes information that makes the pharmacist interrupt his advice and modify it so as to fit the re-evaluated situation. In example (1), the caller also affiliates herself with the pharmacist, by introducing a question sequence ("*Is that good,*" l. 21), instead of just giving an accepting answer such as *okay* or *I'll do that*.

Generally, more than one advice sequence appears in a call. The prototypical position for the main advice is after the interrogative series and the temporary break during which the pharmacist consults a database or archive. In many calls where there is only some risk of poisoning, this is also the place for the first advice. In many calls where the risk of intoxication is greater, a first instance of advice is often positioned in the interrogative series and before the break. Another common advice sequence in the less severe calls (i.e. the S- and N-corpus calls) is the just-in-case advice, where the pharmacist explains what to do if the incident turns out to be more serious than expected. Giving just-in-case advice seems to be a way for the professional and the institution to handle the liability issue. This sequence is normally positioned after the main advice, at the very end of the call, just before the closing sequence.

Calls that are about more severe risks of poisoning contain advice that allow the addressee less freedom of action than do advice for less severe cases (e.g. advice involving verbs like *must* or *have to* instead of verbs like *can, could* or *should*). One reason for this might be that carrying out advice in severe cases tends to demand a greater commitment from the caller than advice in less severe cases (e.g. going to hospital as compared to giving the affected person a glass of milk). Strong advice can then have the effect of reinforcing the need for carrying out the advised action. In calls where the caller does not display the necessary commitment, the pharmacist can interpret the caller's lukewarm response as an indication that the latter will not carry out the advice. In those cases, the pharmacist repeats the advice, often in an aggravated form. Advice that demands a lot of the addressee is also normally motivated with an account for why it is given. In this way, the caller's factual and presumed responses decide the length and shape of the advice sequences.

Notes

1. This organization figure is a slight modification of a figure in Zimmerman (1984:214), which specifically characterized calls to the police, but which also, according to Zimmerman, can characterize a broad class of service calls.

2. The corpus was collected as part of a research project that was the result of a collaboration between FUMS (The Unit for Advanced Studies in Modern Swedish, at the Dept. of Scandinavian Languages at Uppsala University) and The Poison Information Center. The project was funded by The Swedish Counsel for Work Life Research.

3. The transcription conventions that I use in this article follow those originally developed by Gail Jefferson (see Ochs, Schegloff, & Thompson 1996:461–465), with the following exceptions:

(( ))	transcriber's comment
^hh^	laughter
^word^	talk with simultaneous laughter or 'smile voice'
.word	speech during inhalation
eh *or* öh	hesitating sound
%word%	indicates that personal information (i.e. names, addresses, telephone numbers) has been erased from the tape in order to preserve the integrity of the participants.

In the transcribed call fragments, the first line is a transcription of the original stretch of talk. The last line is an idiomatic translation into English, and the middle line is a word by word translation (which is omitted when the idiomatic translation more or less corresponds to the Swedish original).

4. The information that there are 23 first instances of advice in a corpus with a total of 20 calls may seem odd. The explanation is that three calls have been analyzed as containing two first instances of advice each, because the calls have been restarted.

5. In 1995, when these calls were made, the Swedish emergency number was 90 000. Since then the number has been changed to 112, in order to make it more international. In the translated transcription, I have given the US equivalent 9-1-1.

6. The *a* in *ja* is fronted, which gives the word a slightly different meaning than a normal *ja* ('yes'). This *ja* has a connotation of 'got it' or 'I understand, let's move on.' This is the reason for the translation ''kay' instead of 'yes' or 'yeah.'

7. See Aronsson and Larsson (1987:17–23) concerning the same minimal feed-back in doctor–patient talk.

8. See Note 6.

9. Actually, even the caller's following elaborated response ("I will do it," l. 125) is responded to by an aggravated request "Promise me that you'll do it." It seems that the caller's uncooperativeness, which has been consistently displayed throughout the call, has caused the pharmacist to doubt even an elaborated response from the caller.

10. The pause at l. 184 can be explained by the caller's preceding self-interrupted turn-constructional unit, which has caused uncertainty regarding whose turn it presently is. The pause should therefore not be considered a way of delaying the dispreferred response.

References

Agar, M. (1985). 'Institutional discourse'. *Text, 5*, 147–168.

Aronsson, K. & Sätterlund-Larsson, U. (1987). 'Politeness strategies and doctor-patient communication. On the social choreography of collaborative thinking'. *Journal of Language and Social Psychology, 6*, 1–27.

Beach, W. A. (1993). 'Transitional regularities for 'casual' "Okay" usages'. *Journal of Pragmatics, 19*, 325–352.

Börestam Uhlmann, U. (1994). 'Skandinaver samtalar. Språkliga och interaktionella strategier i samtal mellan danskar, norrmän och svenskar'. [Scandinavians in conversation. Linguistic and interactional strategies in conversations between Danish, Norwegian and Swedish speakers.] *Skrifter utgivna av institutionen för nordiska språk vid Uppsala universitet 38*. Uppsala.

Davidson, J. (1984). 'Subsequent versions of invitations, offers, requests, and proposals dealing with potential or actual rejection'. In J. M. Atkinson & J. Heritage (Eds.), *Structures of Social Action. Studies in Conversation Analysis* (pp. 102–128). Cambridge.

Goffman, E. (1974). *Frame Analysis*. New York.

Hakulinen, A. (1993). 'Inandningen som kulturellt interaktionsfenomen'. [Inhalation as a cultural interaction phenomenon.] In Ann-Marie Ivars et al. (Eds.), *Språk och social kontext* (pp. 49–67). (*Meddelanden från institutionen för nordiska språk och nordisk litteratur vid Helsingfors universitet, serie B:15.*) Helsingfors: Helsingfors universitet.

Heritage, J. (1984). 'A change-of-state token and aspects of its sequential placement'. In J. M. Atkinson & J. Heritage (Eds.), *Structures of Social Action. Studies in Conversation Analysis* (pp. 299–345). Cambridge.

Landqvist, H. (2001). 'Råd och ruelse. Moral och samtalsstrategier i Giftinformationscentralens telefonrådgivning'. [Advice and remorse. Morality and conversational strategies in calls to the Swedish Poison Information Centre]. *Skrifter utgivna av institutionen för nordiska språk vid Uppsala universitet 55*. Uppsala.

Levinson, S. C. (1983). *Pragmatics*. New York: Cambridge University Press.

Lindström, A. (1999). 'Language as social action. Grammar, prosody, and interaction in Swedish conversation'. *Skrifter utgivna av institutionen för nordiska språk vid Uppsala universitet 46*. Uppsala.

Ochs, E., Schegloff, E. A., & Thompson, S. A. (Eds.) (1996). *Interaction and Grammar*. Cambridge: Cambridge University Press.

Nordberg, B. (1999). 'On closings in alarm calls'. *Språk och stil, 8* (1998), 65–103.

Persson, H. (1990). 'Giftinformationscentralen – 30 år. Igår: giftkartotek för pediatrikern, idag: "skräddarsydd" rådgivning för alla'. [The Swedish Poison Information Center – 30 years. Yesterday: poison card index for the pediatrician, today: "customized" advice-giving for everybody.] *Läkartidningen, 87*, 1960–1964.

Pomerantz, A. (1984). 'Agreeing and disagreeing with assessments: Some features of preferred/dispreferred turn shapes'. In J. M. Atkinson & J. Heritage (Eds.), *Structures of social action. Studies in conversation analysis* (pp. 57–101). Cambridge.

Schegloff, E. & Sacks, H. (1973). 'Opening up closings'. *Semiotica, 8*, 289–327.

Sjöberg, U. (1999). '"Kan man dö av tjugoåtta tabletter gånger tjugi milligram Cipramil?" Frågor formulerade som generella fall i samtal till Giftinformationscentralen'. ["Will someone die from eating twenty-eight pills of twenty milligrams Cipramil?" Questions formulated as a hypothetical case in calls to a Swedish Poison Information Center.] In P. Linell, L. Ahrenberg, & L. Jönsson (Eds.), *Samtal och språkanvändning i professionerna. Rapport från ASLA:s höstsymposium Linköping*, 6–7 november 1997. (ASLA:s skriftserie 11.) (pp. 235–247). Uppsala.

West, C. (1984). 'Medical misfires: Mishearings, misgivings, and misunderstandings in physician-patient dialogues'. *Discourse Processes, 7*, 107–134.

Zimmerman, D. H. (1984). 'Talk and its occasion: The case of calling the police'. In D. Shiffrin (Ed.), *Meaning, Form and Use in Context: Linguistic Applications* (Georgetown University Round Table on Language and Linguistics) (pp. 210–228). Washington, DC.

Part IV

Consumer assistance

Chapter 10

Opportunities for negotiation at the interface of phone calls and service-counter interaction

A case study

Denise Chappell

1. Introduction

In this study the use of formulations and their sequential organisation have been shown to form a central link to the provision of opportunities for role and goal negotiation. It is perhaps opportune to note that formulations have been variously described as 'summarizing, glossing, or developing the gist of an informant's earlier statement' (Hutchby & Wooffitt 1998) and 'reflexive consultation of the conversation thus far and/or some naturally bounded segment of that conversation' (Heritage & Watson 1979: 128). An important observation is made with concern to the ways that formulations act to 'minimise breaks with respect to the flow of topical talk' (Heritage & Watson 1979: 152). From the perspective of goal outcome Iacobucci assesses their usefulness in terms of ensuring coordination and determination of institutional task goals (Iacobucci 1990).

A feature which has been found to be a useful negotiating resource has been the account. The view taken by Firth (1995) in his study of commodity negotiation is of accounts being 'problem-solving resources' which contribute to the 'overall structure' of the interaction by providing opportunities to negotiate outcomes. Here the 'context-sensitive' nature of accounts is highlighted and he notes one important aspect of context is the 'stance' of the other party (1995: 205). This aspect of accounts is of consequence to outcomes as different degrees of sensitivity will present interactants with varying opportunities for the proactive and creative negotiation of goals and roles in the ongoing talk.

The role that formulations and accounts play in the definition of talk type is in this study specifically related to 'troubles-telling' and 'problem-solving'. In their study of 'Troubles-Telling', Jefferson and Lee (1981) emphasize the problematic nature of distinguishing between these talk types suggesting that the distinction lies in the different focii of attention; in the former case the focus is on the 'teller and his experience', and in the latter it is on the 'problem and its properties' (1981: 141). They note that affiliative responses form an integral part of the 'troubles-telling sequence' in 'ordinary' settings and that affiliation is rare in their data of institutional talk on the phone where focus was on 'the problem and its properties'.

We might surmise, then, that the roles played by interlocutors both shape and define the talk as being one of a 'troubles-telling' or 'problem-solving'. In this paper it has been found that one way in which such shaping and defining of interaction can take place is through the use of formulations and accounts.

2. Aims of the study

The overall aims of this study are to describe the ways in which the Service Provider (SP) and Customer (C) use initial requests, formulations and accounts to accomplish the task-goal of bed-booking and how these set up opportunities to continually realign needs and adjust orientations (Iacobucci 1990).

Of particular concern are questions relating to (i) whether the interactants i.e. the 'interested party', the customer, and the SP, perceive the (one-sided) talk between the Service Provider (SP) and Guest House Proprietor (GSP) in unsuccessful phone calls to be part of a 'Troubles-Telling Sequence' and (ii) the sequential location at which these perceptions change and develop during the course of interaction.

3. The data

The spoken data consisting of three phone-based bed-bookings were selected from a larger corpus of 34 service encounters collected from a number of Tourist Information Centres in Cambridgeshire (Chappell 1993). The data used in this case study approach were recorded at the Tourist Information Centre at Ely in Cambridgeshire. Each of the conversations were transcribed using the transcription conventions shown in the Appendix. The Service Providers

(SPs) and Customers (Cs) were aware that their conversations may be audio-recorded.

It should be noted that the transcribed phone calls are 'one-sided' with the caller's talk having been recorded and transcribed as it was not possible to record the called party's contribution. The lack of the addressee's talk does, however, represent talk as it naturally occurred in these bed-booking service encounters.

The recordings were made during the summer at the Ely TIC which serves a dual role in that the house, now turned Tourist Information Centre was used by Oliver Cromwell. Visitors are able to visit the house which is furnished appropriate to the age with recorded commentaries about its history in each of the rooms. The front office where the bed-bookings were made doubles up as a shop selling a variety of souvenirs. Normally there are two or three SPs on duty at any one time. They serve behind a semi-circular desk located at the side of the shop.

4. Background to the study: Stages of 'Troubles-Telling'

It is proposed that an analysis of this data shows that the 'Troubles-Telling' is organised in two stages; Stage 1 is located at the beginning of the conversations where customers provide initial accounts of their needs and expectations while Stage 2 arises as a consequence of Service Providers' one-sided phone-calls which occur at different locations in the talk and may be 'listened in to' by the 'interested party' waiting at the service counter.

These two stages create opportunities for the negotiation of roles and goals. More specifically, in Stage 1, customers were found to ask questions explicitly related to 'help-seeking' thereby investing the Service Provider with the role of 'problem-solver' or, as Leppanen (2000) observes, ascribe them the role of 'expert' and in Stage 2 there were examples of customers showing evidence of their role as 'troubles-recipients'. (One of the reasons for which these Service Provider/Customer roles may be felt most appropriate could be culturally and experientially influenced.[1])

Stage 2 of the 'Troubles-Telling' in this study is located in positions where customers listen in to unsuccessful phone calls made by Service Providers to Guest Houses and by doing so become the 'Troubles-Recipient'. On completion of the calls, the negotiation of role and task goal via formulations or account-giving appears to play an important part in the mutual understanding of the Service Encounter as a 'Troubles-Telling' sequence. This stage is located at the interface between the end of the successful, or, in many cases, unsuccessful phone calls and ongoing face-to-face conversation. It is important to bear in mind that the length of time the Service Provider spends making these phone calls varies ranging from one successful call to seven calls before eventually booking the accommodation. This interface acts as a Turn-Transition Relevance Point whereby either the Service Provider or Customer are able to select turn. Interesting to note is that in cases where there have been unsuccessful attempts at bed-booking, accounts are provided by the Service Provider of the talk so far, affiliation responses given and other reciprocity work carried out.

We will now explore these two stages further by examining in detail some examples of talk to describe the relationships between the negotiation of SP/C roles and goals and SP/C perceptions of talk sequences as 'troubles-telling' or 'problem-solving' and the role of formulations and account-giving.

5. Data analysis: Example 1

Stage 1: Customer as 'troubles-teller'

In example (1) the customers require accommodation and have a dog. Below is their initial request:

(1)

```
1   SP:  hello can I help?
2   C1:  ( ) a night's accommodation
3   SP:  yes?=
4   C1:  =em(.) we've got a dog
5   SP:  ((laugh)).right.did you want to be in Ely or (.)out( )
```

Here the SP's response to the initial request in line 3 can be seen to be dispreferred as it is a request to provide further information. In line 4, the 'Troubles-Telling' commences with the introduction of the 'trouble' – 'having a dog'. It is interesting to note that the SP seems to view this initial request as both a trouble as well as a service request, firstly, by her use of laughter as an affiliative response and establishment of her role as 'troubles-recipient', and, secondly, by a direct request for their preferred location so downgrading its importance in terms of goal achievement and confirmation of her institutional role of service provider.

After offering the customers the accommodation list and discussing topics concerning location of accommodation, type of dog, and restaurants, the customers direct the SP to act as follows:

(2)

```
42  C2:  ring those guest houses (5.0)
43  SP:  would you like to try that? (2.0)
44  C2:  mm.we have to take it then
45  SP:  not at all no.no what we can do is I can phone around
46       and.erm see who has [vacancies.
47  C2:                      [yeah.
```

The role allocated to the SP by the customers at that point was one of 'secretary' and although they asked her to phone these guest houses they still did not want to be tied into making an immediate decision about the accommodation.

Stage 2: Customer as 'troubles-recipient'

After the first phone call was made, Stage 2 of the 'Troubles-Telling' began with the SP becoming the 'troubles-bearer' and the Cs being cast in the role of 'troubles-recipients'. This is shown in the extract below:

(3) Call 1

```
85  SP:  ↓hello good afternoon it's Alison here from the Tourist
86       Office.any accommodation left please?(2.0)you're full
87       are you?right thank you very much.thank you bye
88       bye.((gap))((looking through accommodation list)))
89  SP:  I think we may have to go out of Ely a little way
90       ( ) Mrs. Levitt who has two twin rooms er: ( )and
91       Mrs. O'Lochan who's just up
92       at the main road and she takes pets by
93       arrangement(2.5)other than that I'm afraid we've to go
94       out of Ely.
95  C1:  ↓no try the ones in Ely.=
96  SP:  =try the ones in Ely.(0.5)you'd like me to try the
97       ones without(.05)en-suites.
```

Here the SP repeats a suggestion about location of accommodation at the interface of the call and the face-to-face interaction in line 89/90 and at the end of her turn in line 94. 'Sandwiched' between these suggestions in lines 90–93 is a section of talk about two other guest houses which may be able to help. This formulation is clearly seen by the SP as having 'resolution potential' (Firth 1995) although it is not used at this stage as the Cs direct the SP to stay with the initial request of being in Ely in line 95 but decide to change their accommodation needs in lines 96/97.

After two further unsuccessful phone calls the customer self-selects at the interface of Call 3 and ongoing talk perhaps expecting some affiliative work as a result of this trouble. As 'troubles-recipient' she continues:

(4) ((end of Call 3))

```
104 C1: last one is it?
105 SP: ((pause))sorry?
106 C1: it's the last one?=
107 SP: ↓=no no(.)we've got a couple more [((laugh))
108 C1:                                   [uh
109 C2: (Tattersall sales today)
110 SP: yes the Tattersall Sales
111 C2: so Newmarket would be quite difficult as well
        ((receptionist makes Call 4))
```

The lack of account-giving and formulation at the end of Call 3 seems to stimulate a need for C1 to self-select (Sacks, Schegloff, & Jefferson 1974) in line 104 and after reformulating the question checks the state of the talk so far which is provided by the SP in line 107 with a confirmation that there are a 'couple more' guest houses to call. By self-selection when seeking an update at the end of the call and referring to another location being 'difficult as well' in line 111 the customers enact their roles as 'troubles-tellers' as well as 'troubles recipients'. Meanwhile the SP continues to reconfirm her role as that of Service Provider and 'problem-solver'.

However, at the sixth attempt there is success in finding accommodation.

(5) Call 6

```
124 SP: ↓hello Mrs.B?he((hh))llo it's A(hh)lison er I'm after erm a
125     double room wondered if you'd got anything left.(2.0)you
126     have a↓twin.I have a couple here with a large friendly
127     labrador as↓ well(5.0)((laugh))(2.0)oh immaculately behaved
128     I'm sure!(1.5) ((laugh))they're nodding enthusiastically
129 C1: actually it's a small friendly one
130 SP: ((laugh))it's a small friendly
131     labrador ((laugh)) (3.5)right.er Mrs.(2.0)sorry?
132     (1.0)you haven't got a dog no.(3.0)oh well
133     labradors are lovely breeds aren't they we we've
134     got a labrador.
```

In this call it becomes increasingly evident that the customers are the 'interested party' listening in to the call being made. This is demonstrated by the interjection in line 129 and shows the customer perceiving herself as having equal participant rights in the call as a co-owner of, and co-participant in, the ongoing talk. It is interesting to note that this interjection could be viewed as an example of what Jefferson terms 'embedded correction' (Jefferson 1987) with utterance X being 'large friendly', (line 126) utterance Y being 'small friendly' (line 129) and the repetition of utterance Y (in line 130). It may be argued that the receptionist also wants to assure the guest house owner that the dog will not be a problem and in order to emphasize this point comments in line 133 on how 'lovely' a breed they are.

Following the customer's interjection by way of embedded correction, the SP provides an account of the talk so far as follows:

(6) ((interrupts Call 6))

```
135 SP: ( )she says she hasn't got a dog of her own but when she had
136     one she use to ( )I'm sure she'll be ( )it's in Fleetwood
137     which is just(.) up here this is Fleetwood here.and if you
138     want to walk into the town tonight there's a little
139     cut↓through here onto the main road so you don't have to
140     actually take your car to get
141 C1: [well that's alright]
142 C2: ( ) ((continues Call 6))
143 SP: ↓yes could I book it for Mr. and Mrs.C? that makes you full
        doesn't it tonight?(2.0) er the name is↑Chalmers.
```

The turn in lines 135–140 takes place while the SP is on the phone to GH owner and enables the interaction to steer once again towards correct orientation of task goal. It also functions to confirm the customer's agreement to take the room which is confirmed in line 141. This turn functions as both a formulation of talk thus far i.e. retrospectively, as well as an indirect check that the customer wants to book the room i.e. in a prospective way. Viewed in its broader context, the appearance of the formulation at this point in the overall structure of the conversation perhaps reflects the initial reluctance shown by Cs to make a firm booking earlier on in this encounter (see line 44 above). The formulation outlines a number of reasons why it is perceived by the SP to be a suitable choice including the fact that the guest house owner likes dogs, is happy to take them and the suitability of location of, and accessiblity to, the guest house. A further 'unpacking' of the formulation is given by suggesting a route to take and emphasizing the guest house's convenience since the car would not be necessary to get to town. The choice of the words such as 'just' in line 137 and 'little' in line 138 serve to 'downgrade' potential problems.

6. Data analysis: Example 2

Stage 1: Customer as 'troubles-teller'

The following example involves two American couples who are seeking accommodation for the night. The immediacy of need is clear in their initial request and places it firmly within what is referred to in this paper as Stage 1 of the 'Troubles-Telling' as shown below:

(7)

```
1   C1: can can we get some rooms(.)for some tired Americans?=
2   SP: =((laugh))yes indeed!how many tired Americans are
3       there=
4   C1: =well we'd need two double rooms(.)and and(.)four people and
        two double rooms(.) and er(.) we'd like baths in the rooms.
```

In this case the SP shows affiliation to the 'trouble' presented by her laughter as well as exact repetition of the customers' physical state, 'tired', and social identity, 'Americans' in line 2. However, the SP, in her institutional capacity and bearing in mind the task goal then requests the number of people who require accommodation. Here then she acts as both 'troubles-recipient' and 'problem-solver'.

It seems from the outset of this interaction that customer and service provider roles became established; the latter as 'helper' for the 'tired Americans' and former 'help-seekers'. This is further exemplified in the following turns:

(8)

```
7   SP: right.did you want a hotel or erm(.)[ a private house
8   C1:                                     [no we'd like we'd
9       like a private house(.) with a bed with breakfast (4.0)
10      and maybe you could pick us something really nice or
11      if you've got any ideas ( )=
12  SP: =yes did you want to be in Ely(.) or out in the
13      villages=
14  C1: =it doesn't matter out in the villages is O.K.wouldn't
15      you say so
16  C2: (.)yes we want I guess we probably need to think about
17      being able to get food tonight. or if ( ) if we stay
18      somewhere.some place that doesn't have one then we need a
19      [suggestion
20  SP: [right.right.
```

This extract indicates the C's perceived role of the SP as one of helper with expertise and knowledge to solve the problem of finding accommodation. This is clearly shown in lines 10 where C1 asks her to 'pick something' for them, and to suggest ideas in lines 11 and 18/19. One might suggest that decisional control is handed over to the SP by these customers through their use of relational talk in their initial request and in providing further information the conversation became more personalised with appeal to their 'troubles-teller' roles. At this point in the conversation they did not know which of the options in line 12/13 would be the less 'troublesome' hence the answer in line 14 of 'it doesn't matter'.

Stage 2: Customer as 'troubles-recipient'

This neutral attitude shown by C1 towards finding accommodation arises at Stage 2 of the 'Troubles-Telling' after a variety of topics have been discussed such as location of restaurants, type of room, booking form completion, cost of room, road information. This point is illustrated in the following extract taken from the interface between the third unsuccessful phone call to guest houses and the ongoing face-to-face interaction and forms the commencement of Stage 2 of the 'Troubles-Telling' in that these customers have been able to listen in to these phone calls, realised there are problems and so reversed their roles from that of 'troubles-teller' to 'troubles-recipient'.

(9) Call 3

```
89  SP: hello is that Mrs.Knott?
90      (1.2)it's Alison here from the Tourist Information
91      Centre in Ely.do you have any accommodation please
92      for ↓tonight.(2.2)I'm after two rooms if possible(8.0)
93      oh alright then O.K.then thank you thanks
94      bye bye.
95  C1: we we don't have to be down there if we can't do
96      anything down↑there=
97  SP: =I think we're working our way slowly back [towards Ely
98  C1:                                            [that's
99      ↓right it.it really makes no difference=
100 SP: =right try ( ) ((gap))
```

Despite this trouble, however, the customer at this point is clearly cooperative and open to negotiation of location. This is signified in lines 95/96 and line 99 where he states it 'really makes no difference'. The SP provides an account of the upshot of the talk so far while commenting on the strategy she's taking in Line 97, hence both reconfirming her role of 'problem-solver' and ascertaining customer's agreement to her action which is given.

After completion of Call 4 the SP provides a part account of Call 3 in the following way:

(10)

```
101 SP: hello Mrs. Jones Alison here from the Tourist
102     Office(.)hello do you have any accommodation free
103     please (1.0)((laugh))you as WELL!((laugh))I think
104     everyone(3.0)oh.(2.0)she's not in ((laugh))(.)
105     never mind(4.0) ah right right O.K.then.not to worry.
106     thanks. bye.
        ((gap))
107 SP: one lady is free but she's (.)not answering the phone
108     so try one a little bit further up in Barlow
109 C2: ( )
110 SP: ((laugh))there
111     are a few more places [to try
112 C1:                       [sure
113 SP: before we (3.0)( )we despair
```

In this example there is reference to Call 4 line 104 where it seems that the guest house proprietor had suggested another guest house which the SP had already tried. The interface between the call and ongoing talk provides an opportunity for the SP to comment on this particular section of the call's talk sequence in line 107 and although confirming this as a 'trouble', nevertheless suggests another solution in line 108 of trying one a 'little bit further up'. The use of the words 'little bit' acted as 'downgraders' of the problem. The SP's repetition of 'try' indicates her role as active 'problem-solver' in her institutional role and the use of relational talk in line 113 where reference is made to their 'despair' reconfirms her solidarity with the customer and the ongoing problem as well as affirming the encounter as a personalised, as well as institutionalised, interaction. It is interesting to note that the word 'despair' reflects the mood first established by the customer's use of 'tired Americans' in the initial request in line 1.

Call 4 brings with it success for this group of tourists and the SP formulates the news for the customers before confirming the booking with the guest house proprietor. The interface between the call and face-to-face talk provides the opportunity for the customers to clarify details of accommodation and negotiate goals if necessary. In this extract clarification of facilities is sorted out before confirmation of the bed-booking is made.

(11) Call 4

```
114 SP: hello.can I speak to Mrs. Cook please?(1.5)yes I
115     will thank you((pause))thank you thank you((pause))hello
        there! it's Alison here from the Tourist Office hello I want
        ( )any accommodation free(2.0)you have I was after both
        rooms(.) just
```

```
119         for one night(4.0) and double(1.0) that's
120         alright is it?(.)oh lovely alright hang on just
121         a sec=
122 SP:  =Mrs. Cook will be able to help ↑you she's got a(1.0)
123         double room with an en-suite and also a(.)[twin=
124 C1:                                               [that's fine
125 SP:  =will that be [alright? you'd like me to book?=
126 C1:                [sure]
127 C1:  =yeah go ahead=
128 C2:  =got a bathroom [in]
129 C1:                  [any] bathrooms in those
130 SP:  this one has yes but this one hasn't the two would
131         share
132 C2:  (3.5)we would share with (.)can [we] do that?
133 C1:                                  [yes]
134 SP:  would that be alright?=
135 C1:  =yeah.yes guess so
136 C2:  yes.not with anyone else just [just]
137 SP:                                [no no]just[the two
138 C1:                                           [yeah yeah]
139         we'll sort it out yeah
140 C3:  we'll do that
         ((Call 4 continues))
141 SP:  right yes can we book them please?(1.0)the name
142         is Mr.↓Laing
```

Perception of this situation as a problem and SP's own role as 'problem-solver' is reinforced by her use of the word 'help' in line 122 and shows SP's shared understanding of the problematic nature of this task, having already had three unsuccessful calls to book two en-suite rooms.

However, problems arise at this embedded stretch of talk where C2 raises the question about bathrooms in the rooms in line 128. In response to this the SP offers a confirming formulation in line 130 which serves to realign task-goals since only one en-suite is available. The clarification request in line 132 and indirect clarification in line 134 confirm this adjustment of task-goal but yet again in lines 136 clarification is sought and provided in line 137 so re-confirming this adjustment. In line 139 C1's role is also reinforced as one who is cooperative and open to negotiation where he says 'we'll sort it out yeah' in regard to any troubles associated with having only one double en-suite as opposed to two doubles en-suite, the initial request.

The final outcome of this service encounter is not what the couples had initially wanted but as increasingly more trouble arose due to unsuccessful calls C1 was willing to negotiate location in lines 99 and 112 and type of accommodation in lines 124; 127; 135; 138/139. Account-giving and formulations played a central part in these negotiations as they worked both retrospectively in terms of bringing the customer/s up to date with the talk so far and prospectively in terms of future action of SP.

Concomitant with these negotiations of goal were negotiation of roles which changed according to the stage of troubles-telling. In Stage 1 the customers took on the role of 'troubles-tellers' while the SP was 'troubles-recipient and problem-solver'. In Stage 2 the customers reversed their role becoming 'troubles-recipients' while the SP was 'troubles-bearer'. The final part of the encounter indicates that the customers had collectively accepted responsiblity for solving any further problems that might arise. The SP's confirmation of bed-booking redefined her institutional role of problem-solver.

7. Data analysis: Example 3

In this example the initial request the C uses endeavours to establish SP role as 'helper' in line 1 but it does not fulfil this function as there is a dispreferred second in line 2 where the less experienced SP1 hands the accommodation list to the customer. She is about to explain what is in it when the customer interrupts in line 3 so problematising the request by adding that they 'don't know anything about the area'. This turn serves to shape her own role as one of 'help-seeker' rather than 'information-seeker'. SP1 then responds by asking which of three types of accommodation is required thereby fulfilling her obligations in her institutional role of 'expert' and 'helper'. The customer indicates her requirement of a bed and breakfast with private facilities and a television set in the room as in extract (12) below.

(12)

```
C:   could you find a nice hotel for us for one night?
SP1: there we are.there [are(.) ((hands customer list))
C:                      [will you do the accommodation we
     don't know anything about this area ((laugh))
SP1: (.)erm d did you want a hotel or a guest house or bed and
     breakfast in [a private home?
C:                [well bed and breakfast but er in a
     nice place=
SP1: =in a nice place(.) ↓alright. what what did you want
     er er how many of just you?
C:   [just my husband and myself.
SP1: [just just two
```

```
13        of you (2.0)
14   C:   it's got to have private [facilities erm
15   SP1:                          [private facilities.↑O.K.
16   C:   and a television set as well.
17   SP1: right erm:t.v.in rooms er t.v.lounge with
18        t.v.((reading from list))
```

This request subsequently developed into a problem as the talk progressed. A large proportion of the 72 turns which followed the initial request were spent discussing which of the guest houses were able to offer the required accommodation with the more experienced SP taking an interest in this problem by contributing suggestions of a number of places which she thought may be suitable. The first call was made in line 101 but this was unsuccessful and the problem was marked by SP1's use of 'oh dear' in line 102, extract (13) below.

Call 2 is then made and a formulation triggers Stage 2 of 'Troubles-Telling' where the customer takes on the role of 'troubles-recipient' at the interface beweeen the call and the face-to-face interaction at line 117. SP1 breaks the news that there is no television but the formulation serves as 'resolution potential' with the customer now having the opportunity to negotiate requirements. In line 121 SP1 offers an account to the guest house proprietor of what is happening and the customer returns with her answer, a rejection, in line 126.

(13)

```
100 SP1: alright are you ready to ((unintell))
101 C:                      [ah ah!
         ((Call 1- no answer))
102 SP1: oh dear they're not in↓erm (4.0)I have it written down but
         I'll try Mrs. Philip?(2.5)( )
104 SP2: yes it's a good idea.((gap))
105 SP1: 6699479

        Call 2
        ((gap))
108 SP1: oh good afternoon this is the Tourist Information Centre(.)
109      erm I'd like to book a room for(.)someone if I may.(.)if
110      you coul- it's for today(2.2)have you got? ↑yes have you
111      got any free?(2.5)yes it's a double room(.)they would like
112      a double room er::rm preferabl- preferably with a a
113      television in the room is that possible?(2.0)haven't you.
114      (1.0)alright but d-do you have a double room erm(.)a double
115      room ↓en-suite(4.0)↑YES(2.0)you've got(4.)O.K. so you would
116      have a double room er-en-suite. I'll just check that(0.5)
117 SP1: erm these people do have a a double room en-suite but
118      no television.
119 C:   I'll just go and ask=
```

```
120 R:    =O.K.(.)
121       ((Call 2 continues)) the lady's just going to check(.)won't
122       be a moment(.)
123 SP1:  erm no there's no television ↑there er(3.0)
124 SP2:  Mrs. Fuller might be able to help(2.0)if erm (1.5) Mrs.
125       Cheese can't (1.2)she's got a ( )
126 C:    no he wants a television
127 SP1:  he does alright
128 SP1:  ((Call 2 continues)) erm no th- they
129       want a television(1.0)alright thank you very much
130       bye↓bye (.)
```

After a further unsuccessful call in line 136, continued discussion takes place between S1 and SP2 and it is significant that in the extract given below (14), SP1 says 'don't worry!' in line 155. This is addressed to the customer who is standing at the counter showing affiliation with her. Call 4 is made in line 158 and appears successful. SP1 indicates her anxiety, however, by asking for confirmation that the booking will 'not be a problem' in line 163, a clear indication that she now perceives herself to be a 'problem-solver'. In line 168 below she provides a formulation which functions in two ways; as final confirmation of booking, and to re-establish SP1's role as service-provider in her institutional role.

(14)

```
155 SP1:  ((laugh))↓don't worry! ((laugh))right so who's this Mrs
156       Peck (1.5)740(2.0)360((gap))
157       ((Call 4))
158       oh good afternoon is that Mrs Peck? this is the
159       Tourist Information Centre at Ely(1.0)er::rm we
160       have a lady that would like to stay with you this
161       evening if you er(.) have a double room(.)available.
162       (1.2)it is a double er en-suite. can you do that?(1.0)
163       that's not a problem.(.)fine and you have a television
164       in the room do you?you do.(.)erm would you hold on a
165       moment?er oh can you tell me confirm the price of the
166       room.(0.8)it's th=
167 SP1:  =thatlady has a double room en-suite with a television
168       and it's thirteen pounds per person is that
169       [alright?]
170 C:    [oh yes!] ((laugh))
171 SP1:  ((laugh))
172       ((Call 4 continues))
173       yes I think that's fine!(0.5)O.K. er whe-will
```

Following this confirmation of booking, there are two embedded sequences; one concerning the time of arrival in lines 174–179, and the other, the clarification of customers' names in lines.

The first embedding is shown in extract (15) below.

(15)

```
174 SP1: when will she expect you?
175 C:   oh I don't know how long it takes to get th[ere.
176 SP1:                                            [oh that
177      that's true.you're going there now are you more or less.
         ((Call 4 continues))
178      er she'll be with you within twenty minutes or so I
179      expect.alright?
```

The customer again appears to be handing back decision control to SP1 with the indirect 'appeal for help' in line 175 and investing the SP1 with the role of 'helper' once more. This dispreferred response brings with it an implied understanding that they will go straight away. In response to this, SP1 offers further opportunity to negotiate the time of arrival by her use of 'more or less' in line 177 to which there is no verbal response. This uncertainty of response is reflected in line 179 where SP1 uses vague language in 'or so I expect' as she has not actually checked with the customer whether this timing will be suitable.

The second of these embedded sequences relates to taking of names a shown in extract (16). Here SP1 attempts to read the name of the customer has written on the booking form. A direct appeal for help is made in line 181 and this forms the first part of a side sequence with the second part found in turn 182 with customer's preferred response. The name is repeated in line 183 and forms the third part of the sequence together with an information check before eventual return to the phone call. The use of the customers' Titles, and Last Name may have been designed to avoid face loss and reinstated SP1's institutional role of 'expert'.

(16)

```
180 SP1  (1.5) the the name(0.5)er::rm
181      how do I how do I pronounce it?=
182 C:   =Joubert=
         ((Call 4 continues))
183      =Joubert(3.0)alright?
184      it's Mr. and Mrs. Joubert.(.) alright?thank you
185      very much.bye bye.
186 SP1: ↓right so that's erm(1.5) you're going to stay there with
         Mrs. Peck O.K?
188 C:   alright.
```

These two embedded sequences precede the final formulation shown in line 186. This outlines the upshot of the talk and confirms customers' accommodation and while doing so reaffirms SP1's institutional role as service provider.

8. Summary and conclusion

By using a case study approach towards three bed-booking conversations it has been possible to track the locations at which formulations and accounts play a role in 'problem-solving' and related negotiation work. The analysis has shown that these features are used as discourse material of, and for, change (Firth 1995) at the interface of call closure and continuing face-to-face interaction i.e. at Transitional Relevance Points and function in ways that help accomplish the task goal by realigning needs and orientations.

With regard to other concerns of this study, namely, the changing 'stance' of negotiators, in this case, Customer and Service Provider, and the location of any changes in negotiators' perspectives, it has been demonstrated that bed-bookings differ in the degree to which they might be termed 'troubles-tellings' and 'problem-solving' encounters. One of the interests of this study has been to identify the location at which interlocutors' perceptions of the ongoing talk changes from being viewed as a 'troubles-telling' to one of 'problem-solving'. From this analysis it appears that the Service Provider starts to view the encounter as a problem-solving task only where difficulties arise when seeking suitable accommodation. For the customer, however, it may continue to be viewed as a trouble in that their role changes to one of 'troubles recipient' when problems arise. It is suggested that it is at this stage in the interaction a need is set up to realign and negotiate task goals. One way in which opportunities for negotiation are provided is through SP's use of account-giving and formulations which help towards the achievement of complementarity of purpose and attainment of the task goal of bed-booking.

Negotiation of roles also appears when customers make initial requests and it seems that the primary purpose of these initial requests are, firstly, to establish interactants' personal and institutional identities and, secondly, to indicate participants' respective levels of 'institutional acculturation' (Iacobucci 1990) or understandings of how the task of bed-booking should carried out. It is proposed that from these findings that the function of negotiating roles differs according to interlocutors' perceptions of whether the interaction is of a 'troubles-telling' or 'problem-solving' nature. In the former case, the function it performs is to establish initial role-relationships between customer/s and service provider while in the latter it acts as a catalyst to negotiating task goal where problems arise.

It is suggested that the negotiation of role can assist in realignment of task goal at points in the conversations where anxiety and concern arises especially at Stage 2 of 'Troubles-Telling'; when there are problems finding suitable accommodation as shown in these extracts. The onset of this anxiety takes place at the stage in the conversation when the service provider has made a number of unsuccessful phone calls to guest houses. In these cases service providers seek to reassure the customer who has been an 'interested party' listening in to the calls and this is done by providing affiliative responses and account-giving.

It is proposed that the ongoing talk consists of 'links' located at the interface between the bed-booking attempts and the face-to-face encounter and often involve the change in SP/C perceptions of the talk-in-action from one seen as 'Troubles-Telling' to one of 'Problem-Solving'. These links function both as 'calming' or 'tempering' measures and demonstrate the service provider's solidarity of purpose through affiliative responses as well as providing opportunities for the interlocutors to realign participant goals. It is further suggested that an additional way in which this takes place is by the SP's use of formulations and account-giving of the phone calls made to the guest house. A consequence of this is the creation of an opportunity for the SP to return to the customer a degree of decisional control lost through the making of unsuccessful calls and in so doing provides opportunities to realign or negotiate task goals.

These data have shown that accounts have played a central part in providing opportunities for negotiation of participant roles and acts of solidarity to take place. In cases where neither occur the Customer was found to self-select. The functions of self selecting include (a) asking for clarification of the upshot of talk so far (Extract 4 line 104) (b) commenting on what s/he perceived the outcome of the calls to have been (Extract 9 line 95) and (c) demonstrating an understanding of this outcome in terms of implications for the fulfilment of the task in hand i.e. to negotiate goal outcome and realign orientations to the task in hand, bed-booking (Extract 11 line 128). This also acted as useful material for negotiating role relationships and often included relational talk which, demanded affiliative work on behalf of the Service Provider. Additionally, self-selection at that stage in the service encounter, allowed clients to topicalise their knowledge of the outcomes of preceding phone calls despite the fact that neither the client nor service provider had made any explicit reference to them. In this way, it is suggested that the client not only indicates knowledge about the talk-so-far but, more specifically, how the 'possible trouble' may have developed into a 'possible obstacle' (Jefferson & Lee 1981).

One might propose that in terms of function, the strategy of offering these ongoing links between the information-holder and information-seeker bears resemblance to those used by helpline operators referring back to customers as they check information on their computer screens. (One major difference, perhaps being, that in these conversations the computer operator often informs the caller of what s/he is doing online as this is 'unseen', 'unshared knowledge'.) This new knowledge is then topicalised and has the potential for providing opportunities for the caller to regain control and realign task goals. In these bed-booking encounters, however, the degree to which the conversations between the service provider and the guest house proprietors are shared, or 'part-owned', with the customer/s themselves has been shown to differ. It is suggested that the degree to which customers see themselves as co-owners of the phone conversation while listening in to it, both reflects upon, and through, their role as 'troubles-recipient' or 'help-seeker'.

Despite this study's limitations in terms of amount of data and areas of interest it is, nevertheless, proposed that this paper adds support to the view that formulaitons are 'socially produced phenomena in their own right' (Psathas 1979:50) and worthy of further research in respect of different types of negotiation discourse. To conclude, helpful links are provided by formulations in service encounters where customers are often required to wait while calls are being made.Further investigation into other types of situations in which similar activity takes place may be useful in ascertaining how customers' needs are best served; for example, where customers anxiously wait at service counters while calls are being made to other establishments, or where they need to 'hold' or 'bear with' the Service Provider in phone-based calls. In both these examples, it would appear that more detailed research into the role played by formulations and accounts would both inform practice and hold wider implications for the training of Service Providers in a variety of institutional settings.

Note

1. In this study nationalities of participants include French, American and Scottish. The institutional, societal and cultural roles played by the Tourist Information Centre and their employees in different communities cannot be assumed. Empirically-based comparative investigation into their role may help assist observer interpretation of this particular talk-in-action. The same could be said for 'secretary roles'.

Appendix: Transcription notation

they:	Underlined fragments indicate speaker emphasis.
(1.0)	The number in brackets indicates a time gap in tenths of a second.
(.)	A dot enclosed in a bracket indicates a pause in the talk of less than two-tenths of a second.
=	The 'equals' sign indicates 'latching' between utterances.
[ ]	Square brackets between adjacent lines of concurrent speech indicate the onset and end of a spate of overlapping talk.
(( ))	A description enclosed in a double bracket indicates a non-verbal activity. Alternatively double brackets may enclose the transcriber's comments on contextual or other features.
?	A question mark indicates a rising inflection. It does not necessarily indicate a question.
( )	Empty parentheses indicate the presence of an unclear fragment on the tape.
.	A full stop indicates a stopping fall in tone. It does not necessarily indicate the end of a sentence.
:	Colons indicate that the speaker has stretched the preceding sound or letter. The more colons the greater the extent of the stretching.
↑↓	Pointed arrows indicate a marked falling or rising intonational shift. They are placed immediately before the onset of the shift.
!	Exclamation marks are used to indicate an animated or emphatic tone.
.hh	A dot before an 'h' indicates speaker in-breath. The more h's, the longer the in-breath.
CAPITALS	Words in capitals mark a section of speech noticeably louder than that surrounding it.

References

Chappell, D. (1993). 'Interactional Opportunities for Language Learning: A Comparative Study of Classroom-Based and Authentic Hotel and Travel Service Encounters'. Unpublished Ph.D. Thesis, University of Southampton.

Firth, A. (1995). 'Accounts in negotiation discourse: A single case analysis'. *Journal of Pragmatics, 23*, 199–226.

Heritage, J. C. & Watson, D. R. (1979). 'Formulations as conversational objects'. In G. Psathas (Ed.), *Everyday Language*. New York: Irvington.

Hutchby, I. & Wooffitt, R. (1998). *Conversation Analysis: Principles, Practice and Applications.* Cambridge: Polity Press.

Iacobucci, C. (1990). 'Accounts, formulations and goal attainment strategies in service encounters'. In K. Tracy & N. Coupland (Eds.), *Multiple Goals in Discourse*. Clevedon: Multilingual Matters.

Jefferson, G. & Lee, J. (1981). 'The rejection of advice: Managing the problematic convergence of a 'troubles-telling' and a 'service encounter'. *Journal of Pragmatics, 5*, 399–422.

Jefferson, G. (1987). 'On exposed and embedded correction in conversation'. In G. Button & J. R. E. Lee (Eds.), *Talk and Social Organisation.* Clevedon: Multilingual Matters.

Leppanen, V. (2000). 'Callers presentations of problems in telephone calls to Swedish primary care'. Paper given at *Symposium 'Calling for Help'*. Aalborg, Denmark, September 2000.

Psathas, G. (Ed.). (1979). *Everyday Language.* New York: Irvington.

Sacks, H., Schegloff, E., & Jefferson, G. (1974). 'A simplest systematics for the organisation of turn-taking for conversation'. *Language, 50*, 696–735.

CHAPTER 11

Institutionality at issue

The helpline call as a 'language game'

Brian Torode

1. Introduction: when institutionality is at issue

In a recent paper (Hester & Francis 2000a) which has provoked further debate (Hester & Francis 2000b; Watson 2000), Stephen Hester and David Francis argue that conversation analysis (hereafter CA) has neglected the specifically institutional character of talk which it analyses, and that while recent CA has attempted to remedy this defect, it has done so in an in inappropriately "linear" manner, seeking to specify distinctive features of turn-taking organisation which constitute certain talk as recognisably "institutional". Instead, Hester and Francis (Hester & Francis 2000a: 405) propose "a reflexive circumstantial mix of analytical concerns" which arises in such a setting.

Sand, gravel, cement, air and water each contribute to the strength of concrete, but we cannot easily identify the separate contribution of each. Hester and Francis similarly suggest that institutional talk is a synthetic product which resists analytic division into distinct components. They particularly resist any attempt to isolate *sequence* from other components, and treat it as privileged.

In contrast to this "concrete mix" approach, I shall show that the organisation of talk and texts into institutionally specific *language games* (Garfinkel 1967: 140ff.; Pitkin 1972; Wittgenstein 1953; Wittgenstein 1958) has priority over other aspects of "institutional" interaction. Empirically, I suggest that a study of institutional talk from an ethnomethodological point of view should seek talk whose institutionality is explicitly warranted by its participants. This requirement may not be best met by contexts where institutionality is securely established and taken for granted. In routine medical consultations (the source of examples (1) to (3) cited by Hester & Francis at pp. 398–401) "institutionality" is not topicalised nor in any way problematic in transactions which focus on pressing personal health issues.[1] Neither, arguably, is it best met by conversations taken from the midst of extended series of talk (e.g example (1) is taken from interaction between a health visitor and the mother of a two week old child, who the authors themselves suggest will have experienced a series of routine encounters with medical personnel throughout her pregnancy and birth; similar considerations arise in examples (2) and (3)). Explicit discussion, and perhaps negotiation, around the institutional context of the interaction is likely to have taken place early in the series and to have been long resolved by two weeks' after the birth.[2]

The parties to a helpline call typically have no prior or subsequent transactions: their entire relationship is constituted there and then within the call. Furthermore, unlike medical consultations, which have been written about since ancient times, telephone helplines are a recent innovation. Callers may be unclear what if any "official" status, professional training, or authority the call taker may possess, or may be prepared to invoke,[3] and as to precisely what – if any – direct assistance with their problem the line may be able to provide.[4] Finally, the helpline adviser may not restrict him or herself to the "professional" advice in which he or she has been explicitly trained. As Sacks (1992: II, 391–395) pointed out, agents of a public authority have an interest in offering advice which members of the public can accept,: anticipating this, Callers can exercise leverage to gain acceptable advice. In pursuit of this, Caller ("CR") requests and Helper ("HR") offers "everyday" advice as to what Caller should "do" about their problem.

2. Language games in Wittgenstein and Garfinkel

The notion of a "language game" was developed by Wittgenstein in the 1930s, but not published until two decades later (Wittgenstein 1958). He proposes (1953: §43, p. 20e) that "for a large class of cases … the meaning of a word is its use in the language". To investigate such uses, he discusses "those games by means of which children learn their native language". Extending the notion to "speak of a primitive language as a language game", he refers to both "language and the actions into which it is woven" as a "language game" (ibid.: §7, p. 5e). The concept seems to be a flexible one and Wittgenstein famously resisted defining it (ibid.: §3, p. 3e), regarding "definition" itself as merely one language game among many (ibid.: §27, p. 13e), which should not be privileged (Wittgenstein 1958: 1ff.).

Other writers are, however, willing to define it. Surprisingly, one of these is Harold Garfinkel, a writer who has vigorously questioned the adequacy of formal definitions both in professional social science and in ordinary everyday life (Garfinkel & Sacks 1970). Garfinkel (1967: 140–141) proposes nine formal features of games. He does so in part in order to identify situations which do *not* conform to the game definition. But he describes many occasions, both simple and complex, which *do* constitute "game-structured episodes" or "game-like" occasions. Briefly, his first six properties are as follows:

1. Definite game completion procedures are known to each player
2. It is always possible to "leave the game"
3. Serious life is suspended to participate in the artificial world of the game
4. Mutual biographies are specific to the game, including repeated plays of the game
5. Each single play of the game is an encapsulated episode
6. Success and failure are decidable within the play of the game, not by outsiders at later times

We may summarise these in two rules:

I. **Each play of the game is episodic, having spatial and temporal boundaries which define a precise beginning (entry point) and end (exit point).**

This incorporates rules 1, 2, 3, 5 above.

II. **Success and failure are determinable within each play of the game, and across a series of plays.**

This incorporates rules 4 and 6.

Garfinkel developed this definition in order to discuss the actions of Agnes, a transsexual person raised as a male, but passing as an adult female when he interviewed her in 1958. The game model accounted for many specific situations which Agnes recounted in interviews, but other occasions failed to satisfy the model.

> One type of such an occasion occurred … frequently: Agnes, by acting in the manner of a "secret apprentice" would learn, as she told it, "to act like a lady".
> (Garfinkel 1967: 146)

For instance her boyfriend Bill would give her "long lectures" on occasions that she or another woman did something he disapproved of. As Garfinkel puts it,

> Agnes was required to live up to … standards of conduct, appearance, skills, feelings, motives and aspirations while simultaneously learning what these

> standards were. ... They had to be learned in situations in which she was treated by others as knowing them in the first place as a matter of course.
> (ibid.: 147)

Both rules broke down here, Garfinkel suggests, because (I) there was no entry or exit point: the task which Agnes had set herself, of "acting like a lady", was lifelong, with no time-out; and (II) success or failure were not clearly determinable.

For Garfinkel, Agnes' project is one of "continuous development" towards being "natural and normal" in her new life in which she will not be "playing games" in a pejorative sense with her boyfriend or with others. Yet like a teenager on a date,[5] or a novice practising the piano,[6] she can only advance this aspiration by finite steps each constituting a specific episode, exhibiting game-like properties of its own. Likewise Garfinkel can only account for her action by narrating distinct stories, each of which itself carries game-like qualities. Rather than escaping the game model, Garfinkel's delicate descriptions enrich our understanding of the resource which that model provides for us, but also the inescapable hold which it has over our activities and the accounts we give of them.

3. Big and little language games in call 01

Consider the following complaint call made to the Office of Consumer Affairs, Dublin, ["OCA"].[7]

Language games in a Consumer Complaint narrative

The call concerns a pair of Bulldog boots which allegedly wore out in four months. (All names of persons and products have been altered.) The call opens as follows:

line *G=Game; P=Play*

Extract 1a. Call Opening as Caller's Story – Preface

0	`CR: (ring)`	G1	P1
1	`HR: He-low(.)Consumer Affairs`	G1	P1
2	`CR: Hell↑o erm my name is er Jenny Wise and`	G1	P2
3	`CR: I was just wondering what are my rights`	G2	P1

Lines 0–2, comprise the call *Opening* game (**G1**),[8] made up of two matching plays (**P1** and **P2**). In **P1**, Adviser pronounces "Hello" and a self-identification.

In **P2**, Caller reciprocates in the same terms.[9] Though they match, these moves are not equal: an institutional/everyday distinction is made by the fact that Helper offers an impersonal organisational identification, whereas Caller names herself.

At line 3, CR identifies her reason for call as a question about her "rights", which initiates her complaint narrative, and which she reprises to conclude that narrative at line 26, below. In so doing she anticipates terms in which HR may be able to advise her. She ascribes knowledge to him of this matter, whereas naturally she claims knowledge of the circumstances of her complaint. "Rights" are a legal object, and the Office of Consumer Affairs (OCA) is a quasi-legal institution – and indeed, later, HR shows himself quite willing to pronounce on this topic. Hence we identify the *Rights* game as **G2**, **played** at line 3 to begin the narrative, and **replayed** at lines 25–26, below, to end it.

In terms of the four-part structure of a narrative story telling identified by Harvey Sacks in lectures originally given in 1970 (Sacks 1992: II, 215–288)[10] **G2 P1** and **P2** respectively formulate the *Preface* and a candidate *Response* to the complaint story.

Extract 1b. Caller's Story – First

CR:	as regards a pair of(.)er(.)Bulldog boots	G3	P1
	I bought last September for my nine year		
	old (0.2)		
CR:	and er in October the:(.)sewing (0.2)	G3	P1
	a:ll (0.2) came=apart and er		
CR:	I got them redone in the (0.5) er	G3	P1
	cobblers > we'll say <		

At line 4, Caller mentions her purchase of a product, and *begins* to document her specific complaint. She does not detail any period of satisfaction with the shoes but describes a problem which arose in the month after she bought them. However she took action to resolve this herself, by way of the "cobblers". It may seem in retrospect that the problem was more serious than it appeared at the time and that by not complaining, Caller showed admirable self-reliance and restraint: practical *competence* to handle everyday problems appropriately and moral *credit* in not being an egregious complainant. I identify lines 4–10 as **P1** of **G3**, the *Cobbler's* game, on this basis. It is an everyday practical game, and its outcome is positive.

Extract 1c. Caller's Story – Then

CR:	and erm then in (0.2) January the soles	G4	P1
	came off them (.)		
CR:	now > I sent them back to the	G4	P1

14		manufacturers <		
15	CR:	but > I have received a letter back from	G4	P1
16		them <(.)saying that erm it wasn't a		
17		genuine cla:im		

At line 11, Caller's story takes a different turn. Three months later, the soles came off the boots. This time she did not approach the cobbler. (Later, lines 70–72 below, she is told that that is what she should do.) She "*sent them back to the manufacturers*" but to no avail. Lines 11–17 comprise **P1** of **G4,** the *Manufacturer's* game. Not only is the outcome of this game negative, but its method, an exchange of letters, is theoretical rather than practical. Whereas the cobbler did what she asked and repaired the boots, accepting a subservient status to herself, the manufacturer asserted a higher status, impersonally rejecting her claim as not genuine: Game 4 is an institutional game. In terms of Sacks' four-part structure, **G3** and **G4** respectively constitute the *First* and *Then* narrative steps of her complaint story proper.

Extract 1d. Caller's Story – Candidate Response

18	CR:	Now the boots are (0.1) cost me	G5	P1
19		twenty=three pounds(.)		
20	CR:	and er (.)I only had them for four	G5	P1
21		months(.)		
22	CR:	and as far as I'm concerned I don't think	G5	P1
23		I got (0.2) value for money		
24	HR:	Rri:ght hhh	G5	P1

Caller provides her own negative assessment of the situation, calculating that she has spent twenty-three pounds for four months' wear.[11] This is play 1 of her *Value* game, **G5**, acknowledged by Helper at line 24. It is her rejoinder to *Manufacturer's* impersonal game, **G4**, but institutionally subservient to it, being expressed as a personal opinion ("as far as I'm concerned").

Thus we see a hierarchy of institutionality in the language games played so far:

INSTITUTIONAL	*Manufacturer*	>	*Value*	>	*Cobbler*	EVERYDAY
	G4		**G5**		**G3**	

Extract 1e. Caller's Story – Response Request

25	CR:	and > I was just wondering <(.)what are	G2	P2
26		my rights		
27	CR:	or what (0.2)	G2	P2
28		[can I do about it?]		

CR invites HR to cap the institutional pecking order by means of a game whose authority will surpass the others and which, she hopes, will rule in her favour. This is play 2 of the *Rights* game, **G2**, with which she introduced her complaint.

INSTITUTIONAL	*Rights*	>	*Manufacturer*	>	*Value*	>	*Cobbler*	EVERYDAY
	G2		**G4**		**G5**		**G3**	

Insertion sequence

In an insertion sequence, Helper takes issue with Caller's account, and especially her claim, above, that she returned the boots directly to the "manufacturer". This elicits a revision of her narrative, according to which she returned them to the shop (also identified as "the store") who returned them to the "wholesaler". The issues are rehearsed several times, one of which is as follows:

(... lines omitted)

Extract 2a

49	HR:	your comeback is against the shop (0.1)	G6	P3
50	HR:	if you purchase goods that turn out to be	G6	P3
51		faulty widin a short period of time		
52	HR:	hhh so you would have to (0.2) go through	G6	P3
53		the sto:re (.) Okay?		
54	CR:	well I(.)I returned the boots to	G6	P2
55		[the store]		
56	HR:	[to them] (.)yeah	G6	P2
57	CR:	and they returned them to the (0.1) to	G4	P5
58		the wholesaler		
59	HR:	okay [e r]	G4	P5

Consider the effect of this revision on the hierarchy of language games which is in process of being constructed by both parties. At HR's insistence, CR has revised her account to say that she dealt with the store, who dealt with the wholesaler. Thus the new *Shop* Game 6 is interpolated between her own assessment and that of the Manufacturer, now reformulated as Wholesaler, Game 4:

INSTITU-	*Rights*	>	*Manufacturer*	>	*Shop*	>	*Value*	>	*Cobbler*	EVERYDAY
TIONAL	**G2**		**G4**		**G6**		**G5**		**G3**	

Game 4 is now replayed in an enhanced version which provides grounds for the rejection of the claim:

Extract 2b

60	HR:	[they] got (0.1) a letter back	G4	P5
61		(0.1) > which I have here in my hand <		
62		(0.1)		

63	HR:	and in their opinion they erm they said	G4	P5
64		that they received a		
65		considerable=amount=of=wear (0.1)	ΦG5	ΦP2
66	CR:	and under ↑ no ↑ circumstance=stances	G4	P5
67		could they consider them to be a genuine		
68		claim.		
69	HR:	O(.)K(.)	G4	P5
70	CR:	and they said that I was to send them	G3	P2
71		back to the shoe-menders and get them re-		
72		soled		
73	HR:	mmh hm=	G3	P2

(The symbol Φ indicates a theme later picked up by a different game.)

The manufacturer advocates that Caller repeat her earlier recourse to the cobbler.

4. Big game, play one

Caller has set out her situation plainly in terms of the language games which have been repeatedly replayed in the call so far. In general terms, she represents her dispute with the Manufacturer/Wholesaler as her reason for calling the OCA. More precisely, she invites the OCA adviser to assert the institutional hegemony of what she anticipates to be his *Rights* game, **G2**, atop the hegemony which the *Manufacturer's* game, **G4**, has asserted over her own *Value* game, **G5**. According to **G4**, her boots problem is not an institutional problem at all but an everyday one, and she should solve it practically as she has done before by recourse to **G3**, the *Cobbler's* game.

Thus the configuration of Little Games is positioned by the conversation so far into a single Big Game, that of INSTITUTIONALITY versus EVERYDAYNESS. The protagonists in the big game are Manufacturer and Caller who are themselves aligned as institutional: everyday, relatively speaking. Manufacturer has recruited Cobbler in support of his case that the matter is open to an everyday solution. To be more precise, the *Manufacturer* Game **G4** exit strategy from the Big Game is an everyday exit via the *Cobbler* Game **G3**. To counter this, Caller's *Value* Game **G5** pursues an INSTITUTIONAL exit from the Big Game via the *Rights* Game **G2**.

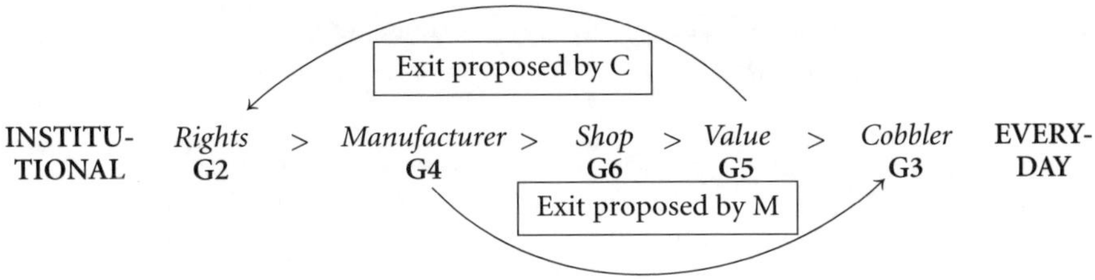

Figure 1. Big game, play one, in progress

Extract 3

CR:	now(.)er (0.2) four(.)months(.)wear(.)you	ΓG5	ΓP2
	know > for a nine=year=old	ΦG8	ΦP1
CR:	< the soles gone	G5	P2
CR:	and(.)they're leaking and=every>thing<	G5	P2
CR:	=I don't think is good (0.5)	G5	P2
	d'you know?		
HR:	OK	G5	P2
HR:	well I'll explain what rights you may	G2	P4
	have in this partic'lar case		

(The symbol Γ indicates a theme earlier addressed in a different game; the symbol Φ indicates a theme later picked up by a different game.)

In **G5 P2**, CR reiterates her own stance in direct response to **G4 P5**, elicited from her by HR's questioning. Her previous value calculation (lines 18–21) was "twenty three pounds for four months". Now she evaluates "four months wear for a nine year old" as not good in direct challenge to Manufacturer's evaluation of a "considerable=amount=of=wear", inviting HR to agree. But rather than directly participating in this game by answering the question, HR promises a response to CR's initial question regarding "Rights".

The Sale of Goods Act and the "short length of time"

In a scripted account (compare that given by Helper as reported in Torode 1995), Helper outlines the provisions of the Sale of Goods Act (**G7**) and, informed by this, returns to the rights issue (**G2**):

(... lines omitted)

Extract 4a

CR:	Now(.)the general rule is(.)	G2	P5
HR:	if the fault is minor(.)		
HR:	your rights(.)are to a=repair		
HR:	(.)hhh where if the fault was major (0.1)		

```
105        > widin a short length of time since purchas:e <
106
107 HR: hhh you could have a right then to
108        a=replacement or to a=refund of money=
```

We have suggested that the EVERYDAY solution to Caller's problem is via the Cobbler: this is encompassed by "repair" at line 103 (though perhaps if fault was accepted, the store would carry the cost of the shoe-mender's work). The alternative "replacement" or "refund" alternatives amount to INSTITUTIONAL remedies. In this sense the two alternatives which arose in the interplay of five games (**G2, G3, G4, G5** and **G6**) are reiterated by HR, and posed as soluble, within **G2**, the *Rights* game, alone.

Extract 4b

```
109 CR: =well four(.)months wha=what would that          G2    P6
110        be?
111 HR: OK=                                              G2    P6
112 CR: =it isn't four months [ (.) is it? ]             G2    P6
```

At lines 109–112, Caller requests calibration of her case in terms of the rule which HR has enunciated. That is, having already received a general opinion on the institutional issue of "rights" – Play 5 of **G2** – her request initiated a more specifically focussed Play 6 of the game. To address this, HR explicitly animates the issue of "wear and tear" as Game 8 in its own right.

Extract 4c

```
113 HR:                    [ h h h     ] well            ΓG8   ΓP1
114        w=what you would have to do here
115        like(.)de short lengt' of time takes into
116        'count the volume of wear(.)and(.)tear
117        hhh the lifespan of the item hhh
```

(The symbol Γ indicates a theme earlier addressed in a different game.)

However this issue is already controversial. In **G4 P5** Manufacturer asserted that wear and tear was "considerable", i.e. too high, whereas in **G5 P2**, Caller asserted that "*four*(.)*months*(.)*wear*" was not good, i.e. too low. HR now launches another language game, **G9**, *Usage*, to measure this, outlining two rival plays of this game, one (**P1**) low the other (**P2**) high. These two positions constitute a scale to be "taken=into='count" as an indication of wear and tear (**G8 P2**).

Extract 4d

```
118 HR: like for an=example one person might buy         G9    P1
119        a pair o'boots hhh and only use them
120        maybe once(.)a(.)fortnight [ hhh   ]
```

```
121 CR:                               [ Yeah ]                  G9    P1
122 HR: hhh where another person might use them                 G9    P2
123     maybe four(.)five(.)times per week
124 CR: Yeah                                                    G9    P2
125 HR: Erm(.)all that has to be                                G8    P2
126     taken=into='count
127     [a s t o ]
```

HR's calibration is technically precise. Of the two positions indicated, the *low* of once-a-fortnight would constitute 8 occasions in the four months; the *high* of four-times-a-week would amount to 64 occasions in the four months. An item used once-only would be off the bottom end of the scale.[12] These three points are powers of 8: $8^0 = 1$, $8^1 = 8$ and $8^2 = 64$.

Caller now submits her daughter's usage for calibration in these terms:

Extract 4e

```
128 CR: [they were] worn now going=to=school mm                 G9    P3
129     Monday=to=Friday (("continuous" sound))
```

Caller specifies five times weekly, amounting to 80 occasions in the four months, clearly off the top end of the scale. On this basis, HR responds:

Extract 4f

```
130 HR: right OK so (0.1) what (0.1) one would                  G8    P3
131     say then that(.)there has been a si=a
132     fair amount of wea:r and tea:r hhh you
133     know involved=
134 CR: =[[yeah ]                                               G8    P3
```

HR resolves the matter, in effect, by recourse to another, micro language game. His correction downgrades "significant amount" – tantamount to "considerable=amount=of=wear", lines 131–132 – to "*fair amount*" in a mini-game which contrasts "some" with "none". In effect he says there has been wear and tear, hence the short length of time is over, the fault if any is minor, and repair is the appropriate redress. He has endorsed the Manufacturer's case, **G4 P5**.

5. Replaying the big game

It is intriguing that whereas HR strongly asserted that Manufacturer has no role in assessing Caller's case (a view which he reiterates later in the call), when he himself assesses Caller's "rights" under the Sale of Goods Act he does so in

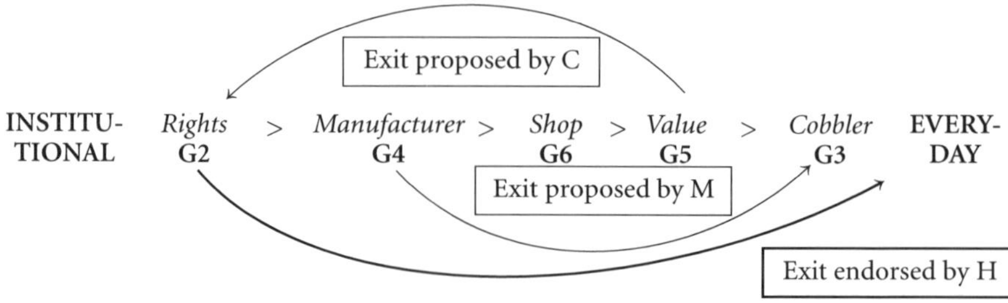

Figure 2. Big game, play one concluded by HR, who invites new play by CR[13]

the same terms, reaching the same result, as did Manufacturer. The language game is the same, and appears reliable in use by different operators.

Caller does not demur from this judgement. She had submitted the boots for inspection having made her own (negative) determination of their value. Rejected by one institutional source she has turned to another with the same result in the same terms. From her viewpoint, it might seem, there is little to chose between the old order (prior to 1980 manufacturers were responsible for faulty products, and were thus in a kind of "*loco parentis*" role in relation both to customer and store) and the new (since 1980, the customer has a contract with the store, to whom any complaint should be made: in the event of dispute, recourse should be made to the Sale of Goods Act (Government of Ireland 1980). A telephone call to the OCA helpline serves as an informal version of such recourse).

Caller provided her own 'calculations' of the value she had received, in language Game 5: "*twenty=three pounds(.)and er (.)I only had them for four months(.)*" (**G5 P1**) and "*four(.)months(.)wear(.)you know > for a nine=year=old*" (**G5 P2**). HR refuted these by reformulating calendar months in terms of wear and tear, and wear and tear in terms of usage, for which a clearly calibrated scale was devised, and in terms of which she herself supplied a precise measurement. At this point, it might seem, HR has not only declined to endorse Caller's case. He has positively *proven* the failure of her case, from the institutional point of view expressed by the language game of *Rights*. And indeed it is the case that the language game of *Rights* – which Caller initiated in line 3 in stating her reason for call, and in which HR has actively participated in detailed discussion of the provisions of the Sale of Goods Act (not quoted here) – makes no further appearance in this call. In its stead, HR launches a new initiative:

Extract 5a

```
135 HR: =[[hhh ] but you still are back to the         G5   P3
136     er difficulty
137 HR: > that you felt that they should have          G5   P3
138     lasted < maybe that(.)bit(.)
139     [ l o n g e r ]
140 CR: [oh certainly ]                                G5   P3
141     = you know                                     G5   P3
142 HR: and that=                                      G5   P3
```

We have seen that HR's *Rights* calculations effectively undermine CR's *Value* calculations and particularly her claim that four months was too little wear to get out of the boots. But despite having invalidated this claim in institutional terms, HR now goes out of his way to acknowledge it in other terms, as something Caller "felt" to be the case. He is crediting her value calculations with some kind of non-institutional, that is, everyday validity. Caller responds early and emphatically, line 140.

Although it refers to "feeling", a term new to this call, which effectively replaces the *Rights* game to which CR aspired, I suggest this is not a new language game so much as a revitalisation by HR of Caller's existing *Value* game, the core of her complaint, which otherwise he could have seemed to have killed off. Thus *feelings* play a similar part in Helper's discourse as *shoe-mender* (cobbler) played in manufacturer's discourse. These are *everyday* games which are recommended to clients by agents when *institutional* games are being withheld.[14] The professional agent thereby helps the client to build an everyday case, when appropriate to do so.

Caller responds strongly to the proposal. From this point on in the call, it is her strength of "feeling" which is her chief resource, according to HR, rather than the "rights" which she phoned to enquire about. Yet it would be misleading to assume this "feeling" is simply *subjective*. In further replays of her *Value* Game 5 she provides new historical documentation to restate her case *objectively*. This makes up a narrative story-telling but in reverse order:

Extract 5b. Caller's New Beginning – Now

```
143 CR: I(.)I(.)I(.)I(.)I(.)they were actually         G5   P4
144     recommended to me to get(.)
145     [you know what] I mean
146 HR: [ O k a y     ]                                G5   P4
```

Extract 5c. Caller's New Beginning – Then

```
147 CR: other than that I would=I was I was            G5   P5
148     always dealing with Bennem's shoes
```

```
149 HR: [and I always had]                                G5    P5
150 HR: [Y e a h         ]                                G5    P5
151 HR: ((telephone rings))
152 CR: better value > out=of=them <                      G5    P5
153 HR: Okay                                              G5    P5
```

In place of her earlier questionable *quantifications* ("23 for 4", "9 for 4") she here makes precise *qualitative* comparisons with (1) what was recently recommended to her (**P4**); and (2) her sustained ("always") prior personal experience with other shoes (**P5**).

6. The footing shift in the replay

The call goes through dramatic falls and rises but I will advance now to the final sequence. At line 168 (not shown here) HR recommended that CR return to store and present her case based on (i) her feeling that the deserved better value, as indicated above; (ii) the Sale of Goods Act; and (iii) negotiating with the store on this basis.[15] No particular result could be guaranteed. At line 210 (not shown here) CR questioned whether she should approach the manufacturer, but this course of action was categorically rejected by HR. At lines 229–231, she then sought explicit confirmation that she should return to store. When he affirmed this at lines 232–248, repeating the (i-ii-iii) sequence just outlined, she expressed doubt whether she would gain anything by doing so (lines 249–250). At lines 254–260, he advised that "if all that fails" she could consult a solicitor, but that this was not justified by the value of the goods in the present.

In the call up to line 134, Caller had sought institutional endorsement for her everyday complaint, which HR withheld.

When the matter was played out through the institutional games of *Rights*, **G2**; *Value*, **G5**; and *Usage*, **G9**, regarded as a decision-making mechanism (which same mechanism, as we noted, had already been played out by manufacturer), the outcome was negative. In short her everyday complaint received neither institutional endorsement nor redress.

From line 135 the boot is on the other foot. Helper offers Caller a way to pursue her complaint which falls short of demanding her "rights". In a sense Caller already prepared for this by asking (lines 27–28 above) not only "what are my rights" but also "what (0.2) can I do about it?". She may have envisaged that even with institutionally agreed "rights" she would still face the everyday problem of "what to do" to actualise those rights. She may not have envisaged

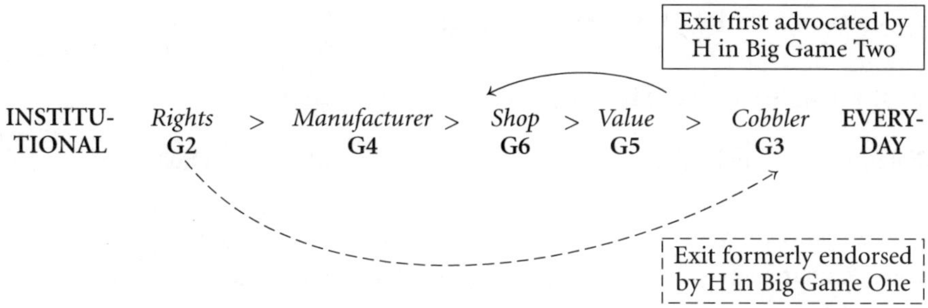

Figure 3. Big game, play two, in progress

that, even without "Rights", as determined by Helper at lines 130–133, there would still be things she could do.

But now it is her turn to withhold endorsement from his proposal that she act on her "feelings" in the absence of his endorsement of her "rights". In other terms, HR recommends that CR try to persuade the store (least institutional) of the merits of a case of which she has failed to persuade either the manufacturer (slightly institutional) or himself (in his fully institutional capacity).

Caller's new ending – Preface

Having expressed doubts, CR now apparently takes up his advice with enthusiasm.

Extract 6

```
261 CR: (0.2) right(.)so (0.2) erm I'll get back          G6   P10
262     onto the the (0.1) owners and I'll see
263     what they (0.2) [say]
264 HR:                 [see] what kind of                G6   P10
265     react[ion you get   ]
```

She launches a new play 10 of the *Shop* game **G6**, and is supported by him in doing so. The previous nine plays each ended negatively. Both CR and HR assure one another that options are open this time. This is recognisably the Precursor to a course of action. Otherwise stated, it serves as **Preface** to a story-telling, although one which, it would seem, cannot be told until a later date. In this story **Preface**, Caller's "try it and see" approach is promptly endorsed by Helper. But she has trapped him into this stance[16] perhaps in the expectation that, being an "arrangement" for future activity,[17] it will be call-closing implicative. But it is not – for she already has empirical evidence to report.

7. Institutionality at issue in a little game

Institutionality at issue in caller's new ending – First

Caller now reveals that she already did just what her **Preface** anticipated. Play 11 of **G6** enacts what occurred. In the prior play 9 of **G6** (lines 249–250, not shown here), Caller had *claimed* she would not get much from the store. Perhaps surprisingly, Helper humorously endorsed this view and they co-produced a negative exit from the game. Her new narrative now *proves* that claim in a serious sense.

Extract 7

```
266 CR:       [when I went in ] on Saturday I just            G6    P11
267      met the girl who=was=working=there
268      you know(.)and she said (0.1)
269 CR: well they're they're not going to give            μG6   μP11
270      you any erm(.)any refund she said(.)and
271      they're not going to do anything for you
272 HR: (.) O [ kay ]                                      μG6   μP11
```

(The symbol μ indicates the game which initially appears to be in play.)

Caller reveals a very recent visit to the store. Lines 269–271 report remarks by "*the girl who=was=working=there*". These remarks, in context, are apparently attributed to the store owners (lines 261–263 above) and institutionally *should* be authored by them, since they are the party responsible for any offer of redress under the Sale of Goods Act, according to Helper at lines 49–53 quoted above.) Consider the institutional: everyday contrast which is achieved in this narrative so far. In her **Preface**, Caller referred to "*the (0.1) owners*" and now she refers to "*the girl*". In relative terms, *they* – if they can be found – should speak with authority, whereas *she* can only express an everyday opinion.[18]

Legally, the store is responsible for offering redress for faulty product. Linguistically, Caller has distinguished the owners from the girl and now quotes the girl's remarks regarding what "*they*" won't do in respect of redress. Listener is led unavoidably to treat that quotation as referring to the store owners, in a continuation of play 11 of game 6. In this narration, the personae of the store have now been divided. Within this division, the girl is represented as a *confidante* of the lady Caller, to whom she has passed inside information – albeit serving the institution's interests, for its upshot is "resistance is futile" – and in this sense shown everyday solidarity with her against the institutional owners, whom Caller had planned to confront.

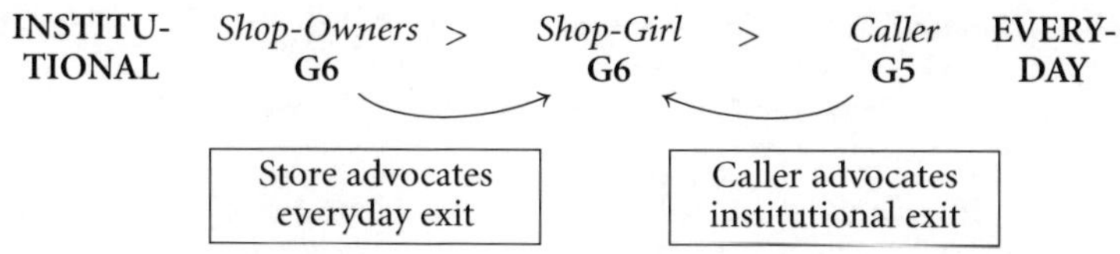

Figure 4. Apparent state of play between CR, shop-owners and shop-girl

Institutionality at issue (ii) in caller's new ending – Then

Caller's story continues at line 273, but because it throws new light on lines 269–271, I re-quote those lines first:

Extract 8

```
269 CR: well they're they're not going to give          μG6   μP11
270     you any erm(.)any refund she said(.)and
271     they're not going to do anything for you
272 HR: (.) O [ kay ]                                   μG6   μP11
                                                        λG4   λP8
273 CR:       [ you ] know and she didn't say            G6    P11
274     that they themselves would do anything or
275     (0.5) you ↑ know?
```

(The symbol μ indicates the game which initially appears to be in play. The symbol λ indicates the game which is later realised to be in play.)

At line 273 Caller, continuing her story, reports a non-continuation of the girl's remarks, i.e. what "*she didn't say*". This formulates a new membership category, "*they themselves*", who are reported as undertaking the exact same action as was attributed to "they" in lines 269–271, namely "not doing anything". The effect of the new report regarding "*they themselves*" (line 274) is to *displace* the earlier referent of "they". At the time of delivery, "they" apparently referred to the store owners, contrasted with the girl, their employee and Caller's confidante. But now "*they themselves*" refers to the store, including both owners and the girl, as contrasted with another party, also familiar enough to CR and HR to be identified simply as "they". This can only be the manufacturers, with whom by her own account – as reported earlier – CR has had direct dealings,[19] unlike the owners with whom she has – as reported here – dealt with only indirectly.

To repeat the point, HR's *institutional* perspective is that CR and MR (Manufacturer) have no business with one another: CR should deal with SR (Store). But CR, in what thereby becomes an *everyday* rebuttal of HR, claims to have dealt directly with MR. Further, she now claims that when she approaches SR,

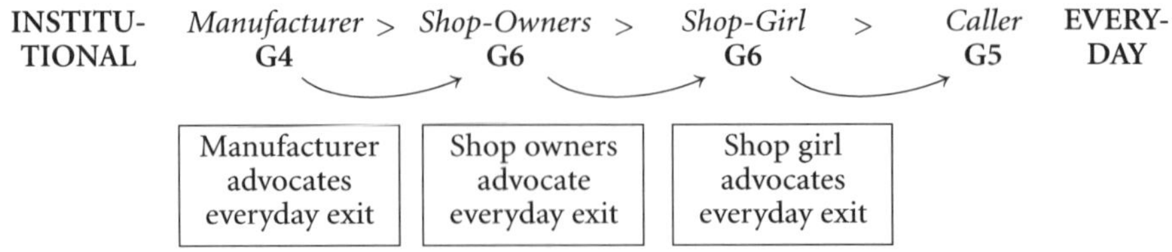

Figure 5. Revealed state of play between CR, Manfacturer, and Shop

they merely refer her to MR. The Manufacturers are now seen to be the ones referred to in lines 269 and 271. Hence that story component is seen to be not play 11 of *Store* game **G6** but play 8 of *Manufacturer* game **G4.** We now see that, contrary to their legal duty, the store are assigning responsibility for redress to the manufacturer, to whom a specific negative decision is attributed: "*they're not going to give you any erm* (.) *any refund*". The store themselves are merely "not saying they will do anything".

In this reformulation, "*they themselves*" includes the girl herself who has switched sides in CR's story. At first she seemed to side with Caller against the impersonal store owners: this account could conform to HR's legal formulation of responsibility in the case. But now she speaks for the store, a junior employee in solidarity with her senior employers who themselves are acting as a lower authority bolstered by the manufacturer as a higher authority against the unfortunate everyday consumer Caller. This hierarchical stacking of three institutional layers – one of which had initially appeared as Caller's personal protector against the other two – is quite contrary to the OCA view, according to which the manufacturer is not involved in the case, has no institutional status, and no higher standing than either the retailer or the customer/Caller.

Caller's new ending – Response

In response to this story, HR appeals to a higher power, the Sale of Goods Act.

Extract 9

```
276 HR: (1.0) yeah well(.)in some situations by           G7   P7
277     maybe discussing the matter(.)by quoting
278     the terms of the Sale O'Goods Act hhh
279     [they may then]
```

This is appropriate since the Act, attributing responsibility for faults to retailer rather than manufacturer, directly challenges the rationalities (games) which, Caller demonstrates, have been used against her. These rationalities (games) were 'authoritarian' in that the decisions to do nothing were attributed to parties to whom the customer had no direct access – the manufacturers, the store owners – while the party to whom the customer did have access (the girl) had no capacity to act in her own right. (As we shall shortly see, in proving his claim, by narrating how Caller should make her case in terms of the Act, HR demonstrates the play of a different game, in which all the relevant components are "on the table" in view and accessible to inspection and other manipulation by both parties to the dispute, namely Caller/Customer and Store.)

However in terms of the Big Game now unfolding he is in a delicate position. In play 1 of the Big Game he himself was in institutional authority over Caller, who must anxiously await his adjudication. But, having rejected her case in play 1 he is now trying to put together an unavoidably weaker case – a "second best", an "ugly double" rather than an "ideal" in Garfinkel and Sacks' (1970) terms, which he nonetheless wants both himself and herself to be able to endorse, in play 2. His weakness seems apparent in the pause, appreciation, and qualification with which he re-introduces the topic.

8. Institutionality at issue in the big game – helper's new ending story

Extract 10a. Helper's New Ending – Preface

```
280 CR: [well what ] ↑ is the Sale Of Goods             G7   P8
281     Act ( ) ?
```

Caller's question highlights Helper's weakness, i.e. the fact that now he has to account to her for the efficacy of his proposed remedy, the Sale of Goods Act language game, in terms of which he already defeated her case once in this conversation. However her question also provides him with an opportunity, an occasion to re-tell that story. The first time around, the *SOGA* game (**G7**) was invoked to answer Caller's *Rights* question (**G2**). Game 7 was made subservient to Game 2, and the final outcome was that, in this particular instance, in view of the *wear and tear* (Game 8) to which the boots had been subjected since purchase, the Caller did not have any remaining "rights" as defined by the Act.[20]

But now, second time around, the game is being played in the context of Caller's *Value* game (**G5**), as re-launched by CR at HR's invitation in terms of "feelings", from lines 135–139 (Play 3). This "output" of **G5** is "input" to **G2**, just as a subroutine result is fed into a computer program, in order to produce a case for redress. Thus *Redress* rather than *Rights* is now the name of the game.

Extract 10b. Helper's New Ending – First

Helper first establishes "quality" as the legal issue

```
282 HR: well(.)the the terms are that the goods        G7   P9
283     must be(.)fit for their purpose
284 CR: yes                                            G7   P9
284 HR: and of merchantable quality                    G7   P9
286 CR: yes                                            G7   P9
287 CR: eh that being that the boots must be up        G7   P9
288     to a reasonable quality
```

Helper's new ending – Then

Now he formulates Caller's "views" in terms of the "quality" issue. He imports, into his replay of the Sale of Goods Act (**G7 P9**) which is currently under way, proven results from Caller's own well-established "value" game (**G5 P5**, lines 147–153 above), now reiterated as **G5 P6.**

Extract 10c. Then-First

```
289 HR: and you obviously express your=views hhh       G5   P6
290     that they didn't hhh er meet that quality
```

He earlier introduced her "feelings" into this game (**G5 P3**, lines 135–142). He now reiterates this concept:

Extract 10d. Then-Then

```
291 HR: hh and= >=accordingly you feel you're          G5   P6
292     entitled to < some(.)type(.)of (.)redress
293 CR: Exac[t l y ]                                   G5   P6
```

CR's "Exactly" endorses this as a precise calculation, along the lines of her earlier "value" arithmetic, enabling the call to end on a collaboratively accomplished positive note. But even as she does so, HR backtracks away from the possible hearing of what they have agreed as an "exact entitlement".

Helper's lines 294–295 revert to an earlier Game 11 (which we have not discussed on this occasion), namely that of Negotiation, whose outcome is uncertain:

Extract 10e. Then-Response

```
294 HR:     [and see] will they concede(.)to that      G11  P3
295     point
296 CR: right (.)                                      G11  P3
297 CR: right (.)                                      G11  P3
```

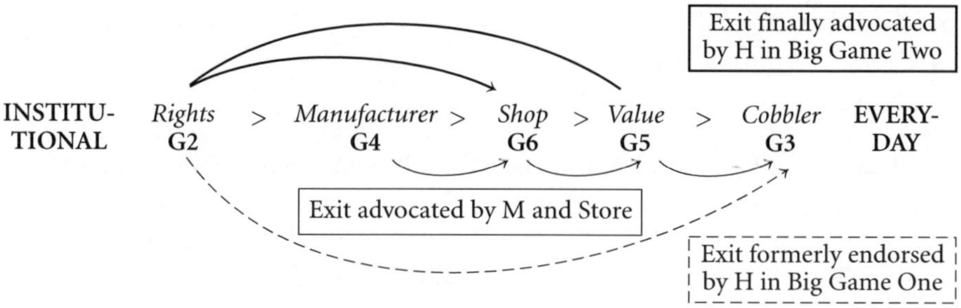

Figure 6. Big game, play two, concluded by CR endorsement, hence call closing

Call closing as caller's story response

There can be no guarantee that this initiative will be effective, but it represents a way forward which marries Caller's conviction of lost value and Helper's concept of entitlement to quality (though not the specific calculations completed by either of them) in an alliance which challenges that of Manufacturer and Store, and which apparently she is happy to accept.

Extract 10f

```
298 CR: (1.0) O↑kay ↑ so (0.2)                    G14  P1
299 HR: Ok↑ay?                                     G14  P1
300 CR: thank you ve[ry much ]                     G14  P2
301 HR:             [rightho ] then                G14  P2
302 CR: bye                                        G14  P3
303 HR: bye bye now                                G14  P3
```

Call completion in two game plays

In coining the neologism, "feeling entitled" (lines 291–292), Helper constructed a new interpolation between the institutional and the everyday, holding out the promise of a result which as he has previously demonstrated (lines 113–134, Games 8 and 9) the institutional approach *sui generis* had rejected, and which as Caller has just demonstrated (lines 269–275, Games 6 and 4) the everyday approach alone was incapable of achieving.

9. Conclusion

Sacks' agent/client game

The helpline call as a whole comprises a single game, played twice. The parameters of this game are indicated by Harvey Sacks' remarks on "Agent/Client Interaction", originally delivered in May 1971 (Sacks 1992: II, 391–395) in which he argued that – by virtue of the record-keeping typical in such organisations – in calls to a bureaucratic organization, agents are motivated not simply to offer advice, but to offer advice that clients will accept. Realizing this, clients can exercise "leverage" over agents to adjust the "disposition of the call". In 1978 Judy Davidson (1978) published a short study of negotiation in the *closing* of a helpline call, which in confirmation of this hypothesis, revealed the client's exercise of leverage in sequential structural terms. Repeatedly, Caller passed over Possible Call Completion points, by pausing, offering weak agreements, or seeking clarification. Davidson does not discuss what Caller achieved by these means.[21]

We have focussed attention on the upshot of advice-giving, and the way this upshot changes as a result of replaying (recycling) the language games, both big and small, which constitute the call. One fundamental change organises the whole call. In the first part of the call, Helper – at Caller's request – is assessing what *Rights* she may have in the case. His assessment is negative. Delivering this judgement in lines 130–133, Helper proceeds uninterruptedly, in overlap with Caller's immediate acknowledgement of it in line 134, to invite her to restate her case in terms of *Feelings*, which she does in an effective manner. The eventual upshot is a positive assessment by both parties of a case which CR can bring to store and which might – or might not – result in redress.

One game or many?

This has been an exploratory and indicative "Single Case Study". We have not discussed every detail of the call but what we have discussed, we have discussed precisely. Our purpose has been to show precisely how institutionality is at issue here. Following Wittgenstein (1953) and Garfinkel (1967: Chapter 5), we regarded the call as comprising many language games.[22] We found that the distinction between institutional and everyday talk arises in relations *between* little language games, e.g. in Caller's initial narration between the *Rights* game **G2** and the *Value* game **G5**, and between the *Manufacturer* game **G4** and the *Cobbler* game **G3**. "Rights" are not in an *absolute* sense institutional, but – as subsequently shown by the comparison between HR's precise calculations at lines 113–133 and CR's vague calculations at lines 18–23, 74–75, and 163–165 – they are *relatively* more institutional than "value".[23] Similarly, such a distinction arose *within* a little game, e.g. in the *Shop* game **G6** (lines 261–275 above), where the store-owners are represented as relatively more institutional (for instance in their remoteness, also in their capacity to offer or withhold redress in response to the complaint) than the shop-girl.

We further found that a Big Game emerged out of the construction and reconstruction, interpolation and extrapolation, of multiple little games, itself organised along the same Institutional/Everyday hierarchy. Its first complete play comprised 8 little games (**G2** to **G9**). Driven by Caller's questions regarding her "rights" at lines 3, 25–26, and 109–112, it pursued a relatively Institutional solution which was discovered to be negative. (A more Institutional solution, employing a Solicitor was subsequently mooted – at lines 251–260, not discussed here – but abandoned.) Its second complete play – driven by Helper's acknowledgement of Caller's "feelings" at lines 137–139 and 291–292 – utilised just 5 of these (**G2** to **G6**). It pursued a compromise Institutional-Everyday solution (based on the neologism, "feeling entitled") which both parties finally endorsed as positive in its evaluation, though uncertain as to its outcome.

The way in which the parties forge a single Big Game, in two plays, out of many little games, confirms Sacks' (1992: II, 391–395) Agent/Client hypothesis. Initially, the two parties define the situation differently – but finally they play the same game. For instance the *Manufacturer* game plays a central part in Caller's account throughout. It would play no part at all in Helper's account but for the fact that he is obliged to acknowledge it in order – in his terms – to correct Caller's account. To eliminate MR, HR must play the MR game. It seems that, interactionally, either party can set such games as they like in motion "on the table" in the clear expectation that the other party will play them, if only to spoil them.[24] As a result, the specific Big Game which emerges is unique to these participants on this particular occasion, even though the components out of which it is built – for instance the *SOGA* game (**G7**), *Shopping* game (**G6**), *wear and tear* game (**G8**), and so on – are familiar from other helpline calls, not to mention the *Opening* (**G1**) and *Closing* games (**G14**) which unavoidably arise in every conversation.

Recently Garfinkel (1992: 182f.) enunciated what he called the "unique adequacy" requirement for understanding practical social action in its specific context. We suggest, however, that any such unique accomplishment of this interaction at this particular time and place is built out of universal components whose properties can be analysed objectively. The work of the conversation analysts from Schegloff (1968) on demonstrates how this can be done. Conversational interaction comprises repeated replays of micro games-within-games, and macro games-atop-games, constituting as their accomplishments what we know as "everyday" and "institutional" social structures, such as "feelings", "entitlements", "rights", and "redress".

Notes

1. Hester and Francis' choice of examples is chiefly dictated by the data reproduced in the introduction to Drew and Heritage (1992), of which their article is a critique.

2. Another context in which institutionality may not be topicalised is the workplace. Lynch (1985) presents data to show that precise technical calibration of laboratory slides is conducted by technicians by means of informal talk including casual chat, swearing, and ribald humour. Kleifgen and Frenz-Belkin (1997) show experienced workers performing painstaking investigation of, and adjustment to, a Surface Mount Technology machine. Apart from the concocted word "Mispick!" designating a machine misbehaviour, the entire conversation is conducted by means of numbers, concluding with a rendition of elementary arithmetic by the senior technician: "Twenty four divided by number eight is three. Yeah, it divides into three."

3. A feature of the tragically unsuccessful Emergency call investigated by Whalen, Zimmerman, and Whalen (1988) was that Caller was passed between separate call taking agents (Desk Operator, Nurse, Supervisor) whose distinct responsibilities, competences and relevancies to his concern were unclear to him.

4. In some circumstances the Dublin Office of Consumer Affairs, discussed below, may be willing to pursue an aspect of a complaint directly through legal channels. Generally Callers either do not know this, or do not know the precise circumstances under which it might occur.

5. Cf. Sacks (1992: I, 782–783).

6. Cf. Sudnow (1993).

7. The OCA telephone helpline has taken calls from the public since the passing of the "Sale of Goods and Supply of Services Act" ["SOGA"] (Government of Ireland 1980). During 1988, in the course of some 250 working days, "nearly 19,000 complaints about unsatisfactory goods and services were received" (Government of Ireland 1990). During Spring 1990, I recorded 228 such calls, representing three days' work. The recorded calls range between 1 and 22 minutes in duration: the median lies a little under 5 minutes The present study examines call OCA01:05 [short name "Call 01"], of 6 minutes' duration.

8. Cf. Schegloff (1968).

9. This two-part language game conforms to Sacks' account of such first exchanges in his first published lecture (Sacks 1992: 3ff.).

10. See fuller discussion of this in Torode (1998).

11. Similarly in their study of market pitchers, who promote their products by elevating the value and lowering the price, Pinch and Clark (1986: 175) point out that when a number is the first element in a contrast with selling price, that number is routinely larger than the number of pounds mentioned as the selling price.

12. This is a common construct in consumer complaints regarding clothes and shoes, cf. Call 07 in which manufacturer met a customer's complaint that a child's dress had fallen apart in the wash by claiming it had only been purchased for a single occasion of use, namely a First Communion:

```
86   CR: now the manufacturer this morning inferred (.) didn't say straight
         ou:t but infe:rred hhh that I bought the suit (.) I used it for the
         confirmation now that I didn't want the suit I was looking for my
         money back
87   HR: mh hm
88   CR: it isn't true hhh ...
```

Helper gave some credence to this claim:

```
99   HR: you know and that because like obviously okay there is p-hoh!-ssibly
         a case where (.) the store would try and say a person goes out hhh
         buys an item (.) uses it and then insists on a refund of money (.)
         you know and has got th-heh! houh!-tfit for the occasion and that's
         i:t
100  CR: yeh
```

13. Games **G7**, **G8** and **G9** are omitted for simplicity's sake.

14. A related practice was reported by Silverman (1987) by a medical consultant dealing with Downs syndrome children suffering a cardiac complaint, which he proposed *not* to treat surgically.

15. Detailed reference is also made to SATRA, a trade organisation which will undertake independent inspection of shoes whose quality is in dispute.

16. Just as, in the section not quoted but summarised above, she led him first in the "manufacturer" option (**G4**), which he terminated, then into the "shop" option (**G6**) which they jointly terminated.

17. Cf discussion of "arrangements" as call-closing implicative by Button (1987, 1990).

18. Wilkie (1994: 500) distinguishes two distinct sales strategies: *expertise* – stereotypically exercised by an older white male who exercises authority over the customer; versus *identification* – perhaps employed by a younger ethnic-minority female who bonds with the customer as an equal. Just such a contrast seems implied here.

19. She initially represented this transactions as direct, lines 13–17 above, though HR later provoked her to repair this to describe it as an indirect transaction, mediated by store, lines 54–58.

20. This was never proposed as an absolute ruling by HR, who is a para-legal rather than a law officer, and is not in the business of pronouncing definitive legal judgements by telephone. At lines 251–260, in the context of Big Game Two, HR mentions the possibility of consulting a solicitor, but rules it out on grounds of expense involved in relation to the price of the shoes.

21. See Torode (1998) for a contribution to such a discussion.

22. See Torode (2003) for an analysis of texts by Wittgenstein and by Garfinkel in these terms.

23. In fact "value for money" is not a legally-defined consumer right.

24. The *Manufacturer* game was spoiled by HR in **G4 P6-P7** (lines 210–228, not discussed here). This did not prevent CR re-launching it in **G4 P8** (lines 269–272, discussed above) but may have contributed to the indirect manner in which she did so (see discussion of lines 273–275, above). A *Solicitor* game, **G13 P1** (lines 253–260, not discussed here) was launched by HR, but was spoiled by him in the same turn, and not resumed in this call.

References

Button, G. (1987). "Moving out of closings." In G. Button & J. Lee (Eds.), *Talk and Social Organisation*. Clevedon: Multilingual Matters.

Button, G. (1990). "On varieties of closings." In G. Psathas (Ed.), *Interaction Competence* (pp. 93–147). Washington, DC: University Press of America.

Davidson, J. (1978). "An instance of negotiation in a call closing." *Sociology, 12*, 123–133.

Drew, P. & Heritage, J. (1992). *Talk at Work*. Cambridge: University Press.

Garfinkel, H. (1967). *Studies in Ethnomethodology*. Cambridge: Polity Press.

Garfinkel, H. & Sacks, H. (1970). "On formal structures of practical action." In S. McKinney & E. Tiryakian (Eds.), *Theoretical Sociology* (pp. 338–366). New York: Appleton-Century-Crofts.

Garfinkel, H. & Wieder, L. D. (1992). "Two incommensurable, asymmetrically alternate technologies of social analysis." In G. Watson & R. Seiler (Eds.), *Text in Context: Contributions to Ethnomethodology* (pp. 175–206). London: Sage.

Government of Ireland (1980). *Sale of Goods and Supply of Services Act*. Dublin, Ireland: Stationery Office.

Government of Ireland (1990). *Annual Report of the Director of Consumer Affairs and Fair Trade for 1988*. Dublin, Ireland: Stationery Office.

Hester, S. & Francis, D. (2000a). "Ethnomethodology, conversation analysis, and 'institutional talk'." *Text, 20*, 373–375.

Hester, S. & Francis, D. (2000b). "Analysing 'institutional talk': A reply to Watson." *Text, 20*, 391–413.

Kleifgen, J. A. & Frenz-Belkin, P. (1997). "Assembling knowledge." *Research on Language and Social Interaction, 30*, 157–192.

Lynch, M. (1985). *Art and Artefact in Laboratory Science*. London: Routledge.

Pinch, T. & Clark, C. (1986). "'Patter merchanting' and the strategic (re-) production and local management of economic reasoning in the sales routines of market pitchers." *Sociology, 20*, 169–191.

Pitkin, H. F. (1972). *Wittgenstein and Justice*. Berkeley, CA: California University Press.

Sacks, H. (1992). *Lectures on Conversation, Vols. I and II*. Oxford, UK/Cambridge, USA: Blackwell.

Schegloff, E. (1968). "Sequencing in conversational openings." *American Anthropologist, 70*, 1075–1095.

Silverman, D. (1987). *Communication and Medical Practice*. London: Sage.

Sudnow, D. (1993). *Ways of the Hand: The Organisation of Improvised Conduct*. Cambridge, MA: The MIT Press.

Torode, B. (1995). "Negotiating 'advice' in a call to a consumer helpline." In A. Firth (Ed.), *The Discourse of Negotiation: Studies of Language in the Workplace* (pp. 345–372). Oxford: Pergamon.

Torode, B. (1998). "Narrative analysis using code-a-text." *Qualitative Health Research, 8*, 414–432.

Torode, B. (2003). "Between uniqueness and universality: An ethnomethodological analysis of language-games." In C. B. Grant (Ed.), *Rethinking Communicative Interaction* (pp. 211–236). Amsterdam: Benjamins.

Watson, R. (2000). "The character of 'institutional talk': A response to Hester and Francis." *Text, 20*, 377–389.

Whalen, J., Zimmerman, D. H., & Whalen, M. (1988). "When words fail: A single case analysis." *Social Problems, 35*, 335–362.

Wilkie, W. L. (1994). *Consumer Behaviour*. New York: Wiley.

Wittgenstein, L. (1953). *Philosophical Investigations*. Oxford: Blackwell.

Wittgenstein, L. (1958). *The Blue and Brown Books*. Oxford: Blackwell.

Part V

Aspects of call management

Chapter 12

Some initial reflections on conversational structures for instruction giving

Ged M. Murtagh

1. Introduction

This chapter is the result of some initial ethnographic investigations into a mobile phone call centre in the UK. After a brief period during the research, the organisation provided me with tapes of recorded calls to the centre (22 calls in all). This afforded me the opportunity to investigate some of the interactional details of these calls and the following discussion reports on that investigation.

If one could be permitted a crude gloss, most of the calls in the data used were brief, consisting of a series of questions and answers producing standard inquiries and standard responses. In this respect they typified interactional exchange in these kinds of settings. On the face of this gloss one might be persuaded of the arguments presented by some schools of sociological thought who describe the structural features of modern day work settings (e.g. technological advance) as determining the character or the experience of the work itself.[1]

However, the aim of this chapter is to transcend arguments of this kind by focussing on how the character of the work (insofar as that concerns one aspect of that work i.e. dealing with customers through the call) is an accomplished phenomenon achieved in and through the structural features of the talk between caller and adviser. That is to say, the evident fact of standard inquiries and standard responses is something that is oriented to and collaboratively produced by caller and adviser in, what might be described as, a "seen but unnoticed" fashion. To explicate those "seen but unnoticed" features of the calls, the discussion will focus on their construction through the situated details of the organisation of speech exchange.

The discussion itself does not state any new findings in relation to the properties of social interaction. Details of these properties have been covered extensively within the ethnomethodological/conversation analytical tradition. Rather, the focus of this endeavour is to point to some of these properties as they were regularly produced and displayed by callers and advisers in and through the talk. The discussion builds upon previous research by addressing single instances or single episodes of interaction as they concern the practical management of the call as an orderly methodical production.

2. Instruction giving and instructional sequences

As an example of organisational telephone talk, one of the striking features of the data corpus is the centrality of sequences of instruction transfer. Further exploration of the conversation analytical literature concerning organisational telephone talk revealed that there are comparatively very few investigations of the phenomenon of instruction transfer within those settings. This is strikingly apparent in the literature concerning calls to emergency services (see for example, Zimmerman 1992a and 1992b; Whalen & Zimmerman 1987).

These studies deal with several aspects of the structure and organisation of calls to emergency services that include, the pre-opening, the opening, information transfer, the delivery of topic, the request for help, interrogative sequences and the reason for the call etc. These studies identify a considerable range of interactional phenomena that are then examined in light of their sequential organisation within the stream of talk between the caller and the person receiving the call.

In one sense one might expect to find instances of instruction transfer in these calls, where, for example, the adviser might instruct the caller as to what to do before the paramedics arrive. Nevertheless, there are sound practical reasons why there are few analysed instances of instruction transfer in these studies. These calls are often short with participants' concentrated focus on the reason for the call and the nature of the emergency often resulting in "institutionally constrained" sequences (Zimmerman 1992a; Whalen & Zimmerman 1987).

Notwithstanding these reasons, instruction giving is, on many occasions, a central part of the organisation of emergency calls, e.g. Zimmerman notes the use of coding and abbreviations used in the dispatch package designed to instruct the particular emergency service of the nature of the situation and the persons involved (Zimmerman 1992b: 423).

3. A system for the transfer of instructions

At the analytical level, the paucity of the analysis of instructional sequences in studies of organisational telephone talk is surprising for another reason. One of the few papers to deal specifically with this topic is one of the very early contributions to the CA corpus that is provided by Jo Ann Goldberg.[2] The article itself is a foundational piece that is also cited in the classic paper by Harvey Sacks, Emanuel Schegloff and Gail Jefferson, 'A Simplest Systematics for the Organisation of Turn-Taking for Conversation'.[3]

Goldberg's discussion appears to be one of the very first empirical investigations of the serial organisation of turns at talk as it concerns the phenomenon of the transfer of instructions in ordinary interaction. Her discussion details the systematic organisation of instruction transfer in ordinary settings. Yet by and large this article seems to have gone unnoticed by the majority of CA practitioners in their investigations of information transfer in different settings. As a consequence a core aspect of help line interaction has largely been overlooked.

The analysis provided in this discussion will utilise Goldberg's analytical framework by focussing on the serial or sequential features of calls to a help line as they produce an "instructional structure" (Goldberg 1975: 269). In so doing it is hoped that a further substantive contribution can be made to our understanding of telephone calls for help and advice. In addition, the discussion aims to highlight one of the key objectives of instruction giving that Suchman's work (1987), in another context, describes.

> An appreciation for what is required in instruction-following makes it easier to understand the problem that the communication of instructions attempts to solve: namely the troubles inherent in turning an instruction into an action.
> (Suchman 1987: 104)

In these calls, solving the problem of turning an instruction into an action through the communication of instructions, as a practical matter, is crucial for the adviser and the caller in order that both parties "recognise what the instructions are definitely talking about" (Garfinkel 1967: 22). The purpose of this discussion is to explicate some of the common sense reasoning procedures used and displayed (in and through the talk) by the participants, as they orient to the request for and receipt of instructions as a social activity. In brief, the focus is on the self-organising features of these calls as instances of social action.

Requesting the telephone number: The basic instructional chain

Typically, the calls collected for analysis can be divided into a series of steps exhibiting a general format that is naturally oriented to by caller and adviser.

- At the beginning of the call the adviser makes a request for the telephone number of the caller and the caller instructs the adviser by providing the information. This can be seen as the first instructional course of the calls.
- Subsequent to this the adviser offers help and the caller provides a description of (for example) technical problems experienced with use of the phone.
- The adviser then engages in delivering a series of instructions to deal with the caller's inquiry. Naturally these instructions vary according to the problem presented.[4] Finally the call is closed often by the adviser formulating what has just occurred.

The first instructional chain can be evidenced in the request for the caller's mobile telephone number. Consider the following sequences.

(1) [CN2: 4-00]

```
1  A:  Pay as you talk Can I take your mobile number ple::ase?
2  C:  Uh (.) xxxxx
3  A:  xxxxx
4  C:  xxx
5  A   xxx::
6  C:  xxx
7  A:  xxx
8  A:  You're through to B can I take your name ple::ase?
```

(2) [CN3: 4-00]

```
1  A:  Welcome to pay as you talk , can I take your mobile number
2      ple::ase?
3  C:  You can just a secon::d=
4  A:  =Okay
5  C:  It's xxxx
6  A:  Yeah
7  C:  Double x xxxxx
8  A:  (0.3) You're through to Y Can I take your name ple::ase?
```

(3) [CN7: 4-00]

```
1  A:  Welcome to pay as you talk can I take your mobile number
       ple::ase?
2  C:  Uh it's xxxxxx
3  A:  (0.3) Yeah
```

```
4   C:  xxxxx
5   A:  You're through to B can I take your name ple:ase?
```

These sequences exhibit the opening and entry into the call. Within the conversation analytic tradition the opening to ordinary and institutional telephone conversations has received considerable analytical attention. (See for example, Zimmerman 1992a and 1992b; Whalen & Zimmerman 1987; Schegloff 1986, 1979.) These studies examine the interactional details of the calls and demonstrate, amongst other things, how participants determine the type of call, the appropriate entry in to the call (Schegloff 1979) and the structure of subsequent interactional sequences.

However, these sequences also exhibit the first instructional episodes within the calls. That is to say, the caller delivers the requisite information over a series of instructional units.[5] The basic unit of this instructional chain consists of the same structure as the adjacency pair (Goldberg 1975). Thus, participants to the call organise their interaction utterance by utterance to manage the sequential implicativeness of each conversational turn (Sacks & Schegloff 1974) over an extended series of instructional steps.

In the opening sequences the caller breaks down the telephone number into two, three or even four component parts. These parts are adjacently paired with the adviser's response. These responses come in the form of either repeating the caller's utterances or producing continuer tokens (Schegloff 1982) such as 'yeah', 'okay' or 'um hmm' displaying to the caller that the talk has been heard and understood.

Given the similarity to the adjacency pair structure, it might be proposed that these first sequences consist of instruction/receipt pairs (Goldberg 1975) where an action (the first instruct) has sequential implications for a next action (its receipt). The continuers provide closure of the instruct/receipt pair providing an opening for the caller's next utterance.

> The Recipient's utterance both completes the utterance pair as well as signals the occasion for generating a next Instruct and Receipt Pair. A Further instruct will not be offered until a Recipient has marked Receipt of the present Instruct. The absence of a Recipient's utterance is consequential to the ongoing course of interaction. (Goldberg 1975:274)

More mundanely, the transfer of instructions in this context also displays the participants' orientation to a feature of this kind of work. That is to say, the delivery of instructions can involve the adviser in some form of activity, i.e. to perform an action pertaining to the instruct. In cases like these the recipient can be found acknowledging receipt of the instruct as well as, for example, repeating and writing down or recording the instruct in some way.

The serial organisation of the utterances facilitates that practical management, allowing for writing or recording time. In this way, both caller and adviser furnish a "texture of relevance" to the call that exhibits one another's competencies in the interactional and practical management of these occasioned activities. Writing down the telephone number, however, isn't just the privilege of the adviser as the next example demonstrates. In this example, the caller is writing down the number as he utters it to the adviser. See lines 2 and 6.

(5) [CN5 4-00]

```
A: () to pay as you talk can I take your mobile number ple:ase
C: iis ju just a second love
A: Okay
C: uhh is xxx (0.2)x
A: [[Yep
C: [[Uh (0.2) just a secu:nd .hhhh (3.0) I'm writin' off by
   heart here you see
A: That's no problem
C: xxx
A: Ya
C: xxxx
A: Okay You're through to B can I take your name ple:ase
```

Departures from the typical structure

As the first basic instructional chain there were occasions where the request for the telephone number went unacknowledged by the caller thus marking a departure from the typical serial organisation of the opening.

(6) [CN1: 4-00]

```
A: Pay as ya talk ( ) your mobile number please?
C: Uh::m good afternoon [Uhm
A:                      [Good afternoon
C: I can't (1.0) register my ehm top up card
A: Ok sir if I can take your mobile number I can put the top on
   for you now,
C: Oh right (0.9) 111
```

In this excerpt the caller proceeds in the second turn by identifying the problem. In so doing the caller disturbs the serial organisation of the beginning of the first instructional sequence. Using the utterance at line 5, the adviser displays the absence of the answer to the question as an accountable feature of the interaction. To render that absence accountable the request is repeated until the answer is obtained. In so doing the adviser explicates or makes visible the necessary practical resources that must be made available to them to successfully complete the task.[6] The significance of this aspect of the identification sequence is also displayed in the next sequence.

(7) [CN6: 4-00]

```
1  A:  () to pay as you talk can I take your mobile number ple::ase?
2  C:  Oh hello I do know it its (0.2) ya self I spoke to earlier
       on(0.1)
3  A:  [[It wasn't ()
4  C:  [[It's Mis (0.2) mister Smith here about my (0.1) daughter's
5      telephone=
6  A:  =Okay can I take you daughter's telephone number please Mr
       Smith=
7  C:  =Ya it's xxx
```

In this instance the caller ignores the initial request for the mobile phone number. His concern is to identify who he is now talking to and whether it was the same person he was talking to previously. He does this by relaying specific features of a previous conversation with an adviser, at lines 2 and 4. At line 3 the adviser denies knowledge of Mr. Smith's previous inquiry and then after the caller informs the adviser that he is calling on behalf of his daughter, the adviser requests his daughter's telephone number. The caller then orients to that request by delivering his daughter's telephone number.

Amongst other things the serial organisation of these instructional sequences provides for the participation rights and obligations of both parties and renders those same rights and obligations accountable features of the interaction. In so far as these instances might be singled out as "deviant cases" marking a departure from the serial organisation of the interaction they can also be seen as what Goldberg (1975) describes as a "structurally violative moves", on the part of the caller, within the instructional chain. Although the adviser makes no explicit reference to the violation of an interaction rule, the repeat of the request is a resource used by the adviser to display that a violation has taken place and repair the serial organisation of the call as it concerns the first instructional sequence is required.

Further instructions

Typically, after the caller has provided the adviser with an account of what they see as the problem, the adviser moves to resolve. The excerpts illustrate that a move to resolve the problem is achieved through short and extended instructional courses. As with the delivery of the telephone number, instruction giving throughout the data is broken down by caller and adviser into a series of component parts. These component parts or instruction/receipt pairs (Goldberg 1975) provide caller and adviser with a "minimal conversational unit" (Goldberg 1975:276) to order and produce a mutually intelligible instruct and receipt sequence regardless of the length of the instructional sequence.

One of the longest sequences in the data provides a good example of how this "minimal conversational unit" is used. It concerns a lengthy sequence of interaction in relation to a problem with call barring on the caller's phone. In addition to the initial opening and a series of questions and answers to determine the precise nature of the problem, instructional sequences can be found in lines: 79–92, 98–115, and 121–150 (see Appendix). The instruction sequences are all broken down into their component parts forming instruct/receipt pairs throughout the interaction. In this respect, "the I/R pair is specifically serviceable as a reusable object, available for Instructions of any duration" (Goldberg 1975:282).

The remainder of the discussion will take a closer inspection of the data utilising what has already been said about the organisation of instruction transfer. This closer inspection will focus specifically on how ambiguities and misunderstandings are managed as part of the interactional work of calling for help.

4. Ambiguity and repair

On any occasion of the delivery and receipt of instructions, participants face the "occasionally relevant" (Coulter 1996) problem of misunderstanding or ambiguity. These are practical issues that have to be locally managed in situ amid the unfolding sequence of interaction. For the purposes of analysis it is worth considering some of the resources available to participants to deal with these matters.

It was mentioned earlier that one of the ways in which the adviser can display confirmation of the delivery of information from the caller is by simply repeating the caller's utterance or alternatively uttering continuation tokens. These resources signal that to the instructor that the instruct has been received and allow for the closure of an I/R pair whilst also leaving the floor open for the next instruction. The following sequence exhibits the use of repetition as a resource for the practical management of the receipt of instructions.

(8) [CN2: 4-00]

```
79  A:  Right let's try a quick (0.5) experiment here to see if this
80      works (0.5) Could you press the hash button for me? Which is
81      the button underneath the number ni::ne
82  C:  (1.0) U:h yeah
83  A:  Then type in x,x,x,
84  C:  x,x(0.5)[x
85  A:          [x Then press the sta::r button
86  C:  Star
87  A:  Then put into the phone x,x,x,x
88  C:  x,x,x,x=
89  A:  =Press the hash button which is underneath the number ni:ne
90  C:  Yeah
91  A:  Then press your se::nd button
92  C:  Says call barred
93  A:  (0.5) It's not allowing you to do that either?=
94  C:  = Na:h
95  A:  (2.0) OK if you can just hold the line for me for a moment
        please
```

At line 85 subsequent to a small pause by the caller, the adviser inserts what Goldberg (1975) describes as a "pre partial repetition". In this instance the caller's pause can be seen to "signal a repair" (Goldberg 1975: 279) of the instruction provided. In accordance with that the adviser repeats the last digit which in this instance is simultaneously uttered by the caller. Again at lines 86 and 88 the caller repeats the adviser's prior instruct of, in the first instance, the instruction to press the star button and in the second, the instruction to input the four digit number.

In relation to this, Goldberg (1975) observes that a common feature of instruction sequences is that, not unlike the basic adjacency pair, issues of repair occur or are inserted between the basic instruct/receipt unit. Indeed several instances in the data corpus exhibit how the repair of an instruction is generated within the structure of the instructional course. Consider the following sequence.

(9) [CN2: 4-00]

```
57  A:  =Could I just ask you to confirm your password for me
        ple::ase
58  C:  ()
59  A:  Your Paa:ssword
60  C:  (0.5) Password
```

61 A: Yeah
62 C: u::h (2.5) what give it to you
63 A: Yes please, it's the password that you have with us when you
64 registered in September

An insertion in between the I/R pair is an interactional resource available designed to deal with ambiguity or misunderstanding immediately before further instruction continues. At line 57 the adviser requests the caller's password, this request is repeated at line 59. At line 60 the caller inserts a repair which is specifically directed to the prior instruct. This occurs again at line 62 where the caller's utterance deals specifically with the prior instruct to submit the password. A noticeable feature of both of these repair insertions is that they occur subsequent to a pause signalling a possible trouble in the receipt of the instruct. Consider the following sequences.

(10) [CN2: 4-00]

101 A: Let's just get that press the ha::sh button form
102 C: Yeah that's the:e
103 A: Th the (0.5) the one that looks like the noughts and crosses
104 bo:ard
105 C: (0.5)Yeah

(11) [CN2: 4-00]

125 A: Then go intu (0.5) security
126 C: (1.0)Yeah
127 A: And there is a section for ba::rring
128 C: (2.0) What is that?
129 A: It's for ba::rring B.A.R.R.I.N.G. Is there a section that
130 says ba::rs or ba::rring

The repairs inserted in these sequences are designed to deal immediately with the prior instruct. However, the design of these utterances (notably the pauses between the utterances) also demonstrates the participants' orientation toward the activities implied by the instruction. As mentioned earlier, the delivery of instructions can involve the caller, or more typically the adviser, in some form of activity, i.e. to perform an action pertaining to the instruct. In cases like these the caller can be found acknowledging receipt of the instruct as well as performing a specific activity (in these instances selecting and activating the appropriate menu options on the telephone). It is with instances like these that the caller and adviser face the practical task of "...turning an instruction into an action" (Suchman 1987: 104). It is imperative, then, that the participants to the talk "recognise what the instructions are definitely talking about" (Garfinkel 1967: 22).

(11) [CN4: 4-00]

```
48 A:  =She needs to pick a bra::nd new pin number it can't be the
49     same as her old one and enter it into the phone now
50 C:  Uhm pick a new pin number=
51 D:  A new one?
52 C:  Ya
53 A:  Put it into the phone no:w
```

In the above sequence there are three parties to the interaction, the adviser, a mother who is calling on behalf of her daughter (D) who is in the background. At line 51 it is the daughter who inserts a repair of the prior instruct to select a new pin number. This type of repair move stands as an alternative to a continuation marker and has been referred to as a "repair slot" (Goldberg 1975). As mentioned, these particular repair moves operate to address the relevance of the immediate prior instruct. As a practical resource to manage the practical contingencies of instructional sequences they are "topically restricted to the locality in which they are initiated" (Goldberg 1975: 281). Thus the repair slot is a device available to both parties to the interaction to determine the structure of the instructional course.

End repairs

The insertion of repair is not always restricted to the slot in between the I/R pair. Repair utterances are also found at the end of instructional sequences and have been referred to as end repairs (Goldberg 1975). The following sequence provides and instance of an end repair.

(12) [CN3: 4-00]

```
30 A:  It's just topping up for you [now
31 C:                               [lovely
32 A:  It's now seven days of service (0.3) and calling credit is 13
33     pounds and 85p
34 C:  Ri::ght
35 A:  Oka:::y
36 C:  .hhh [Now as I say this it wont be getting used again now
37     until me daughter for Christmas .hh will I lose all the
38     credit if it's not used by then?
39 A:  The service credit will go:: down in seven days
40 C:  [Right
41 A:  [A week today. But what will happen is the calling credit
42     will remain their
43 C:  Right
```

44 A: It it uh once you've used your service credit up it gives you
45 () again
46 C: Right
47 A: before the service credit disappears
48 C: Ah that's [lovely

This sequence exhibits two key features. Firstly, at line 35 the adviser initiates closure with the phonetic extension of the utterance okay. At line 36 the caller orients to this as a possible closure by repeating her concern expressed at the start of the call. She thus selects the next slot (at line 36) to insert a repair. In this instance the repair slot is found at the end of the instructional course designed to "target back on a particular prior instruct" (Goldberg 1975: 281).

Secondly, the sequence displays the fact that the end of the instructional sequence is mutually achieved and not merely evident as a "logically findable position" (Goldberg 1975: 282) within the sequence. The phonetic extension of okay is a typical move toward closing the sequence. However, the end of the sequence is always subject to ongoing negotiation between the participants. Thus, alternative moves are always available to initiate repair within an instructional course. Repair can occur immediately after a prior instruct or towards the end of an instructional course. This occurs again in the following sequence.

(13) [CN4: 4-00]

60 C: And press okay (5.0)
61 A: And now should be unblocking itself and that's her new pin
62 number every time she turns the phone on=
63 C: =Ya an now this numbar if it 'appens again can we put this
64 number in again or do we av to ring again=
65 A: =It's the same number you put in again but it's normally best
66 to give us a ring in case thez any problems cos it will only
67 allow you ten times which to put it in before it permanently
68 blocks the pho::ne
69 C: Oh does it=
70 A: =So we advise you if it happens again just to give us a call
71 and we'll go through it with you
72 C: .hh ya okay then
73 A: Oka:::y
74 C: Alright then [thanks ever so mu::ch
75 A: [Thanks for calling xxxxxxx
76 A: [Okay Bye
77 [Bye

The end repair is inserted by the caller at line 63 who targets back on the instructional sequence to check if it is appropriate to insert the same number should the problem occur again. The sequence exhibits not only the alternate repair slot at the instruction end, but also how its structure and organisation is locally produced moment by moment from one turn to the next. In this respect there are (to quote Goldberg 1975) no "intuitively derivable terminators" to the sequence. As a result there is always scope for the participants to open the floor for further instruction. In this respect determining the end of an interaction is strictly a matter for the participants involved.

Formulating instructional courses

A related feature of these instructional sequences is the participants' use of formulations. Formulations are used to sum up, confirm or ratify a gloss proposed by an interlocutor (Watson 1986). In many instances in the calls, the adviser or caller will formulate a summary of the prior talk. Heritage and Watson (1980) detail some of the properties of formulations particularly their placement in conversations. They point out that formulations are one (among many others) of the self descriptive conversational devices through which members "do describing". With regard to the use and placement of formulations they point out that describing is rarely done "for its own sake" but is often tied to some conversational activity.

In several instances, the adviser would formulate or provide the caller with the "upshot" of the interaction or instructional sequence. The following sequence is an example of this.

(14) [CN2: 4-00]

```
172 C:  =I think they've gave this (0.3) number this morning but
173     (2.0) there's nothing happened. They didn't really (0.3)
174     help me at all you know
175 A:  Try them again. Explain to them that you've spoken to
176     xxxxxxx. We've checked there are no bars on your handset at
177     all. Your babysitter lock isn't on (0.3) and (0.2) we can't
178     unbar the phone. They should be able to help you
179 C:   [Alright then=
180 A:  =[Oka::y
181 C:  Cheers [thanks bye
182 A:         [Thank you for calling xxxx [Bye bye
183 C:                                     [Yeah Bye
```

This formulation occurs at the end of a long sequence and several attempts by the adviser to resolve the caller's problem. This formulation is provided as a "proper" gloss of the talk and action engaged in throughout the sequence. In this particular instance, the formulation at line 175 can also be seen to initiate a closing to the interaction where the adviser informs the caller of who to speak to next. This utterance is (to quote Heritage and Watson) a "candidate pre-closing" providing the caller with a signal of the end of the sequence.

> So far as formulations are concerned, the sense of 'all is well' is tied into the ways in which they manifest for members the fact that a conversation is, and has been, a 'self-explicating colloquy' (to use Garfinkel and Sacks' term) – that is an intrinsically self-describing, accountable, observable, reportable phenomenon; this is the explicative work of formulations. (Watson 1986: 105)

Heritage and Watson (1980) also point out that formulations are evident utterance by utterance, related to topic and relating to the overall structure of the conversation. Their use in these different levels of conversational structure are, as they maintain, not isolable from each other. As a consequence of its positioning at the end of the sequence the function of the formulation in example (14) is similar to the end repair. The formulation in this instance, provides caller and adviser with a gloss of their exchange "as an overall unit" (Watson 1986: 108) where that gloss describes the outcome of previous instructional activity.

Formulations are one of the ways in which participants in talk can determine the practical relevance of their exchange. More importantly, as verbal utterances they explicate or make visible an interlocutor's assessment of the other interlocutor's talk.

(15) [CN1: 4-00]

```
25  A:  =555 (1.5) Ok that's for a five pound calling credit only
26      voucher that's just topping up for you no:w
27  C:  Ok thank you=
28  A:  =You now have forty two days of service and calling credit to
29      the value of five pounds
30  C:  Ri:ght thanks[very much
31  A:  [Ok (0.5) Thanks for calling xxxxxx Mr. Smith b:bye
32  C:  [Yep thank you
33  C:  By::e
```

Formulations, such as the one given at line 28 in the above example, were a consistent feature of the calls and frequently operated as "candidate pre-closings". Consider line 44 of example (16) and lines 32–33 and 35–37 of example (17).

(16) [CN3 4-00]

```
44  A:  So you've got nearly three months to tie uhm to top it up
45      (0.2) before the service credit
46  C:  Ah that's [lovely
47  A:  [Oka:::y
48  C:  [That's smashing thanks very much
```

```
49 A: Thanks very much for calling xxxxxxx Ba bye
50 C: Thank you Ba By::e
```

(17) [CN5: 4-00]

```
32 A: (2.0) ka:y that's for five pound calling credit voucher its
33    just topping up for you now Mr. Jones
34 C: Thank you very much=
35 A: =You now have a balance of sixty four days and calling credit
36    for six pounds and twenty eight pence
37 C: [[Thank you very much luv
38 A: [[Oka:::y
39 C: [[Thank you
40 A: [[Thanks for calling xxxxxxx ba bye
41 C: Bye bye
```

These formulations operate to provide the caller with a summary of what their preceding interaction has achieved. Again in these instances the caller confirms the adviser's description of the prior talk. The fact that confirmation of that description is exhibited is consistent with findings from other studies that demonstrate members' preference for confirmation of a formulation first pair part (see Watson 1986). Clearly, however, a formulation in any conversational event is always open disconfirmation.

5. Conclusion

The analysis throughout this paper has focussed on the serial or sequential features of calls to a help line as they relate to the production of an instructional sequence when information is transferred. In so doing it has considered some of the features of a sequentially organised instructional sequence identifying how participants attend to matters of repair and formulation in single episodes of data. One may pursue a further analysis of the data to begin to examine the extent to which these single instances are illustrative of a general pattern of the structure of these interactions.

Utilising some basic premises of past ethnomethodological and conversation analytical work, the discussion has tried to demonstrate how these instructional courses are interactionally produced by caller and adviser as a mutual accomplishment. What is noticeable about these episodes of talk in an institutional context is that they exhibit the basic conversational unit of adjacently paired utterances in the form of instruction and receipt pair parts. From this "conversational unit" other features are "built off" but at the same time sequentially placed in relation to the overall instructional structure. This discussion has briefly attended to two of those features, repair slots and formulations. These features enable participants to structure and organise the transfer of instructions and determine the pattern of the overall instructional structure within telephone calls for help.

Notes

1. The classic statement of this kind of position can be found in Marx's work on the relations of production in capitalist society as resulting from the dynamics of technological advance. For Marx, any technological advance will result in a corresponding set of social relations. See also R. Blauner (1964). *Alienation and Freedom*. Chicago: Chicago University Press.

2. J. A. Goldberg (1975). 'A system for the transfer of instructions in natural settings'. *Semiotica, 14*(3), 269–296.

3. Sacks, H., Schegloff, E. A., & Jefferson, G. (1974). 'A simplest systematics for the organisation of turn taking in conversation'. *Language, 50*(4), 696–735.

4. Most of the advisers reported that on the whole the caller's requests were routine matters, one of the most common being "topping up" the phone with credit. On one occasion a caller actually rang in to ask the time.

5. Goldberg points out the fact that different kinds of instructions come in different arrangements. "For example, seven digit telephone numbers can be delivered in a single Instruct but when parsed to fit into two separate I/R pairs, the first Instruct gets the first three digits and the second Instruct contains the remaining four. If the second four digits of the number are further parsed into two more Instructs, each will carry two of the digits" (Goldberg 1975:285).

6. Melvin Pollner's analysis of what he calls explicative transactions is relevant to this point. See M. Pollner (1979). 'Explicit transactions: making and managing meaning in a traffic court'. In G. Psathas (Ed.), *Everyday Language: Studies in Ethnomethodology*. New York: Irvington.

References

Blauner, R. (1964). *Alienation and Freedom*. Chicago: University of Chicago Press.
Coulter, J. (1996). 'Human practices and the observability of the 'macrosocial'.' *Zeitschrift fur Soziologie, 25*.
Davidson, J. (1978). 'An instance of negotiation in call closing'. *Sociology, 12*, 123–133.
Garfinkel, H. (1967). *Studies in Ethnomethodology*. Englewood Cliffs, NJ: Prentice Hall.
Goldberg, J. (1975). 'A system for the transfer of instructions in natural settings'. *Semiotica, 14*(3), 269–296.

Heritage, J. C. & Watson, D. R. (1980). 'Aspects of the properties of formulations in natural conversations: Some instances analysed'. *Semiotica, 30*(3/4), 245–262.

Pollner, M. (1979). 'Explicative transactions: Making and managing meaning in a traffic court'. In G. Psathas (Ed.), *Everyday Language: Studies in ethnomethodology*. New York: Irvington.

Sacks, H. (1992). 'Lecture 1'. In Gail Jefferson (Ed.), *Harvey Sacks: Lectures on Conversation*, Vol. 1. Oxford: Blackwell.

Sacks, H., Schegloff, E. A., & Jefferson, G. (1974). 'A simplest systematics for the organization of turn taking for conversation'. *Language, 50*(4), 696–735.

Schegloff, E. A. (1979). 'Identification and recognition in telephone conversation openings'. In G. Psathas (Ed.), *Everyday Language: Studies in Ethnomethodology* (pp. 23–78). New York: Irvington.

Schegloff, E. A. (1982). 'Discourse as an interactional achievement: Some uses of "uh huh" and other things that come between sentences'. In D. Tannen (Ed.), *Georgetown University Roundtable on Languages and Linguistics*. Washington: Georgetown University Press.

Schegloff, E. A. (1986). 'The routine as achievement'. *Human Studies, 9*, 111–151.

Schegloff, E. A. (1987). 'Analysing single episodes of interaction: An exercise in conversation analysis'. *Social Psychology Quarterly, 50*(2), 101–114.

Schegloff, E. A. (1988). 'On an actual virtual servo mechanism for guessing bad news: A single case conjecture'. *Social Problems, 35*(4), 442–457.

Schegloff, E. A. & Sacks, H. (1974). 'Opening up closings'. In Roy Turner (Ed.), *Ethnomethodology*. Harmondsmoth: Penguin.

Suchman, L. (1987). *Plans and Situated Actions: The Problem of Human Machine Communication*. Cambridge: Cambridge University Press.

Turner, R. (1972). 'Some formal properties of therapy talk'. In D. Sudnow (Ed.), *Studies in Social Interaction*. New York: Free Press.

Watson, D. R. (1986). 'Doing the organisation's work: An examination of aspects of the operation of a crisis intervention centre'. In S. Fisher & A. Dundas Todd (Eds.), *Discourse and Institutional Authority: Medicine, Law and Education*. New Jersey: Ablex Publishing, Norwood.

Whalen, M. & Zimmerman, D. (1987). 'Sequential and institutional contexts in calls for help'. *Social Psychology Quarterly, 50*, 172–185.

Zimmerman, D. (1992a). 'Achieving context'. In G. Watson & R. M. Seiler (Eds.), *Text in Context*. London: Sage.

Zimmerman, D. (1992b). 'The interactional organisation of calls for emergency assistance'. In J. Heritage & P. Drew (Eds.), *Talk at Work*. Cambridge: Cambridge University Press.

Appendix CN2:4-00

```
A:   Pay as you talk Can I take your mobile number ple::ase?
C:   Uh (.) xxxxx
A:   xxxxx
C:   xxx
```

A: xxx::
C: xxx
A: xxx
A: You're through to B Can I take your name ple::ase?
C: (1.0) U::h Miss (0.5) Jones
A: And how can I he::lp yo::u
C: A::h (1.0) my phone is on call ba::rred (0.5) at the moment, yeah
an' I've been phonin' all these people yeah an' (1.0) they even
like tried to put some credit on me cos I've got my credit with
me yeah 15 pound to put an (1.5) () there but it's still call
ba::rred[a::n
A: [Right it's saying call barred everytime you try to make
an outgoing call
C: Uh yeah an' I'm trying to put some credit on it but then the
people, yeah that I call (0.5) th they tried to put my credit on
it in this one but (1.0) nothin' happenin
A: Right so you're trying to put some credit on your phone and
everytime you try it says call barred
C: E:ya=
A: =What number are you dialling in order to top your phone up
C: What my (0.5) top up card=
A: =No. What telephone number are you dialling
C: Ah (1.0) dee xxxx for the credit line
A: xxxx And are you able to make other calls
C: No I can't do anything
A: You can't do any: you can't make any outgoing calls at a::ll
C: (0.5) No. I can get incoming calls (2.0)
A: Have you put a handset bar on your phone to stop (.) you being
able to make outgoing calls (0.5)
C: M::m
A: () you know of(1.0)
C: I don't
A: OK are you callin' from a phone at the moment=
C: =Yeah I (.) I've been spent all of 5 pound already (2.0)
A: Your calling us from your phone
C: Oh no. I'm calling, I'm in a phone box
A: On the phone box. Can you not dial 100 from your mobile phone
C: (1.0) No I ca: I can't dial anything it just (.) say call barred
A: Right cos I can see your phone's been topped up with 15 pounds
worth of credit toda:y
C: Yeah
A: That's gone on to your pho:ne what make of handset do you ha::ve
C: (0.5) u::h what handset? Do you mean the:e phone
A: Yeah what make of mobile phone
C: xxxx=
A: =xxxx
C: Yeah
A: Oka:y (0.5) u::hm
A: Let's see if we can get it unbarred for you do you have the phone
at the moment
C: Yeah I've got it with me at the moment[yeah
A: [Okay=
A: =Could I just ask you to confirm your password for me ple::ase
C: ()

A: Your Paa:ssword
C: (0.5) Password
A: Yeah
C: u::h (2.5) what give it to you
A: Yes please, it's the password that you have with us when your registered in September
C: u::h (1.5) I changed it actually before
A: (0.5) The one your registered with us (.) your password
C: (4.0) It was xxxx
A: Yap that's correct
C: [Yeah
A: [Okay
C: [But I changed it.
A: (0.5) That's the password that you have with us so that 's fine
C: Yeah
A: U::hm Let's have a quick look to see if Is your phone with you at the moment?
C: Yeah=
A: =Is it switched on
C: Yeah
A: Right let's try a quick (0.5) experiment here to see if this works (0.5) Could you press the hash button for me? Which is the button underneath the number ni::ne
C: (1.0) U:h yeah
A: Then type in x,x,x,
C: x,x[x
A: [x Then press the sta::r button
C: Star
A: Then put into the phone x,x,x,x
C: x,x,x,x=
A: =Press the hash button which is underneath the number ni:ne
C: Yeah
A: Then press your se::nd button
C: Says call barred
A: (0.5) It's not allowing you to do that either?=
C: =Na:h
A: (2.0) OK if you can just hold the line for me for a [moment please
C: Alright yeah
A: Thank you very much for holding for me. Right I've got another code that we can try::
C: Yeah
A: Lets just get that Press the ha::sh button for me
C: Yeah that's the:e
A: Th the (0.5) the one that looks like the noughts and crosses bo:ard
C: (0.5)Yeah
A: Press that button (.) Then you type in x,x,x
C: x,x[x
A: [x press the sta:r button
C: Yeah
A: Then type in x,x,x,x
C: x,x,x,x=
A: =Press the ha::sh button again

C: Yeah
A: =And press send
C: (0.5) Right it's call barred
A: It's still coming up call barred
C: Yeah
A: Let's see if we can find anymore instructions on ba:rring
A: If you don't mind just holding the line for me
C: Na (.) it's alright
A: () Could you go into the me:nu of your phone for me?
C: Yeah
A: Then go into the se:t up
C: () Yeah
A: Then go intu (0.5) security
C: (1.0)Yeah
A: And there is a section for ba::rring
C: (2.0) What is that?
A: It's for ba::rring B.A.R.R.I.N.G. Is there a section that says ba::rs or ba::rring
C: There isn't any barring or anything yeah
A: Could you go through the set up Can you tell me what options you have in security?
C: Ah I've got Baby sitter
A: Yeah
C: Uhh (0.3) (Lock) quick (0.1) dial
A: Yeah
C: Uhh (0.2) Pin set up
A: Yeah
C: A:::nd (3.0) (prevent new sims)=
A: =D'ya want to go into ba::bysitter
C: Yeah
A: (2.0) And if you select babysitter what does it say, does it say on or off
C: (1.0) Uh What turn it on
A: No does it what does it say it's off or does it say it's on
C: It's off
A: It's off okay if you just cancel out of tha:t=
C: (1.0)[Yeah
A: [Come out of that we don't wanna (0.2) to mess with the babysitter lock. I'm afraid I'm going to have to put you on hold cos I'm not quite sure what's the matter with your phone. I'm just gonna see if some one else can..ehm help us Oka::y
C: Alright then
A: Thanks very much for ho::lding for me. I've just checked with one of our uhm managers here today and what he's advised me to do is give you the telephone number for xxx::xxxx to see if they can actually assist in getting your phone unbarred. Cos we honestly don't know. .ehm.have any more information here to help you
C: Yeah=
A: =D'ya have a pen and I can give you their telephone number
C: Yeah
A: Their number is xxxx
C: xxxx
A: Double x

C: Double x
A: Double x
C: Double x
A: Double x
C: Double x=
C: =I think they're gave this (0.3) number this morning but (2.0)
there's nothing happened. They didn't really (0.3) help me at
all you know
A: Try them again. Explain to them that you've spoken to xxxxxxx.
We've checked there are no bars on your handset at all. Your
babysitter lock isn't on (0.3) and (0.2) we can't unbar the
phone. They should be able to help you
C: [Alright then=
A: = [Oka::y
C: Cheers [thanks bye
A: [Thank you for calling xxxx bye bye
C: Yeah bye

Chapter 13

Working a call

Multiparty management and interactional infrastructure in calls for help

Jack Whalen and Don H. Zimmerman

1. Introduction

Telephone calls have usually been viewed as two-party interactions in which the participants have access only to each other's speech. Studies of calls for help (to seek redress for consumer complaints or obtain legal advice, to name only two cases in point) or emergency assistance (to the special 9-1-1 number in the United States, for example, or to Britain's 9-9-9) have largely exploited this methodological advantage, and concerned themselves with how the interactional organization of calls engaging emergency service providers are shaped by particular institutional and organizational agendas.

A notable result of this research is an account of the overall organization of the emergency call (as a species of the call for service), revealing how the structure of these calls represents an adaptation of the resources of common, everyday interaction to accomplish goals particular to those institutional environments: the quick alignment of appropriate identities; the immediate pursuit of the business at hand involving the focused gathering of information necessary to produce the service; and an expeditious closing (Wakin & Zimmerman 1999; M. Whalen & Zimmerman 1987; Zimmerman 1984, 1992a, 1992b; Zimmerman & Romero 2002). Other work proceeding from tape and transcript materials and, in some cases, ethnographic observation has examined a number of further topics such as the role of computer-assisted dispatch in the organization of the call (J. Whalen 1995a, 1995b);[1] the vulnerabilities of emergency telephone calls to "activity contamination," disruption, and misunderstanding (J. Whalen et al. 1988; Garcia & Parmer 1999); issues of "practical epistemology" "relational positioning" and "reasonableness" that attend caller's accounts of the need for a particular service (Meehan 1989; Sharrock & Turner 1978; Tracy & Anderson 1999; Tracy & Agne 2004; M. Whalen & Zimmerman 1990); the management of emotion (Tracy & Tracy 1998), including especially hysteria (J. Whalen & Zimmerman 1998); and interactional problems generated from the conflicting frames of customer service and public service (Tracy 1997).

Still, although it is typically the case that requests for help and emergency assistance are initiated by these phone calls, a great deal more is involved in the regular execution and production of "calls" than the conversational exchanges. It turns out that for most emergency organizations, and perhaps for many helping or troubleshooting (Emerson & Messinger 1977) enterprises as well, the occasion of a "call" is not at all coterminous with what transpires in talk on the telephone but rather is best described as an *ongoing and developing sequence of actions,* actions that may well be initiated in a phone conversation but have to then get systematically formed up, through the closely coordinated work of the organization's staff, into a certified organizational event – into a "call" (see J. Whalen 1995a).

As an example of this coordinated work, bear in mind that while there may commonly be only a single caller and, initially, at the emergency organization's end of the line, a single answerer, there may also be multiple hearers as well as other participants at that organizational end who, while not speaking, may be listening (or are informed through textual communication of the conversation's direction and import) and may then become involved in the management of the call's course; by dispatching field personnel via radio to the scene of the trouble, for instance. And even during the triggering phone conversation itself, more than one staff member may – under certain conditions – become an active, speaking (rather than only listening) participant.

For instance, consider the transcript below – from a call that will figure prominently in our analysis, concerning a complaint about "speeding cars and a loud party" – where only what is hearable on the recording of the telephone conversation, and thus only what is available to the caller, is shown. In this transcript, the caller is designated as C, and the different emergency organization staff members who either speak with the caller or can be heard on the recording are identified as D1, D2, and D3; later transcripts of this call and other events will provide more detailed information about these participants.

Extract 1

D1: County dispatch
C: Hi uh (.) m- mahy name is (Warick) and ah live
over in Rolling Woods on Maywood <u>Way</u> (1.4) and I

got[a
D1: [speeding car-?=
C: =pardon?
D1: Speeding cars and a[:: loud party?
C: [yeah I got a complaint about
your cops were here and they left? (.) and ahm not
getting' thuh protection that mah taxes are goin'
for now (0.6) ah saw a cop come around the corner
of Wellington and drive up on the sidewalk
(0.8)
C: and uh your policeman was here -he left and
there's been - cars are <u>still</u> speeding up and
down you could hide in the bushes here you could-
get fifty <u>D.U.I.'s</u>
D1: Okay-just a moment sir?
(36.9)
((talk already in progress))
C: ('teen minutes)
D2: We've got four units there sir
C: They're not there right now I'm lookin at the
house pal unless they got here in the last
thirty seconds (.) uhh (0.8) ahm getting[just a
D2: [Maybe-
maybe we've got more than one party goin on.
C: Huh?
D2: What's the address of where-where <u>you're</u> looking?
C: Right up from Maywood(.) the next street up from
Maywood, it's not Bellington but it's the next
one.
D1: Is it Lexington? ((in background))
(0.5)
D2: Is it Lexington?
(0.2)
C: yeauh.
D3: They're there?! ((in background))
(1.3)
C: Your cop left here about, uhh oh seven or eight
minutes ago and I'm watching up Bellington I've
still got heavy traffic here and a car
[went arou]nd[the street a few]minutes ago=
D3: [((sneeze))] [((sneeze))]
C: =and went up the sidewalk and[almost got a=
D3: [((sneezes))=
C: =mailbox.
D3: =<u>ma</u>::n they're breakin up the party- that guy's
blind= ((in background))
D2: =We are-we are on scene sir they just told

```
51        me on the radio that they're breakin' up the
52        party.
53        (1.4)
54   C:   okay.
55   D2:  Alrighty?
56   C:   thank-you.
57   D2:  Sure!
```

Notice first that there are three voices on the "dispatch side" of the encounter, two of who speak directly with the caller. Obviously, from this fact alone the caller (and anyone listening to the recording) can ascertain that some sort of "multi-party" handling of his call is taking place, that more than one party is involved in its processing. Moreover, given the mundane manner in which these different staff members become involved or exhibit engagement – particularly the way D2 gets on the phone (line 22) and addresses the caller without special comment or explanation, and the timeliness of the subsequent "background" (to the caller only, of course) contributions of D1 (line 33) and D3 (first at line 38 and later at line 48) – can also reasonably infer that the public safety communications center is organized in such a way that this kind of collaborative processing is commonplace. Indeed, the caller gives no explicit response to the fact that (at least) three staff members are involved. This is not to say he exhibits no recognition of this. For example, with the "switch" to D2 after speaking initially to D1 involving a change from a female to a male staffer, the caller then addresses D2 as "*pal*" (a distinctly male and sometimes fairly hostile term of address) in his first remarks to him, which come in response to D2's immediately prior statement to the caller that there are currently four police cars at the reported location of the party, despite the caller's claim to the contrary. Also, after the first timely utterance at line 38 by D3 to which we referred above, "*They're there*," which is addressed to D2 but hearable by the caller, the caller responds in his next turn with a reiteration of his complaint but only after a silence or pause of slightly more than a second. This pause may reflect the caller's orientation to what is for him a "background" utterance (i.e., hearably addressed to another dispatcher rather than to him) by allowing space for D2 to speak before responding.

We can additionally notice that the assorted "call co-participants" at the communications center appear to form a kind of *team* (cf. Lerner 1993), insofar as their involvement is collaborative and task oriented and their roles – call-taking and dispatching, for instance – appear, based on what is hearable on this recording, to be at least somewhat interdependent. The team relationships and, equally, the team solidarity is certainly evident in the collaborative character of D3's second "background" (again, only to the caller) contribution, at line 48, in which she again states that the police are "*breakin' up the party*" and then adds, "*that guy's blind.*" The form of the reference here nicely displays that it is addressed to someone other than the caller, by referring to him as "*that guy,*" and is immediately followed by D2's restatement of this information to the caller but now tactfully omitting the remark about the caller's eyesight and perhaps marking *this* utterance, and not D3's, as the "official" version.

We want to suggest here that what makes such teamwork possible is not only the structure and interdependency of institutional roles and responsibilities but the social relationships and interactional practices that develop among the participants over the course of *all* their conjoint activities, both the institutionally required or sanctioned work (like processing calls) and the informal conversations or encounters that take place at the work site but that are not, strictly speaking, "work" – that take place in between and across the processing of calls, for instance.

This argument will become central to our discussion of the communications center's organization and functioning below. But at this point we want to just summarize two points: that there are many indications available in the call, both for the caller and overhearing analysts, of multi-party involvement; and that there is a team of dispatchers at work, with interactional practices that surround, infiltrate, and facilitate the talk, *at a given moment*, of a single caller and a single communications center staff person.[2]

Plainly, then, while a research focus on the telephone call itself and its conversational participants has been indispensable for understanding important features of the work of emergency telecommunications, we cannot grasp the actual organization of that work until we take into account this teamwork and its often complex interactional infrastructure of personnel who manage calls both on and off the phone, and who otherwise staff the scene (whether working or not working a call). The audio record of the "Speeding Cars" call by itself does not reveal the full nature and extent of this infrastructure and this teamwork and, especially, how it is mobilized. Accordingly, we will need to draw on additional data for our analysis.

Moreover, following on our observations just above concerning the underpinnings of teamwork, this analysis will also point to the problematic nature of using "work" as a predetermined frame for a human environment and for the analysis of all activities that take place there (as a pre-allocated setting-definition, so to speak; see Sacks et al. 1974). We will show how participants in the setting manage the display and recognition of actions as "work" or "not work" and the transitions between these orientations and states of affairs (or "involvements," in Goffman's 1963 sense of this term). And more important, we will also show how actions and activities that are designed and recognized as "not work," as well as the competencies and orientations developed and supported in those activities, provide the scaffolding upon which "work" activities can be assembled. That is, we propose there is a set of competencies and orientations – what we might term a *sensibility* – that is developed and fostered in this setting, and that its physical arrangement affords, which underlies all action that occurs there, whether "work" or "not work." This will then allow us to situate the work of emergency dispatch with respect to the growing number of ethnomethodological and conversation analytic studies of work activities and places.[3]

Because a general understanding of the physical and operational organization of the setting where we conducted our primary research for this chapter, Central County Dispatch Center, will be necessary for what follows, we begin with a description of that setting and then very briefly review the methods we have employed to collect our data. This is followed by a detailed discussion of the interactional infrastructure – the social competencies and practices – that underlie the recurrent accomplishment of multi-party call management at Central County. We then return to the "Speeding Cars" phone call, only now attending not only to the talk on the phone but also to the activity, conversational and otherwise, that took place throughout the setting immediately prior to that telephone call and during it.

2. Central County Dispatch

Central County Dispatch Center is responsible for all public safety – police, fire, and emergency medical – communications in the unincorporated areas of their county (three incorporated cities in the county have their own public safety departments and dispatch operations). The Central County Sheriff's Department polices that unincorporated region; several departments in the region share fire protection, and both fire departments and a private contractor provide emergency medical assistance. The processing of calls and the radio dispatch of sheriff's deputies, paramedic units and fire equipment from Central County takes place in a single, large room. There are six consoles containing connect buttons for the twelve 9-1-1 emergency lines and public, non-emergency numbers as well as numerous direct lines to other emergency service organizations. Virtually any of these lines may be a conduit for a report of a situation requiring some type of official response. The consoles also contain radio equipment required for dispatching the appropriate emergency units.

The consoles are arrayed along a workbench shaped like an upside-down L, and each console or "station" is dedicated to a particular service, such as police or fire. There are four consoles to the long leg of the L, including the stations dedicated to the dispatch of fire and paramedic services, and two to the short leg, which are both concerned with communications with the many sheriffs' units in the field. In the approximate center of the room and paralleling the leg of the L from which fire and paramedic services are dispatched is a large tub containing cards bearing all street addresses in the county. A smaller worktable parallels the other leg. Several chairs that are often used by visitors to the Center sit opposite the tub (see Figure 1 below).

Phones are answered by pressing the appropriate connect button (the button lights up to indicate that a call is waiting on that line) and picking up a telephone handset. The handsets are all equipped with long coiled cords that permit dispatchers to move with the phone to virtually any other station in the room. In addition to the illumination of the button, a distinctive chime in the case of 9-1-1 calls and the seven-digit non-emergency numbers, and a buzzer for all other lines announce calls. Ordinarily, dispatchers must face the console to answer a call, and to operate the radio.

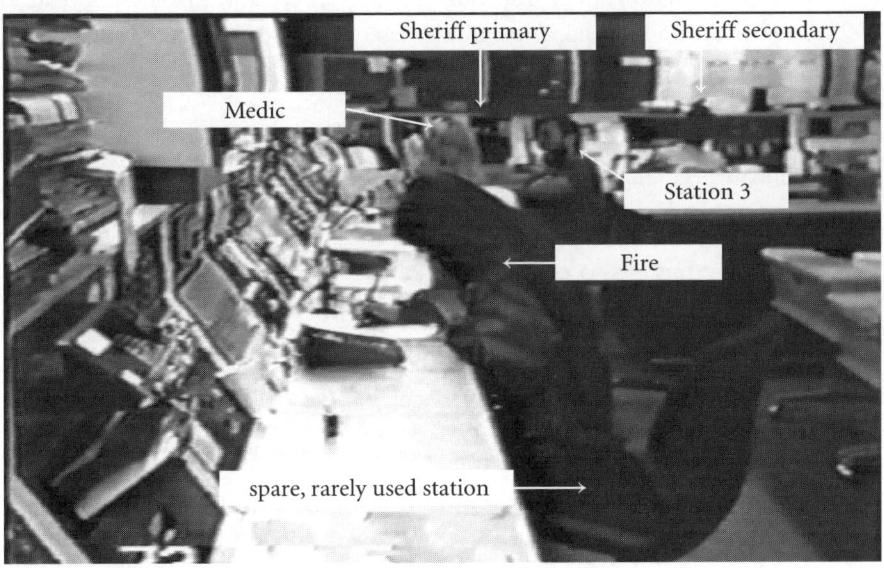

Figure 1. Central County dispatchers and their positions

The consoles are usually identified by function (Figure 1). The first of the four along the long leg is simply a spare station. To its right is "Fire" and next to it "Medic" and, just before the juncture of the two legs, "Station 3." On the other leg is Sheriff's primary or "S1." Next to S1 is Sheriff's secondary or "S2." Staffing is thus also organized by function, and so dispatchers are assigned, for a certain period of time (it could be an entire shift, or only part of a shift) to a particular station, and thus to a particular role, such as "Fire."

There is an established order in which these dispatch functions are to answer the phone lines: first Medic, then Fire, and then "Station 3" (if staffed), which is where shift supervisors usually sit. On the rare occasions when the personnel staffing those positions are all occupied on the phones, the next position in line to answer calls is the secondary Sheriff's position, which is responsible for the operation of the computer that checks on the validity of drivers' licenses and car registration and for the existence of "wants" or arrest warrants for individuals. The primary Sheriff's dispatch position is last in line for calls, because of their responsibility for maintaining radio communication with units in the field at all times.

We can mention here, as a way of introducing some important factors in our analysis, that this kind of division of labor in Central County favors coordinated, multi-party processing of single telephone calls for two reasons.

- Call taking (answering citizen calls for assistance) and dispatching (radio communication with appropriate emergency units) are located together in the same workstation.
- A single team of people performs these tasks, rather than distinctly different and specialized teams, as can be the case in other emergency dispatch organizations. That is to say, the same personnel answer phones, take information, and dispatch by radio.

The physical setting is also configured in a way that facilitates interaction between dispatchers, which then helps support multi-party management.

- There are no barriers – glass partitions, large workstation consoles, or the like – bounding off regions either visually or aurally. Whether operating the phone or radio equipment, then, these dispatchers work side-by-side.
- The absence of such barriers, together with the small size of the setting, makes for comfortable vocal communication and mutual monitoring.
- In addition, the dispatchers all sit on swivel chairs and can regularly swing around to face (and speak with) the other staff (although when they are directly engaged in a phone call or radio transmission, this sort of position-

ing and shift in involvement, even if only momentary, has to be carefully coordinated with the phone and radio exchanges).

However, while these readily observable organizational and architectural features of Central County Dispatch may appreciably assist multi-party management of calls and collaborative task performance, our primary concern in this chapter is the set of *interactional competencies* that are able to develop and are naturally fostered by this sort of environment, and which provides the essentially *social* scaffolding upon which multi-party call processing can routinely and recurrently occur. That is the topic to which we now turn.

3. Interactional scaffolding

Dispatching work, whether at Central County or elsewhere, plainly entails a significant amount of *talking*: on the telephone with callers requesting police, fire or medical assistance; over the radio with units in the field; and among co-workers. Talk of this sort can in a strong sense be described as talk *as* work. But in and around such work, dispatchers also engage each other in conversation (some of which might be described as banter but which may also be matters that emerge in the course of discharging their duties), voiced "noticings" of what their colleagues are up to; and remarks about radio transmissions from deputies in the field, sometimes in a amused or even derisive tone. In fact, there is thus typically a constant buzz of conversation under the chimes announcing in-coming calls, bursts of static and talk from radios, and the occasional clatter of a printer.[4]

In short, there is a lot going on even when the phones are not ringing or the radios are quiet, and a lot of that is interaction between dispatchers that we might initially characterize as talk *at* work. A visit to Central County can thus be counted on to produce a characteristic reaction from first-time viewers. Seen through their untutored eyes, the scene appears remarkably informal. Dispatchers, when observed in all this chatter and other activity, are seen to be casual, perhaps even somewhat nonchalant with respect to the performance of their job. But does this scene reflect a lack of work discipline, or are all the informal, "not work" types of exchanges actually useful – indeed, possibly the elementary scaffolding – for the accomplishment of all cooperative tasks?

We can address this question by first considering some of the social competencies that underlie even these apparently casual exchanges. Any timely contributions – bantering remarks, for example – by a participant in a setting or scene to some on-going talk or activity among others require rather close attention to that talk or activity. Across a set of participants, this will require almost continuous *mutual monitoring*, which can involve both visual and aural attentiveness. It is therefore not a mono-modal activity, nor is it passive. Rather, it is a sustained awareness by each participant of other participants' doings; it involves alertness to audible (speech or other vocal productions) and visible displays (e.g., bodily orientations. including gaze, and changes in both), and it is responsive to activities that are not confined to immediate, particular engagements or engrossments.

To take one example, dispatchers who are not working a call or speaking on the radio listen to calls taken by other dispatchers or at least monitor the latter's side of the conversation. These "eavesdroppers" appeared to be especially attuned to what in the dispatch center were called *key words* like "fire," "gun," and "heart attack," which often occurred in a verification slot post caller's report of trouble. These expressions can also be considered *outlouds* (Goffman 1978: 796; Syzmanski 1999: 19–21) that are available to others in the dispatch office and so can also enable cooperative call processing.[5] As we shall see, other types of *outlouds* – such as imprecations directed at particular parties or actions – may also capture other dispatcher's attentions.[6] Thus, dispatchers *overhear*, or better, *actively monitor* talk and activity in the setting, and as a consequence they are able to start dispatching field units and providing important information to them in a timely manner, sometimes even before the phone call is completed.

The collaborative "working" of a call in this manner, touched off by monitoring activities, can be observed in detail in the following, which is based on a videotape of activity at Central County Dispatch recorded just before and during the processing of a call. In the transcript, the Central County staff members are now identified by their specific dispatch positions: SO1, for Sheriff Primary; SO2, for Sheriff Primary; MD, for the medic position dispatcher; and FD, for fire dispatcher. P1 and P2 refer to police (sheriff's deputies) units in the field who are communicating by radio with SO1.

Extract 2

```
SO1: Diju buy a number Bonnie? ((speaking to SO2))
     (1.0)
SO2: Um: seven six four
SO1: Seven six four I might remember that
     (0.3)
SO2: Heh heh heh
     (2.3)
FD:  I know I won't
     (2.0)
```

MD: Ya won't huh
(0.6)
FD: I'll try though
(0.6)
→ MD: Hi. ((answers a phone line immediately after it rings))
(2.4) ((MD is listening on the phone))
→ MD: Da:d iz with a gun? ((speaking to caller))
(0.8)
→ SO1: ↑NO:::
P1: Thirdy three ((radio traffic for SO1))
→ SO1: ↑NO ↑Gu:n=
((SO1 reaches over and picks up her phone handset,
putting it to her ear.))
FD: =Ah:: WHAT? ((FD picks up his phone, putting
it to his ear))
(0.5)
((SO1 puts the card she had been writing on
in the slot, slaps her hand down on the
tablet of paper in front of her, and listens
on the phone while writing on the tablet.))
SO1: ↑Thirdy three ((responding, over the radio,
to P1))
MD: How old's Patrick ((MD is writing on her
yellow tablet))
(2.3)
MD: Seven fifty five West Main? h:
FD: [What's the (phone=number)? ((speaking to MD))
MD: [What's the (phone number)? ((speaking to
caller))
(1.0)
MD: Uh huh ((to caller; making notes on her
tablet))
(0.7)
P2: Fifdy five thirdy five is clear: (.)
all ( ) ((said over the radio))
MD: An' it wuz Patrick who wuz on ( ) ((to
caller))
(0.7)
MD: O::kay,[thank you (.) bye ((to caller))
FD: [(Patrick's) gunna getta phone call from
(me) ((SO1 hangs up her telephone receiver;
FD picks up his phone and punches in numbers on his
console))
SO1: What's thee address? ((speaking to MD
and/or FD, and overlapping part of FD's
prior utterance))
(0.6)

```
MD:  Sixteen thirty five West Main Patrick Rivas
     advises his ↑Dad has a gu:n an' can't find
     his Mo:m an' if Mom isn't fou:nd the dad's
     gunna kill Patrick
FD:  Line's busy (.) Patrick dead
     ((FD hangs up his receiver as he says
     this))
     (0.4)
     ((During this brief silence, and while the
     next few turns are in progress, the
     supervisor comes out of the back room,stops
     at the desk to the right of SO2,
     picks up three or four bundles of cards,
     walks over to her own desk area, and
     sits down.))
MD:  Line's busy an' I'm makin' a (card)
FD:  ( ) are dead now
SO1: OH:: MA::N
     (2.2)
FD:  Patrick is ah: (.) uncoop'rative (ta )?
     ((to MD))
MD:  Yeah:
SO1: So the father has the gun?
     (0.6)
MD:  The father has the[gun will: (.) shoot*=
FD:                    [Father has the gun hm hm hm
     hm hm ((singsong voice, to the tune of "Farmer
     in the Dell"))
MD:  =Patrick if they don't find the mother.
     (1.2)
SO1: Thirdy two: th[irdy seven* (0.4) possible=
FD:                [Oh:: boy
SO1: =family domestic (.) man with a gun ((to
     officers over radio))
```

As this scene opens, the two police dispatchers, both of whom are seated facing their radios, are talking about whether one of them, SO2 ("Bonnie"), bought a ticket for a chance drawing. SO1 offers some intentionally humorous comments about Bonnie's answer being phrased as the number of her ticket, and Bonnie laughs. The fire and medic dispatchers then make some remarks affiliating with SO1's comments. Note that although FD and MD are on the other side of the work area from SO1 and SO2, they are easily able to overhear and, through their subsequent remarks, join (by self-selecting themselves as participants) the latter's brief conversational exchange, or at least comment on it. This type of event – where we can observe, at various times, all parties present in the setting commenting on, joining in, responding to, or displaying through movement and gesture some interest in or stance toward the in-progress or just completed talk of others – is a routine occurrence at Central County Dispatch; one afforded, as suggested above, by the center's physical layout. A chime indicating an incoming phone call is then responded to by MD, who answers that line. Her answer, "*Hi*" (line 14), is geared to the kind of line it is: a direct phone line to another police agency (see Whalen & Zimmerman 1987). After listening for a few seconds, MD says to the caller (who, it turns out, is a police department dispatcher from the nearby town of Santa Teresa), in a voice loud enough to be overheard by all, "*Da:d iz with a gun?*" (line 16). This brings an almost immediate response from SO1, who – still facing her radio – says "*NO:::*" (line 18) and then "*NO::: Gu:n*" (line 20), both in a very high-pitched, squeaky voice; the pitch keys her remark as slightly mock-serious.

What is most important to notice here is SO1's demonstrable monitoring of talk and action that is taking place behind her back: her "*NO::: Gu:n*" remark, placed just then in the sequence of activity, and expressed in that tone of voice, exhibits for others that she has indeed heard and, perhaps most crucially, understood the ramifications for her responsibilities (and those of her officers) of what she heard.

We can further develop these observations by taking note of what SO1 does next, following her "*NO::: Gu:n*" remark. She picks up the phone receiver at her station to listen-in on the call-in-progress between MD and the Santa Teresa police dispatcher (lines 21–22). Having been alerted and having taken advantage of the overhearing possibilities at Central County, SO1 now moves to obtain information about the incident, and to obtain it as directly and quickly as possible by monitoring not only MD's talk but the two-way phone conversation itself. This gives dispatchers an opportunity to begin their radio dispatch work as soon as possible.

But even more evidence of dispatchers' orientation to mutual monitoring as an appropriate, expected activity in their work can be observed in this activity: the type of turn employed by MD immediately following the initial narrative report from Santa Teresa PD. MD employs (line 16) a verification question – "*Da:d iz with a gun?*" – that includes a repeat of the especially consequential word from the prior report turn (produced by the Santa Teresa dispatcher, but not heard on our video recording, since this record does not include the audio of that phone call). We want to suggest in this regard that this utterance, which so caught SO1's attention, was designed, in how it was delivered and its phrasing, to do just that. And in doing this, it also exhibits an expectation that monitoring by SO1 and others is indeed taking place.

In this case, SO1 is then able to gather some information from her monitoring of the phone conversation – at least she writes on her tablet while listening – but remains possibly uncertain with regard to the address. As soon as MD gets off the phone, SO1 asks for that information (line 53). This brings a response from MD that includes not simply the address, but a brief summary of the situation at that location as well, all delivered in a rapid, somewhat matter-of-fact manner (lines 56–60). Observe that all of this information transfer was accomplished vocally, with some preliminary and additional information obtained by SO1 either indirectly through overhearing MD's first verification query about the gun or directly through monitoring the talk on the phone.

The actions of FD while all this activity by MD and SO1 is taking place are also of interest. After monitoring the phone call from Santa Teresa (starting at about the same point SO1 did) and making a point of asking for the phone number at the incident's location, FD disengages from the line, announces (lines 49–50) he's going to call Patrick – the original caller to Santa Teresa, whose father is the person armed with a gun – and then proceeds to do so.

Taken as a whole, these actions provide additional examples of how Central County Dispatch personnel collaboratively, and in a routine, nothing-special-about-it way "work" a call, with one or more dispatchers who are not themselves the recipient of the phone call or the person responsible for dispatch on the incident, not only monitoring or following the action, but initiating actions relevant to the ongoing processing of the call. This collaboration, like the information transfer described just above, is managed, tracked, and evaluated (as to the outcome of actions) by and through talk.

It should also now be quite evident below that a range of practices can appropriate the local ecology, its furnishings and technological trappings for multi-party management of single telephone calls. With respect to these practices, the problems that participants in *any* multi-activity setting (whether a work environment or not) face in establishing and lapse from a conversational orientation, exhibiting that they are "at work" or "available for participating in 'mundane' conversation," managing the allocation and degree of their involvement in the multiple activities that take place, handling possibly competing claims and obligations (see especially Goffman 1963: 33–79), and interactively organizing attention to the actions or talk of others have been addressed by researchers concerned with the foundations of human social organization (see, e.g., Heath 1986; C. Goodwin 1984, 1986; M. Goodwin 1980–1981; and Zimmerman & Whalen 1987). Because these sorts of problems and the practices members have developed to solve them are so critical at Central County (and other dispatch centers) to what we term *interactional readiness*, which then supports the timely, collaborative working of calls, it will be useful to examine their features in more detail.

4. Mutual monitoring

A critical feature of the Central County Dispatch setting is the processes of engagement, disengagement and re-engagement arising from the fact that there are on-going activities that are distinct from processing calls such as social chit chat and shop talk. Such talk is often part of the process of mutual monitoring that readies participants to assist in call processing. One sort of evidence of this orientation to monitoring the on-going, surrounding activities can be found in those cases in our field notes where, for example, a dispatcher does not exhibit that they have overheard phone activity that has implications for their radio dispatch responsibilities, and another dispatcher then checks to see if they are in fact aware of "what's going on".[7]

This level of awareness is by no means confined to work activities as such. We can illustrate this by considering the following comical exchange, involving everyone in the center at that moment: both police dispatchers (SO 1 and SO2), the dispatcher staffing station 3 (ST3) and her visiting spouse (identified as "SP" in the transcript), and finally the fire dispatcher (FD).

Extract 3

```
1   SO1: (Y'know) King One's goin' out there
2   SO2: Where?
3   ST3: HHHH ((exaggerated in-breath))
4   SO1: I dunno
5   SO2: Wha'd I miss?
6   SP:  They're taking him out there to take pictures.
7   SO2: What happened?
8   ST3: You didn't hear?
9   ST3: Three dead. (.)
10  ?    (                 ) serious
11  SP:  Three walking wounded
12  ST3: Three homeless.
12  SP:  Two stretcher cases
13  ST3: Twelve starving
14  SO1: He's en route
15  ST3: And a partridge in a pear tree
16  FD:  ((sings)) And a[partridge in a pear tree
17                      [((chime))
```

Note that ST3 and Spouse produce an extended, fully collaborative and (eventually) song-like response to SO2's "*What'd I miss?*" and "*What happened?*" questions, which were prompted by ST3 and SP's quick-witted setup, suggesting that something dramatically significant had taken place. Even FD, who had not been an active part of the interaction to that point, joins in at the end with a clear demonstration of her grasp of the structure of that collaborative production. (Of course, FD could have only attended to ST3's utterance in line 15; nevertheless, even suc a minimal hearing speaks to FD's attentiveness to talk in the dispatch center.)

Such sequences are seemingly extraneous to the work of emergency dispatching. Nevertheless, by virtue of being alert to what is going on in the center, dispatchers are ready to pick up on, and possibly join in, *both* work-oriented and touched-off, non-work activities in the setting. In a very real sense, then, a key element of the dispatcher's job is to overhear, to listen in, to be aware of *whatever* activities – including then talk of all sorts – in which others are engaged.

5. Transitions between work and not-work activities

Given that the dispatchers monitor a wide range of activities, and that work and non-work talk and activity intermingle, one task plainly confronting dispatchers is the management of transitions between types of involvement. Such transitions are visible and observed in the setting, and are themselves a feature of the scene being monitored. A useful framework for looking at these kinds of transitions is provided by Szymanski's (1999) study of practices for transitioning from silence to talk, and talk to silence among students participating in reading and writing groups in an elementary school classroom.

Syzmanski examined students' practices in re-engaging and disengaging talk with fellow group members in the course of pursuing their task. She shows how re-engagement is achieved through the use of *questions*; *noticings*; *announcements*; and *outlouds* (recall these practices are also seen at County Dispatch). As Szymanski (1999:7) observes, "[n]oticings are produced as observations of the 'passing world' while *announcements* are produced to highlight an action as noteworthy." Announcements can attract attention, identify something as worthwhile or problematic, and like noticings, can also direct performance of some action (e.g., to look at something).

Participants' *questions* were derived from the nature of the task activity itself, available to all present. At County Dispatch, the relevant issue is the transition between what we earlier termed "talk *at* work" and "talk *as* work." For example, in the extract shown below, a dispatcher who has been involved in an extended casual exchange from which her co-participant has just disengaged (to take an incoming call) directs a question ("*Find anyone?*") to her supervisor that is derived from her awareness of the latter's search for someone to substitute for a dispatcher with family problems.

Extract 5

```
       SO:  Oh I remember that last year that was a riot=Yeah she
            was drunk wasn't she
       ?:   Really
       SO:  That was two years ago wasn't it.Yeah (.) It wasn't
            last year
       MD:  Everybody was racked. These guys were idiots. I said I
            gotta get out of this place
       SO1: That's why I decided not to ((chime)) go this year
       MD:  I don't w- I don't like dealing with these
            ((turns head to console and reaches over pushes button))
            people anyway. Why would I want to deal with them
            drunk=
       FD:  =[((softly)) heh heh heh
       SO1: =[((looking toward MD)) heh ((looks down, then back
            toward center of room)) ( )
       MD:  =[((quickly picks up handset and puts it to her ear))
            County Dispatch.
  →    SO1: ((to Supervisor)) Find anybody?
```

Outlouds are another device available for re-engagement. These are "utterances that are not produced for any particular recipient, yet are available for all the co-present participants" (Szymanski 1999: 19). The outloud can do several kinds of work. First, it can occur with and over the course of continuing activity. In a given instance, it may not be specifically designed to re-engage talk, but can provide a source for future talk. Responsive next actions may transform the outloud into an occasion for re-engagement. (A particularly dramatic instance of the role of an outloud in mobilizing collaborative call management will be addressed below.)

Noticing involves remarking on some feature, appearance, event or activity in the environment, including in the latter the parties present. Beyond occasioning responses, noticings may require that a recipient perform some action such as looking, explaining, and so forth. And like questions, an important source of noticings is the task activity underway.

Visually and aurally available events can set the stage for re-engaging talk in and for some activity; that is, there are occurrences which are perhaps best described as "pre-re-engagements," either discontinuing work activity or producing physical displays that indicate availability, like leaning away from the table or other participants (Szymanski 1999: 8). Pre-re-engagement activities, like the practices of re-engagement, are closely tied to and indeed shaped by the ecology of the setting and the nature of task activities within it. As we shall see, in County Dispatch there are pre-*dis*engagement activities as well.

It is important to observe that questions, noticings and announcements are all "first" or initiating actions. Syzmanski's classroom study's (1999) was concerned with the re-engagement of turn-by-turn talk. These practices are governed by the turn-taking organization itself but are enabled by sequence organization. The responses are projected by some first which yield at least a two (and possibly a three) turn sequence, thus providing for entry into turn by turn talk. Her point is that the devices for re-engaging talk are quite generic mechanisms of talk-in-interaction brought to bear in setting specific configurations to manage the contingencies of a particular task or set of tasks.

Syzmanski's observations, then, can be readily extended to the case where engagement issues concern the type of turn-by-turn talk (and other activities) and the *transitions* between them. We will return to a consideration of re-engagement practices later.

Methods for disengaging talk

Like re-engagement, the disengagement of turn-by-turn talk relies on the organization of employs sequences of action:

> [A] sequence-initiating turn elicits a range of responsive actions and thereby effectively re-engages talk. The antithesis of this is producing a sequence-completing action that does *not* elicit a next action, and instead leaves open the possibility that a next action will not be produced. In the absence of a next action, a lapse in conversation will occur.
>
> (Szymanski 1999: 10; emphasis in original)

The key here is the analyzable completion of some sequence that does not project some further vocal action. This is not to say that some next talk/action might not be forthcoming, but rather that a junction has been reached where no particular next action is called for or is appropriate. This provides for the possible disengagement of turn-by-turn talk. A "return to task" is one strategy for disengaging talk in that such a return, in this case, is to reading or writing, which are silent, individual activities. Talk can be extended, of course, through such devices as repair, or post-expansion sequences which involve rounds of talk across participants.

It is important to note here that County Dispatch differs from Szymanski's third-grade reading groups in at least one important respect. In the elementary school groups, we might characterize the relevant *transitions* as SILENCE/TALK/SILENCE whereas for the Dispatch Center, the transitions are TALK AT WORK/TALK AS WORK/TALK AT WORK. That is, as Szymanski noted, the issue for the student task group is re-engagement and disengagement from turn-by-turn talk, whereas for dispatchers the issue is re-engagement and disengagement from different types of talk. The matter is more complicated still. While there are periods of silence in at County Dispatch, it is not unusual for there to be concurrent and different streams of talk and activity on-going at a particular moment.

Thus, while dispatchers may be oriented to action in the room, and as a consequence, toward each other in varying degrees, this does not mean that there is a common focus of interaction among all dispatchers at any moment. It will be useful in this context to consider the full Speeding Cars transcript [not shown]. The transcript indiscriminately, and in serial fashion, represents what was picked up by microphones in the room. To an overhearer (such as the transcriber), this can be a confusing cacophony. To knowledgeable participants, the streams of talk, chimes, buzzers, radio traffic and printers are organized in terms of foregrounded (or backgrounded, as the case may be) sequences of recognizable actions appropriate to the setting.

Not surprisingly, then, from the "inside," embodied activities, including talk between and by dispatchers, constitute the familiar, mutually accomplished environment within which, and in terms of which, further activities are launched. An examination of some ways in which focused interaction is achieved, and transitions between one focus and another, one type of talk and the other, and one activity and another, will illuminate the role of the organization of talk-in-interaction, mutual monitoring, and readiness to interact in producing and navigating the action environment of County Dispatch, and thereby how certain aspects the work of dispatching are accomplished.

Re-engagement displays and fitting

Dispatchers' management of transitions, and of the disengagements and re-engagements these transitions entail, is remarkable for how these are *fitted to* on-going interactions. Part of the work of fitting hinges on the recognizability in the setting of the implications of particular actions in the setting: talking on the phone or radio, writing an incident card. Involvement in writing an incident card (usually the terminal element of a call or radio transaction) does not preclude some level of monitoring of surrounding talk. Consider the following extract, which represents the talk that occurs just before the fragment shown previously.

Extract 6

```
((FD and MD at right angles to console; FD is rubbing MD's back and
shoulders. SO1 is facing the console, writing on an incident card.
There has been prior talk concerning an annual professional meeting.
The current talk has been occasioned by a query from ? if ? is going
to some professional meeting this year.))

MD:  I went up there one year on my motorcycle and
     everybody was staggering around drunk. Harvey was (.)
     trying to grab Mary and I- oh Jesus I could not get out
     of there fast enough.
SO1: ((finishing writing and filing an incident card,
     turning to center))
     Where was this.
MD:  At the (PSA)
SO1: Oh I remember that last year that was a riot=Yeah she
     was drunk wasn't she
```

Notice the fit of SO1's query to the preceding talk in which MD recounts traveling to the meeting. SO1 has moved from involvement in work to re-engagement in a state of talk currently underway via a question shaped to specify the nature of the setting being described. (The "where" in SO1's question is treated by MD's reply and accepted by SO1 as a query concerning the *occasion* rather than the literal location.) After MD's response in line 8, SO1 produces a change-of-state token and offers a recollection of the event that affiliates with MD's account. Our interest here goes to the fact that more than one stream of activity may be attended to by participants, resulting in the *lamination* of work and non-work activities in the setting.

Disengagement displays and fitting

It was noted above that dispatchers often face to the center of the room or at least away from the console in order to monitor the others present or engage in conversation with them. When an incoming call is announced by a chime or buzzer, or radio traffic occurs, they then have to turn to the console to respond to it. These summonses call them from one participation status to another, namely, *from* more casual engagement with colleagues (e.g., discussing the missed roll call) to attention *to* engagement with a caller or a unit in the field. This transition is particularly marked when it involves withdrawal or disengagement from the speaker role or that of addressed recipient in an on-going interaction as opposed to withdrawal from listenership.

One mode of managing disengagement is moving to respond to a summons (partial swivel to console, picking up a hand set, poising a finger to depress a button) while completing an utterance. Note here that the delay to complete an utterance or action appears oriented to *completion* and to *transition* as further consideration of an extract used above will demonstrate.

Extract 7

```
SO:  Oh I remember that last year that was a riot=Yeah she
     was drunk wasn't she
?:   Really
SO:  That was two years ago wasn't it. Yeah (.) It wasn't
     last year
MD:  Everybody was racked. These guys were idiots. I said I
     gotta get out of this place
SO1: That's why I decided not to ((chime)) go this year
MD:  I don't w- I don't like dealing with these
     ((has turned head to console, reaches over to push
     button)) people anyway ((pushes button)). Why would I
     want to deal with them drunk=
FD:  =[((softly)) heh heh heh
SO1: =[((looking toward MD)) heh ((looks down, then back
     toward center of room)) (          )
MD:  =[((quickly picks up handset and puts it to his ear))
     County Dispatch.
SO1: ((to Supervisor)) Find anybody?
```

As Szymanski (1999:2) has argued, the issue for disengagement is producing an utterance that closes a sequence without projecting further talk (in this case, further talk addressed to the dispatcher answering the call). Of course, completion of a sequence does not preclude a participant for further talk; conclusiveness is an interactional accomplishment. In the extract of concern here, MD has been involved in a telling in which SO1 has joined as a recipient and contributor. Notice that MD's account has reached the point where he describes his desire to leave the scene of drunken behavior, and SO1 has contributed an utterance in line 16 tied to MD's characterization of the scene. The chime announcing an incoming call occurs toward the end of SO1's utterance. MD does not disengage directly upon hearing the chime. In line 17 he begins an utterance "I don't w-" but cuts it off and starts a new (or at least differently packaged) utterance at the same time as he turns his head and moves his arm to the console to connect the call (the button is pushed as he utters "anyway"). He picks up the handset as he begins his second utterance, moving it toward his head, and placing it to his ear just as he utters "County Dispatch." (Although the call has been connected, the caller cannot hear MD until he releases that button on the handset.)

The two utterances together form a premise and an inference from it: "I don't like dealing with these people" [so] "why would I want to deal with them drunk." These utterances perhaps stand as MD's summary assessment of the occasion and its personnel, that is, it is a kind of conclusion to *his* telling. If such a summary remark could, however project further response, such as affiliation (e.g., "Yeah, why would you?"). Another kind of response is laughter, keying to the irony of the contrast which is contributed by FD in line 21 and SO1 in line 22. The occurrence of laughter in this position (post summary assessment) and in this context (MD answering a call) provides a response without implicating further talk from MD. MD has transitioned to take the call, disengaging (at the very last moment) from interaction with colleagues. SO1 subsequently displays disengagement, looking toward Supervisor who is approaching her position and addressing her with a question "Find anybody?" that is tied, as explained earelier, to the Supervisor's search for someone to substitute for a dispatcher unable to come in due to a family emergency. Supervisor's overheard telephone calls on this subject provide anchor for the query, environmental doings of all sort being recoverable as means of initiating or contributing to talk.

We should clear that we are not asserting that the practices described above are unique to County Dispatch, or to institutional settings more generally. We would be surprised if we could not find comparable instances in non-institutional data. Our point, however, is that County Dispatch operations, as an activity type (Levinson) or system (Goffman), is an environment in which such issues are routine features of a day's work, and that management of these issues is part of the tasks of mutual monitoring and cooperative task accomplishment.

It is to the latter that we know focus our attention.

6. "Speeding cars and a loud party" redux

As noted above, dispatchers often join the center formation when they are not otherwise occupied; they also disengage and re-engage as incoming calls or radio traffic command their attention. Here, we examine a particular stretch of interaction leading up to the development of an instance of cooperative call processing, an instance of multi-party management of a single telephone call.

A convenient starting point is S01's utterance in line 1 of the Speeding Cars Transcript (see Extract 8). FD, MD, ST3, and SO2 are arrayed in the *center formation*: seated, facing in toward each other, some with their backs to the console (ST3, MD, SO2) or at a right angle to it (FD). SO1 is standing near ST3, facing away from the console. Spouse (SP), who is visiting, is seated in a chair opposite ST3.

Approximately 17 seconds prior to this point, SO1, who has just concluded telling a story about a bogus computer name check (for "Germaine Shepard") swivels away from the center formation to deal with radio traffic. Just after this ST3, observing SO2 eat popcorn as she works, remarks: "She's got to learn how to breathe and eat." SO2 turns toward him and away from the console and mimics the activity of breathing and eating simultaneously, then laughs. Several seconds later, SO1 rises to move to the NAS console to her left. The point we wish to stress here is that the activities at County Dispatch involve numerous transitions, entailing shifts of focus and alignment of participants. When SO1, the previous story teller disengages from the center formation to respond to radio traffic, ST3 makes a remark targeting SO2 but addressed to the formation. SO2's response is matched to the remark. As will become evident, another transition, and new engagements, is incipient in SO1's movement to the NAS console. Our focus here is on the engagement between SO1, ST3 and MD. The interactions between other participants are indented.

Extract 8

SO1: Did you turn this thing down. ((*gaze toward ST3*))
(0.8)
ST3: ((*gaze toward SO1*)) Now- ((*gaze shifts to M*)) Now
she's yellin' at me for turnin it dow[::n
NAS: [(Kahema)
(1.1)
SO1: We::ll they probly missed us.
NAS: Hello kahema
SO2: (honey)
(1.8)
SO1: Kahema. (That's us.)
ST3: Hell:loh: your mama
(1.2)
RT: (521 traffic)
(0.8)
SO2: Uhfirmative. Eleven twenty nine on
a valid class three

```
18  F:   Did y'say honey=
19  MD:  =I betcha he recalled (us) already even=
20  SO2: =I said honey=
21  SO1: =Pacific State, Central County
22  SO2: Uh huh huh (.) I wasn't talking on
23       thuh radio
24       ((chime))
25  SO1: ((gaze toward ST3)) Turkey.
```

At the beginning of this segment, attention is organized around an element of the work routine in the setting, namely, the answering of a roll call for the National Alert Network. A few minutes earlier, a warning buzzer had sounded; SO1 (the dispatcher responsible for responding to the system) waved both arms in the air, and ST3 rose from his seat and turned the volume down. As the current sequence starts, S01 rises from her chair to approach the console containing the alert radio equipment and picks up a handset; as she does so, she turns her body and her gaze to ST3 with the utterance in line 01: "*Did you turn this thing down*". The first thing to note is that the grammatical shape of SO1's utterance is that of a yes-no question. However, this is not the only import of such an utterance. ST3 treats SO1's utterance as a complaint, responding with a (mock?) counter-complaint: "*Now- Now she's yellin' at me for turnin' it dow::n*". Initially, the first element of ST3's utterance is directed (by gaze) at S01 ("*Now-*") and then by a shift in gaze, the restart of ST3,s utterance is directed to MD ("*Now's she's yelling'*…"). ST3's re-directed utterance receives no observable receipt or response from MD (and hence, she does not ratify her proposed participation in this particular piece of by-play). After the 1.1 second gap in line 05, S01's justification for her complaint in line 06, the initial component of which, "*We:ll,*" secures ST3's gaze, but also does not receive a response from MD who shifts her gaze away from S01 shortly after the completion of the utterance.

This brief engagement is managed with respect to differentiation of target (S01) and recipient (MD) by ST3's rapid shift of gaze and by then MD's lack of affiliation with either the (mock) complaint (line 03) or S01's justification in line 06 (the latter managed by M remaining silent and by shifting her gaze away from S01). Thus, there is a brief, bantering engagement, touched off by SO1's query/complaint, followed by ST3s counter-complaint, and bounded for the moment off by MDs non-response.

It is in point to note here that SO1 is engaged in a work activity (answering the NAS roll call). ST3 are MD are not involved in the task but their presence on scene, their participation (or lack thereof), prior interactions, personalities, and the like enter as contingencies of the moment into the performance of work and the clusters of talk and action incidental to it. In short, the line between talk as work and talk at work is often indistinct, not by virtue of inherent lack of distinctness, but as the consequence of the continuity and contiguity of interaction.

Note, moreover, that a particular episode of work (including the task and the incidentals that surround it) can form the history of subsequent activities, and indeed, the interaction initiated by the NAS alert continues past the exchange between ST3 and SO1. ST3 in line 29 begins a telling touched-off by the "missed" NAS roll call, "We missed it one night". Shortly before ST3's utterance, MD responds to a chime announcing an in-coming call by swiveling to the console and answering the call (line 30). F also picks up a handset to listen in; ST3, SO1 and SO2 and S (a visitor) do not listen.

Extract 9

```
29  ST3: We missed [it one night an ] an he called up
30  MD:            [County dispatch ]
31: ST3: here[ (.) an' he screamed (.) goddamit Central County
32  SO1:      [heh heh heh
33  ST3: (.) get yer shit together
34  ?:   (Goddam calling anyway)=
35  MD:  =Speeding car-
36  ?:   Nuclear holocast[(           )]
37  MD:                   [Speeding] cars an' uh (.) loud party?
38  ST3: Speeding car:s.
```

ST3's touched-off telling of a previous incident in which an NAS roll-call had been missed, addressed (by gaze) SO1, SO2 and Spouse. Spouse rises from his seat just as ST3 utters "*Central County*" and walks over to a wastebasket to ST3's left while ST3 completes his utterance in line 33. ST3 glances at Spouse as he walks over, and then re-directs his gaze to SO1 and SO2 whose bodies are oriented toward him. Just as Spouse bends down to deposit something in trash basket MD, who, as noted above, has answered an incoming call, utters "*Speeding car-*" (line 35) and while Spouse is still there, "*Speeding cars an' uh* (.) *loud party*?" in line 37.

In this case, MD's remark, produced in the verification slot, is a display of independent knowledge designed, possibly, to exhibit to caller that the Sheriff's Department already knows of the situation. ST3 does a noticing of the incident in line 36 ("Speeding car:s") but does not mark it as requiring attention.

Extract 10

```
35  MD:  =Speeding car-
36  ?:   Nuclear holocast[(            )]
37  MD:                  [Speeding] cars an' uh (.) loud party?
38  ST3: Speeding car:s.
39  RT:  =[Thirty five forty two two six] dee
40  ?:   =[Kansas City ( )]
41  ST3: Kansas City's[( )]
42  SO1               [ Ten four ]
43  FD:               [(Go)      ]fuck off::
44  SO1:              [Get uh job]
45  ST3: They all go home thuh day after
46  FD:  Tell 'im to call his mother
47  RT:  Thirty nine delta four possible eleven five, ((static))
48  ST3: What?
49  SO2: Possible?
50  SO1: Delta four
51  FD:  He's not getting thuh pro[tection he's payin' his taxes
52  RT:                           [Two five Ida David Mary
53  FD:  for]
53       [( )] South of Fairfield
54  ST3: [Let me talk to'm
```

FD, however, has been listening in, and at approximately the point at which caller is complaining that his taxes are not providing the service he expects (see Extract [1], lines 10–14), she hangs up and produces an outloud in line 42. "*Oh fuck off*" (which appears to be a dismissive response to the caller's complaint) and then disconnects. Although it may be have been designed to call attention to this call,[8] there is no discernable uptake of this remark by other dispatchers. As she lays the handset on the counter of the console she produces a second outloud in line 46, "*Tell 'm to call his mother*" (a mock, derisive remedy for the complaint). This remark draws the gaze of Spouse who looks in her direction just as he returns to his chair. And just at this point radio traffic begins in line 47 ("*Thirty nine delta four possible eleven five,* ((*static*))") which occasions the turn of both SO1 and SO2 away from the center and back to the console. ST3 then returns his gaze to Spouse to find him looking in the direction of FD.

Perhaps having half-registered FD's remarks in line 42 and 46, or seeking to locate what has drawn S's attention, he also looks in FD's direction and says "*What?*" in line 48. Note here that ST3's query is directed in toward FD (or at least that section of the room), and employs what Drew (1997) has described as an open-class repair initiator. The sequential environment of its use here is different than those addressed by Drew. It is neither a topically disjunctive or inapposite utterance *within* an on-going conversation (FD. was listening in on the call since line 23 in Extract [6], and not involved in the prior talk between ST3 and S).[9] Rather, it is responsive to an outloud, an *open remark* potentially available for anyone[10] within hearing to attend or respond to. Further, it may be an instance of an unregistered or half-heard remark made retrospectively relevant by another's (S's) interest in it (as displayed in gaze directed toward its source). Thus, ST3 may have heard *something*, or upon seeing that S's attention has been drawn to another, he seeks to clarify what is going on.

FD's response in line 49 however is not a repeat of her last outloud, but rather a statement of the issue to which it was addressed: "*He's not getting thuh protection he's payin' his taxes for*" (line 51/53).[11] As she says this, she pushes back from the console for a clear line of sight to ST3. Notice that there is no vocalized prior referent for "he." However, ST3 hears FDs utterance as a quotation (with transformed personal pronouns) of the current *caller's* complaint; ST3's uptake of this is displayed as he glances backward at an even more extreme angle to MD who is on the phone to the caller (FD and MD routinely listen in on calls received by each station).That is, ST3 treats FD's utterance as work-relevant, as is evident in ST3's quick swivel to the console, his utterance (line 52) "*Let me talk to 'm.*" and his grasping the handset and putting it to his ear. He also leans back in his chair with his gaze on MD. His proposed involvement in the call has come about by monitoring another party monitor an environment of activities with which all parties are familiar and where various actions, including talk, are to some degree transparent in their sense and import.

We also note here that this outloud and its subsequent expansions do not employ a term such as "gun" (as in the "Patrick" call discussed earlier) or some other "hot button" term to mobilize collaborative call management. Nevertheless, FD's reply to ST3's repair initiation is understood as calling attention to a feature of the current call that invites (or perhaps commands) multi-party involvement ("*Let me talk to 'm*"). The words or utterances that can do this work are not confined to a bounded list, nor are the issues that give rise to collaborative call management necessarily those of threat to life or property.

Thus far, five parties have become involved in this call: the caller, MD, who first answered the call, FD who listened in and commented on the caller's complaint, SP who "notices" FD's outloud and whose gaze direction provides the pivot for ST3's "open class repair" on FD's remarks. Two other dispatchers will become involved, one actively (SO1), the other (SO2) passively.

Extract 11

```
56   ST3: [Let me talk to'm]
57   SO1: We just cleared a call[( )
56   M:                         [Just uh moment sir?]
```

SO1: southbound south of Fairfield
ST3: Want me tuh talk to 'm?
(0.9)
F: Is this the Maywood [thing?
RT: [South of Roberts
M: [Have they-]Yeah Have
they cleared?
(0.2)
F: I just had a lady an I [threw thuh card in] thuh
SO1: [South of Roberts]
F: garbage, I didn't even go,
SO1: [Two eight two] Ida David Mary
ST3: [You had uh lady?]
(2.0)
M: This guy: This guy is: (.) complaining too.
ST3: ( ) what, a party?
M: He's not gettin' satisfaction from his tax
dol[lars
?: [( )
F: [Till they tell us what's goin' on I'm not
goin'=
M: =Are they still there?
SO2: Stop paying his tax(h)es.
(1.3)
SO1: Where?
ST3: Ye(h)eah. [Sir, if you're] dissatisfied with
M: [M- Maywood]
ST3: [the service stop paying your taxes]
SO1: [<u>Yeah</u>. They're all] out there.
M: They're still there?
SO1: Yeah.
F: They're not giving him thuh protection he needs.
SO1: Ohh ye(h)ah.
ST3: They are still there Lizzy?
((chime))
SO1: They're sti(h)ll there.
(1.6)
SO1: His tax [dollars are hard at work]
ST3: [Sir apparently our] units are
still on thuh scene at that (0.3) problem.
SO1: I've got (.) <u>four</u> units out there.
(0.2)
ST3: We've got four units there sir.
(5.0) ((printer sounds))
M: Maybe its a different house
SO1: Well
ST3: Maybe we've got more than [one party] going on?

```
105  SO1:                              [He's blind]
106  RT:  Forty two thirty three
105  ST3: What's thee address[(right where)         ]
106  SO1:                      [Forty two thirty three]
107  ST3  you're looking
108       (1.4)
109  RT:  We're ten eight with a twenty five (0.8) six
110       sixty two
111       (0.8)
112  M:   Is it Lexington?
113  ST3: Is it Lex[ington?
114  SO1:           [ Ten four ]
115  M:             [They're there]
116  F:   sign a ( )
117  SO1: Thirty nine delta four,
118       (1.0)
119  ?:   (They have to have a bedtime[( )
120  RT:                               [Thirty nine delta
121       four( ) breakin up (( static burst ))
122  ?:   ( in their house on Cleariew)
123  SO1: ((sneezes twice))
124: SO2: Bless you=
125  F:   =What's your address?
126  SO1: He's the(h)re an' they're breakin' up thuh party
127       and that guy's blind
128  ST3: We are: (.) We are on thuh scene [sir. They just=
129  SO1:                                  [ Ten four
130  ST3  =told me on thuh radio that they're breakin' up
131       [thee party
132  F:   [Which way is it coming from?
133       (1.0)
134  SO1: Aw shit ( )
135  ST3: Alright? (.) Sure
```

As ST3 leans back in his chair, handset to ear, MD turns to him, and as she does she says to caller "*Just uh moment sir?*" and puts him on hold. ST3 again offers to speak to the caller (line 59), which is followed by FD's query addressed to this issue and MD begins a interrogatively shaped utterance ("*Have they-*"[12]) which she cuts, shifting her gaze to FD and responding to her ("*Yeah*"). MD then returns to the question that she was on the way to producing: "*Have they cleared?*". After some discussion with FD of the of the disposition of earlier calls on this incident, MD turns from the console and gets up, saying: "*This guy: This guy is:* (.) *complaining too*" (line 72). ST3, who is tracking M as she moves to the area where "cleared" incident cards are stored, asks "*About what, a party?*" (line 73) to which MD replies, reprising the form of the complaint: "*He's not getting*' *satisfaction from his tax dollars*" (lines 74–75). As MD passes behind SO1 she asks (line 79) "*Are they still there:*" (i.e., are the sheriff's deputies still on the scene.") SO1 asks "*Where?*" (line 82) and is informed by MD of its location (line 84). SO1 then confirms that units are still on the scene, which MD follows up with a confirmatory question on the presence of units, which SO1 affirms (lines 86–88). MD continues her movement to the area in which "cleared" incident cards are kept, where she shuffles through them and then begins to return to her station.

As MD interacts with SO1, SO2 joins in with a mock remedy to the caller's complaint, "*Stop paying his taxes*" (line 80), which ST3 picks up on: "*Ye(h) eah. Sir, if you're dissatisfied with the service stop paying your taxes*" (line 83, 85). FD enters in again (line 89) after SO1's affirmation that deputies are still on scene with the remark "*They're not giving him the protection he needs.*" SO1 responds with a sarcastic "*Oh ye(h) ah*" (line 90). This byplay among the dispatchers, primarily among those who are, at the moment, not directly engaged in some aspect of managing the call, subjects the caller's complaint to ridicule and affirms a shared orientation to it (which, as will be seen below, is resumed after the call's conclusion).

ST3 now asks (line 91) SO1 if deputies are present on the scene, and as he does so, MD signals him with a hand gesture to take the call. SO1 confirms (for the fifth time) that deputies are on scene, adding that the caller's "*tax dollars are hard at work*" (lines 93, 95). ST3 speaks to the caller at this point, informing him that there are units on the scene[13] (lines 96–97). SO1, who has picked up her handset to listen in specifies her earlier report to ST3 "*I've got (.) four units out there*" (line 98), which ST3 repeats to the caller.

At this point, the call is being managed by ST3 and SO1. MD, however, has picked up the phone, and hearing the caller dispute the report that there are deputies on the scene (Extract [1], lines 24–27) suggests (to ST3) that it might be "*a different house*" (line 102), which ST3 relays to the caller as the possibility that there is "*more than one party going on*" (line 104). To the caller's "*Huh?*", a type of repair initiation that suggests that caller has found ST3's suggestion disjunctive or inapposite (Drew 1997), ST3 does not repeat his utterance but instead pursues the issue it raised by asking for the address of the problematic party ("*where you're looking*", lines 105–107). The caller is uncertain about the cross street and MD, still listening in, prompts ST3 to ask "*Is it Maywood*" (line 113). After the caller agrees that it is, SO1 (also listening in) can be heard over the phone line to say "*They're there*" (line 116).

Following a 1.3 second silence, caller pursues his complaint that the disturbance is not being dealt with (Extract [1], lines 40–45). SO1 can then be heard on the line with her comment that the deputies are "*breakin' up the party- that guy's blind*" (lines 127–128). This is followed immediately by ST3's restating this information, omitting the remark about the eyesight of the caller (lines 129, 131–132). The call closing follows,[14] after which MD, SO1, SO2 and ST3 engage in a series of remarks continuing to question the caller's report of the problem, suggesting that special powers or equipment would have been necessary for him to accurately assess what was happening at the scene of the party (lines 137–156).

Extract 12

```
137  M:   If he's two streets [away how can he see Lexington
138  F:                       [Thuh music.
139  SO1: Really.
140       (0.6)
141  SO2: Xray vision?
142       (0.9)
143  ST3: He's got mirrors
144  SO1: hem hem hem
145  F:   Folsom's Ranch
146  S:   Could we just tell 'm he was bli[ind?
147  SO2:                                 [He has uh
148       periscope that goes up above his house and he
149       aims it where he wantsit to be heh heh
150  SO1: 's that Mister Hare?
151  ?:   hhh hhh heh heh huh huh ehk .ehk ((could be more
152       than one person))
153  M:   No::
154  ST3: 's awfully clo(h)se
155  ?:   .ehk .ehk
156  SO2: [( ) it's uh magnifying periscope
157  ?:   [(                                )
```

Several matters bear mention at this point. First, the occasion of this interaction is *work*: the matter of managing a caller's complaint that a loud party was not being effectively attended to by the Sheriff's Department. Such work hinges on the mutual monitoring and consequent interactional readiness of participants: how they attend to activities, include talk, in the setting, and involve each other in the course of managing calls. In this instance, the caller's complaint, and the circumstances it targets is thoroughly *socialized*: it is managed in a way closely attentive other participants whose initiating and responding actions co-construct, moment by moment, the County Dispatch center as an achieved work environment *within which the parties function as alert, mutually involved social actors*. Moreover, the coda to the call is an expression of solidarity among the team of dispatcher's handling a call that was as much a complaint about *them* as about the disturbance they were responsible to deal with.

7. Conclusion

We have argued in this paper that the work of calls to emergency services is often not restricted to talk between a single caller and a single call-taker. Rather, calls can involve the closely coordinated work of the organization's staff who may listen in, or simply attend to the talk of other call-takers. The point we stress is that such multi-party involvement is a mode and method of call processing. Thus, we reiterate that however productive research on the telephone call itself has been, we cannot grasp the actual organization of the work of emergency telecommunications until we take into account this teamwork and its often complex interactional infrastructure of personnel who manage calls both on and off the phone.

Multi-party management of single calls rests on an interactional scaffolding that involves require almost continuous *mutual monitoring*, both visually and aurally. It is a mode of active attentiveness to what is going on in the setting, sustained in part by serial interactional engagements that may or may not directly involve work. As we have shown, dispatchers are sensitive to the talk of colleagues, and are sensitive to the occurrence of key or work-relevant terms like "fire," "gun," and "heart attack," which often occurred in a verification slot post caller's report of trouble. When not engaged themselves, dispatchers *overhear*, or *actively monitor* talk and activity in the setting. This enables them to initiate the dispatch of appropriate units in a timely manner, sometimes even before the phone call is completed.

We characterized this active monitoring as *interactional readiness*, and described how the processes of engagement, disengagement and re-engagement organized managed transitions from casual talk to work and back to casual talk. Casual talk, which entails interactional readiness and social presence, is part of the process of mutual monitoring, and readies participants to assist in call processing. The observations reported in this paper, then, suggest that common-sense notions of what it means to be "at work", and what it means to be engaged in task activity, may mislead us into thinking that a good deal of what goes on in a site such as Central County is unrelated to the job that dispatchers are paid to perform. The requirement to be aware of what others are saying and doing in the setting requires – or at least is partially dependent upon – being engaged with others in the setting. Engagement, in turn, requires some degree of mutual monitoring. It is out of the awareness this brings that dispatchers can be ready to pick up on indications that some call might require joint processing. We note that when dispatchers are not talking to caller or to units in the field over radio, they are likely to be talking to one another rather than, for example, reading a book.[15] Thus, while these engagements may appear to be peripheral to the job, or even subversive of it, they may be nicely suited to the interactional demands of call processing in Central County.

Notes

1. In a strong sense, CAD is actually another "player" in the interaction that is usually not transparent to a transcript.

2. Similar evidence of multiparty involvement has been apparent in transcripts of calls analyzed elsewhere. The infamous call to the Dallas Fire Department (Whalen et al. 1988) is opened by a switchboard screening by an operator who transfers the call to a Nurse-Dispatcher. There is the explicit involvement of a second official (the Nurse-Dispatcher's supervisor), and the caller's roommate. The latter is an unratified participant in the first call, perhaps consequentially so, as he makes the successful second call for help possibly primed as to what to say by his overhearing of the unfortunate course of the first call. The hysteria calls also exhibit multiparty involvement, as well as a number of Western and Lane County calls.

3. For recent collections, see Luff et al. (2000), and the 2002 special issue of the *British Journal of Sociology*.

4. In the late-night/early morning shifts when there is very little phone or radio activity, dispatchers may read, knit, or tend to various chores incidental to the job. It is interesting that more interaction between dispatchers occurs when the center is busy with incoming calls and radio traffic.

5. It is interesting to speculate that "verification" is the secondary rather than primary function of the utterance that encompasses two recipients – caller *and* whoever's attention is captured in the dispatch center.

6. These task-relevant terms may also occur in utterances designed to verify a information about the problem obtained from sources outside the current call. In the "Patrick" example discussed in the text, a dispatcher calls the residence for which he has received a report about a domestic dispute involving a father and his son from another law enforcement jurisdiction. The son answers the phone:

D: This iz thuh sheriffs department Patrick
(0.3)
C: Yes
D: Whats goin' on?
(0.5)

C: I jes' went somewhere to get thih (res)
D: What's goin' o:n >Patrick<?
C: I ahreedy (.)g-ive that same information[to-
D: [PATrick
we're on thih wa:y, you answer my questions
(0.8)
What iz, going o:n?
C: I cannot answer yer question right now °sir
D: Iz yer dad there?
C: Yes sir
(1.1)
→ D: Duz he have ah gun?
C: Yes sir

7. These observations are based on our fieldwork at Central County.

8. While we cannot be sure that the remark was designed to capture the attention of others, it was not uttered *sotto voce*, and as noted in the text, it was followed up by a second outloud. Note also that the first outloud, while dismissive, is not informative as to the offense that prompted it, whereas the second outloud provides an allusive specification. FD's reponse to ST3's "*What?*" appears to be the third in a progressively more specific characterization of what FD took issue with in the call.

9. Although FD was not proximately involved in talk with ST3, she is one of a coterie ratified interactants (i.e., a fellow dispatcher) and hence, what ever it is that she said, it can be viewed by ST3 as potentially relevant *for him*. See also Note 15.

10. "Anyone" in this case, being primarily those ratified participants in the activity of the setting, i.e., other dispatchers. Note that while S participates in conversation and banter with the dispatchers in those moments when they are not engaged in on the phone or on the radio, his response to FD's outloud is merely a shift in his gaze toward her. It is ST3 who, following S's gaze, initiates an action that proves consequential for the subsequent management of the call.

11. A common response to a repeat request is to repeat all or part of what was previously said. The prior utterance, however, was an outloud, touched off by what the caller had said, and if designed at all, would be designed for a hearer to inquire into the occasion for the comment. Whether or not ST3 actually heard what FD said, his query in this sequential environment is treated as a request for an explication of what occasioned the remark, not the remark itself.

12. It is likely that MD is in the process of addressing the query "*Have they cleared?*" (i.e., resolved the incident) to ST3, but re-addresses it to FD who has raised the question of whether this call is about the same incident as the previous call. If this is so, MD's truncated query to ST3 is an aborted initiation of an insertion sequence aimed at gathering more information concerning the incident. MD's subsequent activities leading up to her hand gesture releasing ST3 to take the call can then be viewed as addressing the issue that motivated the abandoned assertion, a matter to be settled before resuming contact with the caller.

13. For reasons unclear to us, ST3's utterance at this point does not appear on the tape of the call itself.

14. ST3's closing implicative report that deputies are present on the scene and are breaking up the party is, of course, contrary to the stance that the caller has taken, i.e., inaction by the authorities. The caller's reluctant acceptance of this outcome of the call is seen in the 1.4 second silence, with "Okay" which merely acknowledges ST3's closing move but withholds endorsement of it (as would be done via "thank yopu" as a service receipt). ST3 does not treat does not treat "okay" as a sufficient receipt; he pursues with the query "*Alright?*" that elicits a "thank you" (a service receipt) and rapid call closure occurs (cf. Zimmerman & Romero 2002).

15. An exception to this is in the late-night/early morning shifts when there is very little phone or radio activity. At these times, dispatchers may read, knit, or tend to various chores incidental to the job. It is interesting that the more interaction between dispatchers occurs when the center is busy with incoming calls and radio traffic.

References

Drew, P. & Heritage, J. (1992). 'Analyzing talk at work: An introduction'. In P. Drew & J. Heritage (Eds.), *Talk at Work: Interaction in Institutional Settings* (pp. 3–65). Cambridge: Cambridge University Press.

Emerson, R. & Messinger, S. (1977). "The mico-politics of trouble." *Social Problems, 25*, 121–134.

Garcia, A. C. & Parmer, P. A. (1999). 'Misplaced mistrust: The collaborative construction of doubt in 911 emergency calls.' *Symbolic Interaction, 22*(4), 297–324.

Goffman, E. (1963). *Behavior in Public Places: Notes on the Social Organization of Gatherings*. New York: Free Press.

Goffman, E. (1978). 'Response cries.' *Language, 54*, 787–815.

Goffman, E. (1981). *Forms of Talk*. Oxford: Basil Blackwell.

Goodwin, C. (1984). 'Notes on story structure and the organization of participation'. In J. M. Atkinson & J. C. Heritage (Eds.), *Structures of Social Action: Studies in Conversation Analysis* (pp. 225–246). Cambridge, UK: Cambridge University Press.

Goodwin, M. H. (1980–1981). 'Processes of mutual monitoring implicated in the production of description sequences.' *Sociological Inquiry, 50*, 303–317.

Heath, C. (1986). *Body Movement and Speech in Medical Interaction*. Cambridge, UK: Cambridge University Press.

Lerner, G. H. (1993). 'Collectivities in action: Eastablishing the relevance of cojoined participation in conversation.' *Text, 13*, 213–245.

Levinson, S. C. (1992). 'Activity types and language'. In P. Drew & J. Heritage (Eds.), *Talk at Work: Interaction in Institutional Settings* (pp. 66–100). Cambridge: Cambridge University Press.

Luff, P. K., Hindmarsh, J., & Heath, C. (2000). *Workplace Studies: Recovering Work Practices and Informing Systems Design*. Cambridge: Cambridge University Press.

Meehan, A. J. (n.d.). 'Assessing the police-worthiness of citizen's complaints to the police: Accountability and the negotiation of 'facts''. Unpublished paper.

Meehan, A. J. (1989). 'Assessing the "policeworthiness" of citizen's complaints to the police: accountability and the negotiation of "facts"'. In D. T. Helm, W. T. Anderson, A. J. Meehan, & A. W. Rawls (Eds.), *The Interactional Order: New Directions in the Study of Social Order* (pp. 116–140). New York: Irvington.

Sacks, H., Schegloff, E. A., & Jefferson, G. (1974). 'A simplest systematics for the organization of turn taking for conversation'. *Language, 50*, 696–735.

Sharrock, W. W. & Turner, R. (1978). 'On a conversational environment for equivocality'. In J. Schenkein (Ed.), *Studies in the Organization of Conversational Interaction* (pp. 173–197). New York, NY: Academic Press.

Tracy, K. (1997). 'Interactional trouble in emergency service requests: A problem of frames.' *Research on Language and Social Interaction, 30*, 315–343.

Tracy, K. & Agne, R. R. (2004). 'Sensitivities in calls to the police about domestic disputes.' In J. Cotterill (Ed.), *Language in the Legal Process* (pp. 75–89). Brunel, UK: Palgrave Publishers.

Tracy, K. & Tracy, S. J. (1998). 'Rudeness at 911: Reconceptualizing face and face attack.' *Human Communication Research, 25*, 225–251.

Tracy, S. J. & Tracy, K. (1998). 'Emotion labor at 911: A case study and theoretical critique.' *Journal of Aoolied Communication Research, 26*, 390–411.

Tracy, K. & Anderson, D. L. (1999). 'Relational positioning strategies in police calls: A dilemma.' *Discourse Studies, 1*, 201–225.

Syzmanski, M. (1999). 'Re-engaging and dis-engaging talk in activity.' *Language in Society, 28*, 1–23.

Wakin, M. & Zimmerman, D. H. (1999). 'Reduction and specialization in emergency and directory assistance calls.' *Research on Language and Social Interaction, 32*, 409–437.

Whalen, J. (1995a). 'A technology of order production: Computer-adied dispatch in public safety communication'. In P. ten Have & G. Psathas (Eds.), *Situated order: Studies in the Social Organization Talk and Embodied Activities* (pp. 187–230). Washington, DC: International Institute for Ethnomethodology and Conversation Analysis and University Press of America.

Whalen, J. (1995b). 'Expert systems versus systems for experts: Computer-aided dispatch in public safety communication'. In P. Thomas (Ed.), *The Social and Interactional Dimensions of Human-computer Interaction* (pp. 161–183). Cambridge: Cambridge University Press.

Whalen, J., Zimmerman, D. H., & Whalen, M. (1988). 'When words fail: A single case analysis'. *Social Problems, 35*(4), 335–362.

Whalen, J. & Zimmerman, D. H. (1998). 'Observations on the display and managment of emotion in naturally occurring activities: The case of "hysteria" in calls to 9-1-1'. *Social Psychological Quarterly, 61*, 141–159.

Whalen, M. & Zimmerman, D. H. (1987). 'Sequential and institutional contexts in calls for help'. *Social Psychology Quarterly, 50*, 465–492.

Whalen, M. & Zimmerma, D. H. (1990). 'Describing trouble: Practical epistemology in citizen calls to the police'. *Language in Society, 19*, 465–492.

Zimmerman, D. H. (1984). 'Talk and its occasion: The case of calling the police'. In D. Schiffrin (Ed.), *Meaning, Form, and Use in Context: Linguistic Applications* (pp. 210–228). Washington, DC: Georgetown University Roundtable on Language and Linguistics.

Zimmerman, D. H. (1992a). 'Achieving context: Openings in emergency calls'. In G. Watson & R. Seiler (Eds.), *Text in Context Contributions to Ethnomethodology*. Sage: Newbury Park.

Zimmerman, D. H. (1992b). 'The interactional organization of calls for emergency assistance.' In P. Drew & J. Heritage (Eds.), *Talk at Work: Interaction in Institutional Settings* (pp. 418–469). Cambridge: Cambridge University Press.

Zimmerman, D. H. & Romero, D. (2002). 'The interactional basis of differential deployment of "thank you's" in closing sequences of emergency calls'. Paper presented at the *National Communication Association Meetings*, New Orleans.

Name index

Subject index

In the *Pragmatics & Beyond New Series* the following titles have been published thus far or are scheduled for publication:

143 **BAKER, Carolyn D., Michael EMMISON and Alan FIRTH (eds.):** Calling for Help. Language and social interaction in telephone helplines. 2005. xvii, 351 pp.

142 **SIDNELL, Jack:** Talk and Practical Epistemology. The social life of knowledge in a Caribbean community. xv, 255 pp. *Expected October 2005*

141 **ZHU, Yunxia:** Written Communication across Cultures. A sociocognitive perspective on business genres. xviii, 208 pp. + index. *Expected November 2005*

140 **BUTLER, Christopher S., María de los Ángeles GÓMEZ-GONZÁLEZ and Susana M. DOVAL-SUÁREZ (eds.):** The Dynamics of Language Use. Functional and contrastive perspectives. 2005. xvi, 413 pp.

139 **LAKOFF, Robin T. and Sachiko IDE (eds.):** Broadening the Horizon of Linguistic Politeness. xii, 342 pp. *Expected October 2005*

138 **MÜLLER, Simone:** Discourse Markers in Native and Non-native English Discourse. ix, 282 pp. + index. *Expected November 2005*

137 **MORITA, Emi:** Negotiation of Contingent Talk. The Japanese interactional particles *ne* and *sa*. 2005. xvi, 240 pp.

136 **SASSEN, Claudia:** Linguistic Dimensions of Crisis Talk. Formalising structures in a controlled language. 2005. ix, 230 pp.

135 **ARCHER, Dawn:** Questions and Answers in the English Courtroom (1640–1760). A sociopragmatic analysis. 2005. xiv, 374 pp.

134 **SKAFFARI, Janne, Matti PEIKOLA, Ruth CARROLL, Risto HILTUNEN and Brita WÅRVIK (eds.):** Opening Windows on Texts and Discourses of the Past. 2005. x, 418 pp.

133 **MARNETTE, Sophie:** Speech and Thought Presentation in French. Concepts and strategies. 2005. xiv, 379 pp.

132 **ONODERA, Noriko O.:** Japanese Discourse Markers. Synchronic and diachronic discourse analysis. 2004. xiv, 253 pp.

131 **JANOSCHKA, Anja:** Web Advertising. New forms of communication on the Internet. 2004. xiv, 230 pp.

130 **HALMARI, Helena and Tuija VIRTANEN (eds.):** Persuasion Across Genres. A linguistic approach. 2005. x, 257 pp.

129 **TABOADA, María Teresa:** Building Coherence and Cohesion. Task-oriented dialogue in English and Spanish. 2004. xvii, 264 pp.

128 **CORDELLA, Marisa:** The Dynamic Consultation. A discourse analytical study of doctor–patient communication. 2004. xvi, 254 pp.

127 **BRISARD, Frank, Michael MEEUWIS and Bart VANDENABEELE (eds.):** Seduction, Community, Speech. A Festschrift for Herman Parret. 2004. vi, 202 pp.

126 **WU, Yi'an:** Spatial Demonstratives in English and Chinese. Text and Cognition. 2004. xviii, 236 pp.

125 **LERNER, Gene H. (ed.):** Conversation Analysis. Studies from the first generation. 2004. x, 302 pp.

124 **VINE, Bernadette:** Getting Things Done at Work. The discourse of power in workplace interaction. 2004. x, 278 pp.

123 **MÁRQUEZ REITER, Rosina and María Elena PLACENCIA (eds.):** Current Trends in the Pragmatics of Spanish. 2004. xvi, 383 pp.

122 **GONZÁLEZ, Montserrat:** Pragmatic Markers in Oral Narrative. The case of English and Catalan. 2004. xvi, 410 pp.

121 **FETZER, Anita:** Recontextualizing Context. Grammaticality meets appropriateness. 2004. x, 272 pp.

120 **AIJMER, Karin and Anna-Brita STENSTRÖM (eds.):** Discourse Patterns in Spoken and Written Corpora. 2004. viii, 279 pp.

119 **HILTUNEN, Risto and Janne SKAFFARI (eds.):** Discourse Perspectives on English. Medieval to modern. 2003. viii, 243 pp.

118 **CHENG, Winnie:** Intercultural Conversation. 2003. xii, 279 pp.

117 **WU, Ruey-Jiuan Regina:** Stance in Talk. A conversation analysis of Mandarin final particles. 2004. xvi, 260 pp.

116 **GRANT, Colin B. (ed.):** Rethinking Communicative Interaction. New interdisciplinary horizons. 2003. viii, 330 pp.

115 **KÄRKKÄINEN, Elise:** Epistemic Stance in English Conversation. A description of its interactional functions, with a focus on *I think*. 2003. xii, 213 pp.

114 **KÜHNLEIN, Peter, Hannes RIESER and Henk ZEEVAT (eds.):** Perspectives on Dialogue in the New Millennium. 2003. xii, 400 pp.

113 **PANTHER, Klaus-Uwe and Linda L. THORNBURG (eds.):** Metonymy and Pragmatic Inferencing. 2003. xii, 285 pp.

112 **LENZ, Friedrich (ed.):** Deictic Conceptualisation of Space, Time and Person. 2003. xiv, 279 pp.

111 **ENSINK, Titus and Christoph SAUER (eds.):** Framing and Perspectivising in Discourse. 2003. viii, 227 pp.

110 **ANDROUTSOPOULOS, Jannis K. and Alexandra GEORGAKOPOULOU (eds.):** Discourse Constructions of Youth Identities. 2003. viii, 343 pp.

109 **MAYES, Patricia:** Language, Social Structure, and Culture. A genre analysis of cooking classes in Japan and America. 2003. xiv, 228 pp.

108 **BARRON, Anne:** Acquisition in Interlanguage Pragmatics. Learning how to do things with words in a study abroad context. 2003. xviii, 403 pp.

107 **TAAVITSAINEN, Irma and Andreas H. JUCKER (eds.):** Diachronic Perspectives on Address Term Systems. 2003. viii, 446 pp.

106 **BUSSE, Ulrich:** Linguistic Variation in the Shakespeare Corpus. Morpho-syntactic variability of second person pronouns. 2002. xiv, 344 pp.

105 **BLACKWELL, Sarah E.:** Implicatures in Discourse. The case of Spanish NP anaphora. 2003. xvi, 303 pp.

104 **BEECHING, Kate:** Gender, Politeness and Pragmatic Particles in French. 2002. x, 251 pp.

103 **FETZER, Anita and Christiane MEIERKORD (eds.):** Rethinking Sequentiality. Linguistics meets conversational interaction. 2002. vi, 300 pp.

102 **LEAFGREN, John:** Degrees of Explicitness. Information structure and the packaging of Bulgarian subjects and objects. 2002. xii, 252 pp.

101 **LUKE, Kang Kwong and Theodossia-Soula PAVLIDOU (eds.):** Telephone Calls. Unity and diversity in conversational structure across languages and cultures. 2002. x, 295 pp.

100 **JASZCZOLT, Katarzyna M. and Ken TURNER (eds.):** Meaning Through Language Contrast. Volume 2. 2003. viii, 496 pp.

99 **JASZCZOLT, Katarzyna M. and Ken TURNER (eds.):** Meaning Through Language Contrast. Volume 1. 2003. xii, 388 pp.

98 **DUSZAK, Anna (ed.):** Us and Others. Social identities across languages, discourses and cultures. 2002. viii, 522 pp.

97 **MAYNARD, Senko K.:** Linguistic Emotivity. Centrality of place, the topic-comment dynamic, and an ideology of *pathos* in Japanese discourse. 2002. xiv, 481 pp.

96 **HAVERKATE, Henk:** The Syntax, Semantics and Pragmatics of Spanish Mood. 2002. vi, 241 pp.

95 **FITZMAURICE, Susan M.:** The Familiar Letter in Early Modern English. A pragmatic approach. 2002. viii, 263 pp.

94 **McILVENNY, Paul (ed.):** Talking Gender and Sexuality. 2002. x, 332 pp.

93 **BARON, Bettina and Helga KOTTHOFF (eds.):** Gender in Interaction. Perspectives on femininity and masculinity in ethnography and discourse. 2002. xxiv, 357 pp.

92 **GARDNER, Rod:** When Listeners Talk. Response tokens and listener stance. 2001. xxii, 281 pp.

91 **GROSS, Joan:** Speaking in Other Voices. An ethnography of Walloon puppet theaters. 2001. xxviii, 341 pp.

90 **KENESEI, István and Robert M. HARNISH (eds.):** Perspectives on Semantics, Pragmatics, and Discourse. A Festschrift for Ferenc Kiefer. 2001. xxii, 352 pp.

89 **ITAKURA, Hiroko:** Conversational Dominance and Gender. A study of Japanese speakers in first and second language contexts. 2001. xviii, 231 pp.

88 **BAYRAKTAROĞLU, Arın and Maria SIFIANOU (eds.):** Linguistic Politeness Across Boundaries. The case of Greek and Turkish. 2001. xiv, 439 pp.

87 **MUSHIN, Ilana:** Evidentiality and Epistemological Stance. Narrative Retelling. 2001. xviii, 244 pp.

86 **IFANTIDOU, Elly:** Evidentials and Relevance. 2001. xii, 225 pp.

85 **COLLINS, Daniel E.:** Reanimated Voices. Speech reporting in a historical-pragmatic perspective. 2001. xx, 384 pp.

84 **ANDERSEN, Gisle:** Pragmatic Markers and Sociolinguistic Variation. A relevance-theoretic approach to the language of adolescents. 2001. ix, 352 pp.

83 **MÁRQUEZ REITER, Rosina:** Linguistic Politeness in Britain and Uruguay. A contrastive study of requests and apologies. 2000. xviii, 225 pp.

82 **KHALIL, Esam N.:** Grounding in English and Arabic News Discourse. 2000. x, 274 pp.

81 **DI LUZIO, Aldo, Susanne GÜNTHNER and Franca ORLETTI (eds.):** Culture in Communication. Analyses of intercultural situations. 2001. xvi, 341 pp.

80 **UNGERER, Friedrich (ed.):** English Media Texts – Past and Present. Language and textual structure. 2000. xiv, 286 pp.

A complete list of titles in this series can be found on the publishers' website, *www.benjamins.com*